life in moving fluids

THE PHYSICAL BIOLOGY OF FLOW

life in moving fluids

THE PHYSICAL BIOLOGY OF FLOW

Steven Vogel
DUKE UNIVERSITY

illustrated by

Sally A. Schrohenloher

PRINCETON UNIVERSITY PRESS

Published by Princeton University Press, 41 William Street,
Princeton, New Jersey 08540
In the United Kingdom: Princeton University Press,
Guildford, Surrey

First Princeton Paperback printing, 1983

LCC 83-60465
ISBN 0-691-02378-6 pbk.

Reprinted by arrangement with Willard Grant Press

Clothbound editions of Princeton University Press books are printed on acid-free paper, and binding materials are chosen for strength and durability. Paperbacks, while satisfactory for personal collections, are not usually suitable for library rebinding.

Text design by Katherine Townes. Composed in Garamond No. 3 by Caron LaVallie, PWS Production Department.

Printed in the United States of America by
Princeton University Press,
Princeton, New Jersey

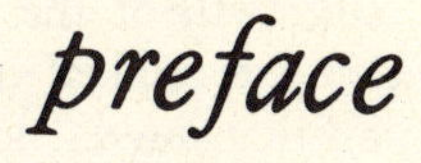

FLUID FLOW IS NOT CURRENTLY in the mainstream of biology, but it has its place. At one time I wanted to be a neurophysiologist; as it turned out, my first attempt at science showed that nerves were irrelevant to the phenomenon in question— I was looking at an odd aerodynamic quirk. One project led to another: flight of small insects, convective cooling in leaves in slow winds, induced flows in burrows and sponges. The interface between biology and fluid mechanics appeared to be a broad, intriguing field, rife with possibilities.

Just ten years ago, I began to give a course on the biological applications of low speed fluid mechanics. It was in part a way for me to learn a little about flow, and student reports on applications were something of an afterthought. The physical part of the course is still being taught, with some permutations of title and content; excerpts have been given at Woods Hole and at the University of California at Irvine. It's now part of what we call "biomechanics" here at Duke and at Friday Harbor, Washington. Since the course began, fluid mechanics has figured substantially in a series of theses in my department, theses done by students other than my own and on topics of which I never would have thought. On occasion the course has generated published products such as my favorite, the Rubenstein-Koehl paper on filter feeding theory.

This book is a direct outgrowth of the course, with commentary on biological applications added to the original notes on physical processes. It reflects my increasing conviction that interest in this material might be more than a local aberration. The book has two missions. First, I want to promote the idea that adaptation to fluid flow underlies much biological design. Second, I hope to provide some guidance on the nature of the physical phenomena, what has been done, and what can be done. It is intended for limnologists, marine biologists, natural historians, interested paleontologists, ecologists of diverse persuasions, comparative and environmental physiologists, and all biologists who study organisms that either contain or are exposed to fluid motion.

The intention is not to convert unsuspecting biologists into full-fledged biofluidmechanicians; rather, my objective is to provide an easy introduction to the subject for those who suspect that it might bear relevance to their projects but whose time and vigor is unequal to the full rigor of engineering and applied mathematics. The book also provides an introduction to some literature in biology; although biofluidmechanics is a relatively unrecognized field, some good research has already been done and much relevant material can be found, scattered among a diversity of journals and subject headings. I usually come away from reading reviews with the impression of doors closing rather than swinging open; most of my harsher comments on existing work are attempts to elicit a strong sense of "doors ajar."

There exists a great misapprehension that fluid mechanics and complex mathematics must walk hand in hand. Admittedly, a great deal of fancy mathematics underlies the simple equations presented in this book. And, so I'm creditably informed, the more mathematics one can manage, the more problems one can solve. But it has long seemed to me that a purely mathematical and non-experimental approach to biologically interesting questions has serious drawbacks and that experimental work does not demand great mathematical sophistication. The real *sine qua non* in the fluids game is insight into the operative physical processes. We must second-guess Nature in an area where physical complexity presents her endless opportunities to be clever. Complex systems are a biologist's stock in trade: the simplest of our systems makes the most complex type of flow look like child's play. For some of us, equations and derivations produce the requisite insight; for others it is words, graphs, and diagrams; still others must stain fingers with dye in flow tanks. As a member of these latter groups, naturally I am writing for other such people. But I make no judgment: in the opportunistic game of science, anything is legitimate if it produces the insights that underlie the hypotheses which explain the organisms.

As the outgrowth of a course, this book is more text than treatise. The subject matter generated a more appealing course than I had originally anticipated, and the course has drawn undergraduates, graduates, and postdoctoral people, premedical students, botanists, ecologists, cell biologists, and paleontologists as well as a few biomedical engineers. In the happy event that it finds use as a textbook, a review of my teaching experiences may be beneficial.

In the initial version of the course, I spent the first half of a semester lecturing on the physical phenomena, and the students spent the second half reporting to each other on biological applications. Each report formed the basis of a term paper, and some developed into experimental projects pursued independently in the following semester. Little effort was made to assign or even suggest topics, and the topics chosen by the students make a persuasive argument that no area of biology lacks flow-related problems. Mention of a few will illustrate the diversity: shape and flow in fish, gliding in reptiles, the flow of water through membranes, blood flow through arteries, flow and the shapes of aquatic plants, cytoplasmic flow, air flow in forests, undulating locomotion in microorganisms, wave action and shore zonation, thermal soaring, flow in burrows, drag minimization, convective cooling in lizards, rheology of small-vessel circulation, filter-feeding, ciliary propulsion, interstitial flow in beaches, flow in the small airways of the lungs, hovering flight, flow in estuaries, and people-powered aircraft.

In its present form at Duke and at Friday Harbor, the course combines fluid and solid mechanics, with about half a semester given to each. This combination works well since both build on elementary mechanics. Furthermore, in many situations, fluid motion causes the stresses with which solid structures must contend. Thus we extend consideration of the shapes of organisms (the main concern here) to a full treatment of structure and deformation. For solid mechanics, Wainwright et al. (1976) serves as text and Gordon (1978) as light background; for fluids, Shapiro (1961) provides background. In the total-immersion, summer course, fluids and solids are studied concurrently, with a morning and afternoon lecture each day; in three weeks of maniacal activity, we cover a semester of material and leave two weeks for even more maniacal work on projects. In the complete semester, fluids and solids are studied sequentially, and projects, as mentioned, are postponed to the next semester. We intend to formalize the follow-up projects as a laboratory course; as now envisioned, a series of specific exercises designed to introduce the relevant technology will be followed by individual projects.

To complement lectures, I have assembled extensive bibliographies, assigned a limited amount of reading of primary literature, and designed quantitative problems which focus on biological situations. Such problems, periodically assigned, are of great value, since physical and quantitative material has a dismaying tendency to pass unimpeded from ear to ear. The sophistication of student reports rose substantially when I began to assign problems. In addition, I do a few demonstrations in the lectures and illustrate some of the physical phenomena with a splendid set of film-loops (SFM series, from Encyclopaedia Britannica Educational Corp., 425 North Michigan Avenue, Chicago, Illinois 60611.)

In practice, this book contains a comfortable semester's worth of material; if combined with solid mechanics, some omissions will be necessary. Each of the fourteen chapters should take about a week, perhaps with the first two combined. Lectures may either follow the text or focus on other applications of the material to complement the interests of particular instructors. Anyone contemplating use of the book as the core of a course may find useful my accumulated teaching resources, including problems (and answers), a list of favorite film-loops, and various handouts. A letter to me (Department of Zoology, Duke University, Durham, NC 27706) will secure this material. If you specifically ask, I will send no other copy to the same institution, to avoid distributing solutions to the problems.

Aside from any practical use in a course or for personal education, I hope the book provides the reader with a measure of amusement. Everyone enjoys the unexpected phenomenon, the bizarre and the counterintuitive, the unusual elements of commonality between overtly unrelated systems. The visceral biologist revels in the cleverness of the living world. I find this interface between fluid flow and the functioning of organisms particularly rich in these pleasures; perhaps the reader will also.

While the content and arrangement of the book can be blamed on no one besides me, practical assistance, constructive criticism and a good bit of gratuitous goading have been generously provided by others. Stephen Wainwright and Jerry Lyons have urged the undertaking, although I doubt whether the result is what either had in mind. Vance Tucker, Knut Schmidt-Nielsen, Michael LaBarbera, and Mimi Koehl have given useful advice and guidance for circumnavigation of pitfalls. I'm particularly appreciative of the efforts of faithful readers Thomas Daniel and Robert Zaret to weed out misstatements and ambiguities in the physics and those of Diana Wheeler to do the same to verbal explanations. Kathi Townes gamefully put up with the pun-ishing prose, deftly deleting the most grotesque and graceless, and Jean-François Vilain swept aside all minor obstacles. Michael Corcoran built the room in which the writing was done. Duke University provided a few months of sabbatical leave from the usual fragmentation of the academic circus together with the books and personnel of a truly splendid library system. The writing was supervised by Coriander Cardamom Cat while I considered the aerodynamics of tapering vibrissae. My family, Jane and Roger, knew when to laugh and when to be scarce, for both of which I am grateful.

acknowledgements

Permission from the copyright holders is gratefully acknowledged for the following figures, redrawn from previously published material.

Figure 4.7: Academic Press, Inc. (*Journal of Theoretical Biology*); 6.1: Cornell University Press (*Aerodynamics* by T. von Karman); 6.12, 12.2: The Company of Biologists, Ltd. (*Journal of Experimental Biology*); 9.1a, 9.1b: Birkhäuser Verlag (*Swiss Journal of Hydrology*); 7.1c: William Collins Sons and Co. Ltd. (*Living Marine Molluscs* by C.M. Yonge and T.E. Thompson); 9.1d, 9.7b, 9.7c, 11.13: John Wiley and Sons, Inc. (*Fresh Water Invertebrates of the United States* by R.W. Pennak); 9.6: Cambridge University Press (*The Invertebrata* by L.A. Borradaile et al.); 9.8b: National Research Council of Canada (*Canadian Journal of Botany*); 10.4, 10.8: United Engineering Trustees, Inc. (*Fluid Mechanics for Hydraulic Engineers* by H. Rouse); 10.6b: Holt, Rinehart, and Winston and Mrs. Alfred S. Romer (*The Vertebrate Body* by A.S. Romer); 10.7a: Cambridge University Press (*Journal of the Marine Biological Association of the United Kingdom*); 10.7b: Marine Biological Laboratory (*Biological Bulletin*); 11.10: Springer Verlag (*Oecologia*); 11.11, 13.2: Springer Verlag (*Journal of Comparative Physiology*); 12.9: Cambridge University Press (*Biological Reviews*); 14.3: Wadsworth Publishing Co. (*An Evolutionary Survey of the Plant Kingdom* by R.F. Scagel et al.)

contents

CHAPTER 13

flow at very low Reynolds numbers, 241

CHAPTER 14

a few final phenomena, 267

APPENDIX 1

making fluids flow, 289

APPENDIX 2

visualization and measurement of flow, 303

To my principal teachers,

H. Francis Henderson, ecologist

the late Kenneth D. Roeder, neurophysiologist

Carroll M. Williams, developmental biologist

No student could want better. Each in his own way set an indelible example of both personal and intellectual excellence.

CHAPTER 1

remarks at the start

WITH THE RATIFICATION OF long tradition, the biologist goes forth, thermometer in hand, and measures the effects of temperature on every parameter of life. Lack of sophistication poses no barrier; heat storage and exchange may be ignored or Arrhenius abused; but temperature is, after time, our favorite abscissa. One doesn't have to be a card-carrying thermodynamicist to wield a thermometer.

By contrast, few of us measure the rate at which fluids flow, however potent the effects of winds and currents on our experimental systems. Fluid mechanics is intimidating, with courses and texts designed for practicing engineers, masters of vector calculus and similar arcane arts. So the effects of flow are too commonly ignored or relegated to parenthetical mention, and we have developed no comfortably familiar and appropriate technology to produce and measure the flow of fluids under biologically interesting circumstances.

By fluids, we mainly mean air and water. The rules for dealing with the two are surprisingly similar and the differences less overwhelming than one might suspect offhand. A life immersed in air or water is, of course, quite unexceptional among organisms. Almost as commonly, the organism and the fluid medium move with respect to one another, either through locomotion, as winds or currents across some sedentary creature, or via the passage of fluid through internal conduits. Fluid movement, then, is something with which many organisms must contend. Therefore it should be a factor to which the design of organisms reflects adaptation.

It is this particular set of phenomena—the adaptations of organisms to moving fluids—that is the topic of this book. Its main points are that such adaptations are of considerable potential interest, even if as yet poorly known and commonly overlooked, and that fluid flow need not be viewed with fear or alarm. Flow may be one of the messier aspects of the physical world. Still, most of the messiness can be explained in words and graphs. It holds considerable intrinsic fascination and gives the evolutionary process endless opportunities to be clever. At the same time, this complexity greatly restricts the possibilities of exact mathematical solutions or simulations; usually the investigator must fall back on a world of physical models and direct experimentation, a world in which the biologist can feel quite at home.

A review of what is known of the interrelationships of the movements of fluids and the design or organisms is not a primary concern here. The intent, rather, is more prospective than retrospective. I take the view that with some minimal sense of the mechanics of fluids the biologist is likely to notice these sorts of relationships. Without that sense, hydromechanical hypotheses will not be put to test, and for the most part, the relevant phenomena will continue to escape scrutiny. The main objective, then, is an attempt to imbue the reader with some intuitive feeling for the behavior of fluids under biologically interesting circumstances, to supply some of the contemptuous familiarity with fluid motion that most of us have for solids.

One could organize a book such as this with either the biology or the physics as a framework. The physical phenomena, though, fall more easily into an orderly and useful sequence; my attempts to make good order of the biological topics inevitably have an air of artificiality. So the physics of flow will provide the skeleton, fleshed out, as it were, by consideration of the bioportentousness of each topic in turn. Where relevance to a particularly large segment of biology wants examination, as when we consider velocity gradients or drag, frankly digressive chapters are interjected. I'm assuming that the typical reader is a biologist seeking an easy entry into fluids rather than a fluid mechanist entering biology. The reader may be vague on the distinctions between work and power, stress and strain, but is assumed to be quite sound on vertebrates and invertebrates, anemones and sea anemones, and can take them in any order. Since the framework is physical rather than biological, topics such as seed dispersal and filter feeding will wander in and out; where a particular biological topic illustrates several applications of fluid mechanics a little redundancy has been inevitable.

Biological examples will vary from well-established through half-baked to wildly speculative; I'll try to indicate the degree of confidence with which each should be vested. But speculation is quite reasonable at this stage. If anyone feels strongly enough for or against some assertion to give the matter a decent investigation, then we'll all be a bit further ahead. In any case it seems an entirely practical procedure to fix on some physical phenomenon and then go looking to see how organisms have responded to it. I've occasionally compared the scheme to shooting at a wall and drawing targets around the bullet holes; the point is that science is an opportunistic affair, not a sporting proposition.

It is intended that the reader work through in sequence at least the basic material on viscosity, the principle of continuity, Bernoulli's equation, the Reynolds number, and the characteristics of velocity gradients. The book can be used out of order or with parts omitted, but certain pitfalls may then lurk. Biologists have been all too willing to grab equations and use them with no more than a guileless glance to see that the right variables were represented. As a result, much mischief has been perpetrated. Bernoulli's equation doesn't work well in boundary layers, Poiseuille's law presumes one is far from the entrance to a pipe, and Stokes' law works (in general) for small spheres, not large ones. Yes, there are equations in these pages; but they can as easily be found elsewhere. More important are the discussions of how and where they apply—when to rush in and when to fear to tread. In this latter matter, the sequence takes on some significance, especially in the first five chapters.

inclusions and exclusions

Some idea of exactly what is covered in this book is apparent from the table of contents. The omissions are naturally less evident, but they deserve mention. Fluid mechanics ramifies in many directions; much of it carries the unmistakable odor of our technological concerns and has little relevance to biology. Thus we can safely ignore high speed (compressible) flows and the flow of rarefied gases, to give a few examples. For reasons which flatter neither author nor readers, the mathematical niceties of fluid mechanics will not loom large here. Furthermore, the interface between fluids and organisms is remarkably extensive, and some rather arbitrary exclusions have been made in order to focus on organisms and small assemblages of organisms which are faced with relatively general problems. Only a little will be said about temporally non-steady flows, the statics of fluids, biometeorology, and the flow of substances which are incompletely fluid, such as whole blood or the contents of cells. Thus I'll largely eschew the treacherous quagmires of mitosis and cyclosis and the features of gas-liquid interfaces such as surface waves, surface tension, ocean currents, wind-driven circulation in lakes, as well as the hydraulics of streams and rivers, and flow through porous media. Many of these topics are treated well elsewhere, and sources of enlightenment will be mentioned at appropriate points.

More extensive and detailed information on fluid dynamics may be obtained from conventional engineering textbooks, such as Streeter and Wylie (1975). The basic processes which will be of concern here were about as well understood forty or fifty years ago as they are today, so the age of the source is immaterial. Indeed, the older generation of fluid dynamicists worried more than their successors about low-speed phenomena and other items of biological relevance; and several inexpensive reprints of old texts are particularly handy. Prandtl and Tietjens' (1934) *Applied Hydro- and Aeromechanics* is particularly good on boundary layers and drag. *Fluid Mechanics for Hydraulic Engineers* (Rouse, 1938) contains a fine treatment of dimensional analysis and the origin of common dimensionless indices. Goldstein's (1938) two volumes, *Modern Developments in Fluid Dynamics,* has nice verbal descriptions of phenomena in between the equations. Mises (1945) gives excellent explanations of how wings and propellers work in *Theory of Flight.* With several of these books at hand one can usually find an understandable and intuitively reasonable explanation of any puzzling aspect of flow.

Popular accounts are surprisingly scarce, with no treatment of fluid mechanics coming anywhere near the breadth and elegance that Gordon's (1978) *Structures* brings to solid mechanics. My favorite three paperbacks are von Karman's (1954) *Aerodynamics,* Sutton's (1949) *The Science of Flight,* and Shapiro's (1961) *Shape and Flow.* The first is particularly witty, the second is clear and honest, and the third, *mirabile dictu,* is about the sort of fluid phenomena that biologists should encounter.

Three other sources deserve mention. *Fluid Behavior in Biological Systems* by Leyton (1975) is nearest to this book in intent. It gives less attention to drag, boundary layers, and propulsion but more to flow in porous media, heat transfer, non-Newtonian fluids, thermodynamics, and micrometeorology. Bascom's (1980) *Waves and Beaches* deserves the attention of anyone working along the edge of the ocean. And the privately published compendium of Hoerner (1965), *Fluid-Dynamic Drag,* is filled with data for simple objects to which analogous data for organisms might be compared.

technology

The how-to-do-it aspect of biological fluid mechanics presents special snares. High technology is certainly no stranger to the microscopist, biochemist or neurophysiologist, but few of us are facile in making and measuring fluid flows. Faced with some upcoming investigation one's first impulse is to seek out a friendly engineer who then prescribes a hot-wire or laser anemometer, and the problem is reduced to a search for kilobucks. But there is, as well, a second and less ordinary problem. The technology used by the engineers is a product of, by, and for engineers. The range of magnitude of the flows we need to produce and of the forces we want to measure is rather different, and engineering technology is often as inappropriate as it is expensive.

In twenty years of facing problems of flow, I've found that the devices I've had to use were (compared to those of my colleagues in more conventional fields) rather cheap. On the other hand, they have rarely been available prepackaged and precalibrated. Flow tanks and wind tunnels, flow meters, anemometers, and force meters have simply been built in my laboratory as needed. I've developed a fair contempt for fancy commercial gear except for those of very general applicability—voltmeters, power supplies, potentiometric chart recorders, variable speed motors, gears and pulleys, an electronic stroboscope, and so forth. The single most valuable tool has been a combination lathe, milling machine, and drill press. Since this book is intended to reduce the activation energy needed for a biologist to work on fluid problems, I've incorporated some general lore in the text and more specific guidance and designs for equipment in appendices. The reader should recognize that the writer is a member of a lineage of quite unreconstructed primitivists. Some of us take perverse pride in the passion of the impecunious immersed in the jetsam of an industrial age!

dimensions and units

Fluid mechanics makes use of an array of different parameters, some familiar (density, viscosity, lift, and drag) and others (circulation, friction factor, and pressure coefficient) out of the biologist's normal menagerie of terms. A list of symbols and quantities used in this book precedes the index. Some simplification will be achieved if the reader tries to keep in mind the dimensions which attach to each quantity.

By dimensions I don't mean units. Thus velocity always has *dimensions* of length per unit time, whether given in units of meters per second, miles per hour, or furlongs per fortnight. Dimensions take on rather special significance in fluid mechanics as a result of the general messiness of the subject and a trivial-sounding but surprisingly potent condition. *For an equation to have any applicability to the real world, not only must the two sides be numerically equal, but they must also be dimensionally equal.* When theory, memory, or intuition fails, this injunction can go a long way toward indicating the form of an appropriate equation.

An example will show how one can proceed. Assume you want an equation relating the tension in the wall of a sphere or cylinder to the pressure inside. Tension has dimensions of force per unit length and pressure has dimensions of force per unit area. (You can obtain the dimensions from common units; pressure can be measured in pounds of force per square inch of area.) To relate pressure to tension, clearly pressure must be multiplied by some linear dimension of the sphere or cylinder, so the equation will be of the form

$$\text{tension} = \text{constant} \times \text{radius} \times \text{pressure}$$

where we know nothing about the constant except that it is dimensionless. Evidently, it takes a larger pressure to generate a given tension when the radius is small than when the radius is large. Even without further information about the constant we can see why it is hard to *start* blowing up a balloon (we are applying pressure in order to produce tension); why a small plant cell can withstand many atmospheres of internal pressure with only thin walls; and why arterioles can get by with much thinner walls than that of the aorta when both are subjected to the same internal pressure.

The easiest way to compare the dimensions of different parameters is to reduce them to combinations of a few so-called "fundamental dimensions." We will usually require only three such dimensions here: *length, mass,* and *time* (and, rarely, *temperature*). This reduction is accomplished by use of definitions or previously memorized functions. For instance, force is mass times acceleration, and therefore force has fundamental dimensions of MLT^{-2}. Work or energy is force times distance and thus ML^2T^{-2}. With such simple manipulations the dimensional equivalence of each term in almost any equation may be verified. Considerable use will be made of this sort of "dimensional analysis" in forthcoming chapters. Also, just because two quantities have the same dimensions doesn't mean that they are synonymous; thus pressure and stress are both given as force per unit area.

Incidentally, not only constants but also variables may be dimensionless and still have meaning in the physical world. Any number which is the ratio of two quantities measured in the same dimensions will be dimensionless. A fairly common example is *strain* as it is used in solid mechanics. Strain is the ratio of the extension in length of a stressed object to its original, unstressed length. Dimensionless numbers are, of course, automatically unitless and quite indifferent to the system of units in use. They find wide application in fluid mechanics.

Units are quite another matter. To eliminate awkward constants, all variables in a dimensionally proper equation should be measured in a consistent set of units. Biologists have usually used something approaching the physicist's centimeter-gram-second (CGS) system, with a few oddities such as calories and millimeters of mercury. We are in the midst of a transition to a common system for all science, the SI or "Systême Internationale," which will generally be used here. Fundamental units (for the fundamental dimensions) are kilograms (mass), meters (length), and seconds (time). It should be emphasized that use of the metric system is not sufficient; rather some more specific scheme such as SI is required. SI allows the common prefixes (mega-, milli-, micro-, and so forth) to be attached to any fundamental or derived unit, but only a single prefix may be used with a single unit, and the prefix must attach to the numerator. Thus meganewtons per square meter is legitimate but newtons per square millimeter is not. It must be admitted that the SI units can get mildly ludicrous in the context of organisms. The drag of a cruising fruit fly (Vogel, 1966a) is about three micronewtons. And Wainwright et al. (1976) give the strength of spider silk as about 10^9 newtons per square meter of cross section; it would take one hundred billion (US) strands to get that combined area. Table 1.1 gives a list of quantities with their fundamental dimensions and SI units. For further introduction to SI units and conventions, see the National Bureau of Standards (US) Special Publication 330 (1977), *The International System of Units,* or Mechtly's (1973) pamphlet of the same title. For dealing with different systems of units for the quantities of interest here, Pennycuick's (1974) booklet, *Handy Matrices of Unit Conversion Factors for Biology and Mechanics,* is an absolute godsend. It's also cheap enough to scatter a few copies around home, office, and laboratory.

It might be noted that this tidy system of consistent dimensions and units falters when applied to empirical equations fitted to data. These typically have odd exponents ("metabolic rate is proportional to body mass to the 0.75 power") and a constant of proportionality accorded less attention than the exponent. All of the unpleasantness is heaped upon the constant. If the expres-

TABLE 1.1

Common quantities, fundamental dimensions, and SI units

quantities	*dimensions*	*SI units*
Length, distance	L	meter (m)
Area, surface	L^2	square meter (m^2)
Volume	L^3	cubic meter (m^3)
Time	T	second (s)
Velocity, speed	LT^{-1}	meter per second ($m\ s^{-1}$)
Acceleration	LT^{-2}	meter per second squared ($m\ s^{-2}$)
Mass	M	kilogram (kg)
Force	MLT^{-2}	newton (N or $kg\ m\ s^{-2}$)
Density	ML^{-3}	kilogram per cubic meter ($kg\ m^{-3}$)
Work	ML^2T^{-2}	joule (J or N m)
Power	ML^2T^{-3}	watt (W or $J\ s^{-1}$)
Pressure, shear stress	$ML^{-1}T^{-2}$	pascal (Pa or $N\ m^{-2}$)
Dynamic viscosity	$ML^{-1}T^{-1}$	pascal second (Pa s or $N\ m^{-2}\ s$)
Kinematic viscosity	L^2T^{-1}	square meter per second ($m^2\ s^{-1}$)

sion is written as an equation rather than a proportionality, the abuse of the constant is concealed by merely specifying the set of units which must be used with the expression. Here the practice among fluid mechanists and biologists is about the same, both being practical people with scattered data, inadequate theories, regression equations and the like.

measurements and accuracy

First, a few words about what is meant by *accuracy*. As Eisenhart (1968) has pointed out, lack of accuracy reflects two distinct components. There is, to begin with, *imprecision*, or lack of repeatability of a series of determinations. And there is, as well, *systematic error*, or bias—the gap between the measured value and some "true value," or the tendency to measure something other than what we intended. In the kinds of problems we'll discuss, unavoidable imprecision is usually vast. Thus it is rarely worth great attention to matters such as fine standards for calibration if it merely gets the systematic errors far below the practical imprecision.

Some quantities such as viscosity and density can be measured very precisely. But the inevitable irregularities in flow streams, the large effects of small variations in surface qualities, the nature of vortices and turbulence, all these and other phenomena severely limit the precision with which the behavior of moving fluids is worth measuring. The drag of an object measured in one wind tunnel will often differ considerably from that measured in another, while a third figure will result from towing the object through otherwise still air. If a figure of, say, 1 m s^{-1} is cited as the transition point from laminar to turbulent flow in some pipe, that figure should not be interpreted as 1 ± 0.01 or even ± 0.1. With extreme care it may be possible to postpone the transition to 10 m s^{-1} or more. Some empirical formulas, especially those for convective flows, use constants given in standard works to three significant figures. My own experience suggests that such numbers should be viewed with enormous skepticism.

The development of hand-held calculators has had a pernicious effect on our notions of necessary precision. The art of rounding off and guessing just how much precision is minimally necessary has suffered from the demise of the slide rule. As a practical matter, the flow of fluids, even without organisms, is a subject which enjoys only a slide rule level of precision—rarely better than one percent and often much worse.

CHAPTER 2

what is a fluid and how much so

A FLUID IS EITHER A GAS or a liquid; it is not a solid. This bald, inscrutable statement begs a number of questions. Why do we find it convenient to lump the gaseous and liquid states even though we know of no intermediates? Why do we persist in making a general distinction between liquids and solids when, in practice, we're made up of and surrounded by substances of intermediate properties? And, finally, by what criterion can we qualitatively distinguish fluids from solids?

It turns out that a simple experiment can be used to tell a fluid from a solid, an experiment which makes no particular distinction between gases and liquids. Consider an apparatus consisting of two concentric cylinders of ordinary size such as that shown on the right in Figure 2.1. Although the outer one is fixed, the inner cylinder may be rotated. Between the cylinders is a gap which we can fill with any material of any state; the material adheres to or "wets" the walls of the cylinders. The rotation of the inner cylinder is resisted only by forces caused by the fixed outer wall and transmitted through the material being tested.

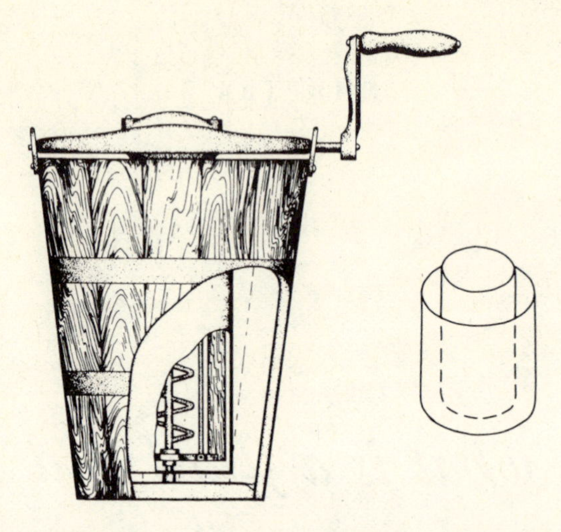

A pair of concentric cylinders (right); if the space between them is filled with a fluid, then the force which tends to rotate the outer cylinder will depend on the rate of rotation of the inner one. A more useful version is the ice cream freezer on the left. The experienced operator monitors the increase in resistance to shear rate in this practical viscometer by noting the increasing difficulty of turning the handle while not turning the bucket.

FIGURE 2.1

One naturally expects that the inner cylinder will turn more or less easily, depending on the material in the gap; and, indeed, that's just what happens; the force required varies over an enormous range. But another distinction emerges. For some materials, the force necessary to turn the inner cylinder depends on *how far* we already turned it from "rest." For others, the force is quite independent of how far we've distorted the material and depends only on *how fast* we turn the cylinder. In each case we have imposed a "shearing" load on the materials; that is, we've attempted to distort them by sliding one surface relative to another, parallel one.

Solids resist shear deformation—they care how *far* they are deformed. In general, the further one wishes to deform them, the more force is required. To put the matter quantitatively, if θ is the angle by which the material is deformed and S the surface area over which the force is applied, then

$$\frac{F}{S} = E\theta \qquad (2.1)$$

E is called the *elastic modulus* or the *shear modulus* and corresponds to our common notion of stiffness. The equation can also be read as stress (F/S) is proportional to strain (θ): Hooke's law. If E stays constant as stress and strain vary, the material is spoken of as "Hookean."

Proper fluids, by contrast, completely lack an elastic modulus because they are infinitely distortable and magnificently oblivious to their shape, present or past. E is not zero; E does not exist. What fluids do care about, though, is how *rapidly* they are deformed. Thus, for fluids,

$$\frac{F}{S} = \frac{\mu\theta}{t} \tag{2.2}$$

where θ/t is the rate of shear and μ is a different property which we call the *dynamic viscosity.* The latter indicates how much a fluid resists not shear but rate of shear. If μ remains constant as stress and shear rate vary, the fluid is said to be "Newtonian." For different fluids, though, the dynamic viscosity may vary greatly from low values for gases to exceedingly high values for substances such as tar or glass which, over long time periods, behave as fluids. Sometimes it is useful to assign values of viscosity to glaciers or to the mantle of the earth.

the no-slip condition

The more disputatious reader may have detected a peculiar assumption in our demonstration of viscosity: the fluid had to stick to the walls of the outer and inner cylinders in order for the fluid to shear rather than slide along the walls. A fluid does tend to stick to itself. If one element of a fluid moves it tends to carry other elements with it—that tendency is what viscosity measures. Less obviously, fluids stick to solids quite as well as they stick to themselves. As nearly as we can tell from the very best measurements, the velocity of a fluid *at* the interface with a solid is always the same as that of the solid. This last statement expresses something called the "no-slip" condition: *fluids do not slip with respect to solids.* It is one of many counter-intuitive concepts in fluid mechanics; indeed, the dubious may be comforted to know that the existence and universality of the no-slip condition was heatedly disputed through most of the nineteenth century. Goldstein (1938) devotes a special section at the end of his book to the controversy. The only known exception is the case of very rarefied gases, where molecules encounter one another too rarely for viscosity to mean much.

Yet another peculiarity is that the nature of the solid surface makes very little difference. If water is flowing over a solid without an air-water interface to complicate matters, the no-slip condition holds whether the solid is hydrophilic or hydrophobic, rough or smooth, greasy or clean.

The no-slip condition has a number of important ramifications. In particular, it implies that velocity gradients are developed entirely within fluids rather than between fluids and solids. If the fluid is homogeneous, the velocity must smoothly approach zero as the surface is approached; there can be no discontinuity within the fluid. Furthermore, the velocity cannot asymptotically approach zero, for that would require a variable viscosity. In practice, the no-slip condition explains (in part) why dust and grime accumulate on fan blades, why pipes scale up in use rather than wearing thin, and why a bit of rock is needed in water for the latter to become effectively erosive.

The no-slip condition is as easy to demonstrate as it is hard to prove. Simply fill a circular basin with water, stir the water into smooth circuitous motion, and inject a small bit of dye at the bottom or side wall. The last layer of dye at the wall will remain there despite quite a lot more stirring. Eventually, of coruse, diffusion will move most of the dye into the moving stream, but diffusion in liquids isn't very rapid. Alternatively, just consider why dishcloths and mops are so much more effective for cleaning surfaces than any mere rinse.

assumptions and conventions

Before proceeding further, I'd like to establish a series of assumptions which will tacitly underlie the forthcoming chapters unless one or another of them is explicitly relaxed. These are mildly preposterous; but, like protoplasm and point masses, they're convenient fictions.

1. *Fluids are "Newtonian."* There exist, as mentioned, many substances with properties of both fluids and solids. But we will disallow any trace of solidity as previously defined—our fluids will have no memory of previous shape or any elasticity. In effect, a line is drawn between molasses (in) and "Jello" (out). As Shakespeare says (King Lear), "Out, vile jelly." Many biological materials are in this complex, multidimensional continuum of non-Newtonian fluids and viscoelastic solids—whole blood, synovial fluid, mucuses of various consistencies—but we'll largely ignore them and, in doing so, avoid perhaps the messiest and least understood branch of fluid mechanics, rheology.

As mentioned, for a fluid to be Newtonian it must exhibit a direct (linear) proportionality between the applied shear stress and the resulting rate of deformation. If so, a value of viscosity may be obtained which is independent of the specific test conditions. Air and water, the fluids of main concern here, are virtually perfect Newtonian fluids.

2. *Fluids are continua.* This assumption presupposes that fluids are non-particulate and infinitely distortable and divisible. We're going to turn the picture of Democritus to the wall and deny molecules. But, consider: is there anything in your everyday world which requires that you assume matter is made of molecules? Why should cheese, sliced thinly enough, ultimately stop being cheese? Molecules are apparently necessary to explain the physical basis of viscosity; but once viscosity is accepted, there is no further need for them. A "fluid particle" will be just a linguistic convenience for specifying an arbitrarily small element of a moving fluid for our particular scrutiny.

3. *Fluids are incompressible.* This does real violence to the experience of anyone who has wielded a bicycle pump. Gases compress easily, liquids only with difficulty. You should put aside these subversive thoughts. It may be easy to compress some fluids with static devices; it's no small matter, however, to get compression through flow. The fluid behind pushes on the fluid in front, and the latter flows rather than squashes. The compressibility of fluids by virtue of their motion through pipes, along walls, and past obstacles is negligible up to velocities comparable to that of sound (340 m s^{-1} in air, 1500 m s^{-1} in water). A report that a deer fly achieved 350 m s^{-1} was carefully demolished by Langmuir (1938) who showed that the fly would either be crushed by its drag or consume its own weight as fuel each second. So we're on safe ground in asserting that biologically important flows are thoroughly subsonic and that the assumption of incompressibility is justifiable. It effects a hugh simplification compared to the world with which airplane designers must contend.

4. *Fluids make no interfaces with other fluids.* Gases, of course, will not discretely interface with each other, but liquid-liquid and liquid-gas interfaces (especially the latter) are commonplace wherever gravity is appreciable. At such fluid-fluid interfaces we have phenomena like surface tension, capillarity, and surface waves. Little attention will be given to these, mainly for reasons of space and not for lack of biological relevance. Thus, unless otherwise specified, bodies of fluid will be considered unbounded except by solids.

One might note that with assumptions (3) and (4) the differences between liquids and gases become quantitative rather than qualitative, and there is no easy absolute test left to tell whether a substance is a gas or a liquid.

In addition to these four assumptions, we'll usually assume that our bodies of fluid are of uniform temperature and that the solids bounding or immersed in them are perfectly rigid.

properties of fluids

Density. For fluids, which continually move past our frame of reference, the notion of mass is somewhat awkward. In practice it is replaced by *density*, or mass per unit volume. The symbol used is ρ (rho), the fundamental dimensions are ML^{-3}, and the SI unit is the kg m^{-3}. The CGS unit, in this unusual case 10^3 times larger, is the g cm^{-3}. Some contrary sources use specific volume instead of density; one is the reciprocal of the other. Since we're dealing with incompressible fluids, density is constant within any particular field of flow.

In both air and water, at ordinary temperatures density drops as temperature increases, but only in air is the effect appreciable. In air the density is inversely proportional to the absolute temperature. In fresh water, density is maximal at about 4 °C; drops slightly when the freezing point is approached; and drops further with the formation of ice. The biological consequences of these minor density differences are profound and are discussed in any book on limnology. In sea water, -3.5 °C is the temperature of maximum density, a temperature at which ice crystals have ordinarily already begun to form. Representative values of density are given in Table 2.1.

Dynamic viscosity. If one views the shearing of a fluid as the sliding of a large number of very thin sheets across one another, then *dynamic viscosity* represents the stickiness or friction between these sheets. For a quantitative view of viscosity, it is helpful to consider a simple (if practically preposterous) experiment. Imagine two negligibly thick parallel flat plates of the same shape and area, separated by a fluid-filled space (Figure 2.2). The lower plate is fixed, and we push on the upper with a steady force, F, causing it to move in its plane of orientation at a uniform velocity, U. How is the relationship between F and U affected by the nature of the fluid and the geometry of the system? It turns out that for Newtonian fluids a uniform velocity gradient develops between the plates, just as if the space were filled with layered sheets of uniform stickiness. The force required for the upper plate to achieve a given velocity will be the product of that velocity, the area of a plate (S), and the property which we are

TABLE 2.1

SI values of some physical parameters at atmospheric pressure

		dynamic viscosity ($kg\ m^{-1} s^{-1}$)	*density* ($kg\ m^{-3}$)	*kinematic viscosity* ($m^2 s^{-1}$)
Air	0 °C	17.09×10^{-6}	1.293	13.22×10^{-6}
	20 °C	18.08×10^{-6}	1.205	15.00×10^{-6}
	40 °C	19.04×10^{-6}	1.128	16.88×10^{-6}
Fresh water	0 °C	1.787×10^{-3}	1.000×10^3	1.787×10^{-6}
	20 °C	1.002×10^{-3}	0.998×10^3	1.004×10^{-6}
	40 °C	0.653×10^{-3}	0.992×10^3	0.658×10^{-6}
Sea water*	20 °C	1.072×10^{-3}	1.024×10^3	1.047×10^{-6}
Acetone	20 °C	0.326×10^{-3}	0.792×10^3	0.412×10^{-6}
Glycerin	20 °C	1.490	1.261×10^3	1.182×10^{-3}
Glycerin (90% aq.)	20 °C	0.219	1.235×10^3	0.177×10^{-3}
Mercury	20 °C	1.554×10^{-3}	13.546×10^3	0.115×10^{-6}

*Sea water of salinity 34.84%. The salinity of sea water varies somewhat from place to place. At 10 °C the dynamic viscosity of sea water is 1.391×10^{-3} and at 30 °C it is 0.868×10^{-3}.

calling the dynamic viscosity (μ or mu), divided by the distance (l) between the plates:

$$F = \frac{US\mu}{l} \quad \text{or} \quad \mu = \frac{Fl}{US} \qquad (2.3)$$

Dynamic viscosity will thus have fundamental dimensions of $ML^{-1}t^{-1}$ or, less fundamentally, (force)(time)/(area). Since the SI scheme uses the pascal for force per area, the SI unit of dynamic viscosity is the "pascal second," Pa s; there is no specific name for the unit. Much of the tabulated data available is given in "poises," the comparable CGS unit equivalent to dyne seconds per square centimeter, and a unit ten times smaller than the pascal second.

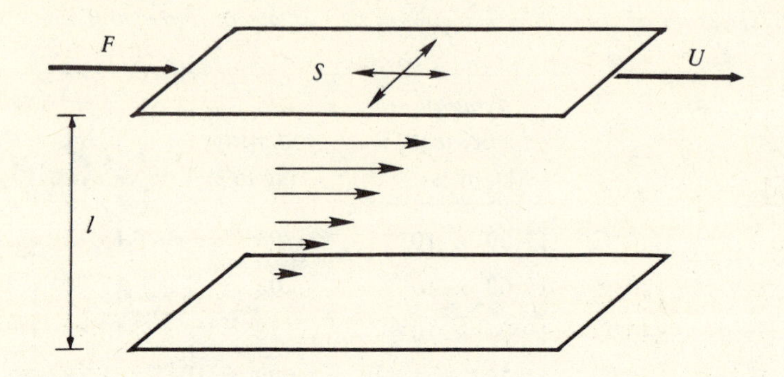

An ideal pair of thin, flat plates. If the lower plate is fixed, it takes a force to keep the upper one moving. The magnitude of that force is proportional to the dynamic viscosity of the fluid between them. The length of each horizontal arrow between the plates is proportional to the local flow speed.

FIGURE 2.2

Dynamic viscosity is commonly called simply "viscosity," although the former designation explicitly distinguishes it from kinematic viscosity (page 19), also sometimes called simply viscosity. Dynamic viscosity is occasionally referred to as "molecular viscosity" to distinguish it from "eddy viscosity," a larger-scale parameter with the same dimensions used mainly by oceanographers (Sverdrup et al., 1942). Physical chemists traditionally use the symbol η (eta) instead of μ for dynamic viscosity.

Note in Table 2.1 that the dynamic viscosity of air increases slightly with an increase in temperature. In water, though, viscosity drops dramatically as temperature rises. Sea water and fresh water behave in a similar manner, with the small differences in viscosity closely proportional to the concentration of salts. Few physical properties have as extreme a temperature coefficient as does the viscosity of ordinary liquids.

Dynamic viscosity in either air or water is substantially independent of the value of density, a great convenience. For water, density varies only slightly with pressure, so the matter is rarely of concern. The oddity of what variation does occur is treated by Stanley and Batten (1968, 1969) and may have some relevance to very deep sea organisms or extremely eurybaric species.

If a more formal definition of dynamic viscosity is preferred, it may be considered the coefficient which relates the shear stress (τ or tau, a force per area) to the local velocity gradient or shear rate (*dU/dl*, with dimensions of time^{-1}):

$$\tau = \mu \frac{dU}{dl} \qquad (2.4)$$

Kinematic viscosity. In reading about fluid flow you may encounter the symbol υ (nu) for viscosity; you are entirely justified in asking "what's υ?" It is, in fact, the so-called ***kinematic viscosity,*** which is nothing more than the ratio of dynamic viscosity to density:

$$\upsilon = \frac{\mu}{\rho} \qquad (2.5)$$

Kinematic viscosity has fundamental dimensions of L^2T^{-1}; the SI unit is the m^2s^{-1}. The CGS unit, the cm^2s^{-1} (10^4 times smaller), is given the name "stokes" (not "stoke"!) or St.

Why devise such an obviously derived unit? It happens that in most situations the character of a flow depends on this particular ratio. It determines the practical "gooiness" of a fluid—how easily it flows, how likely it is to break out in a rash of vortices, how steep will be the velocity gradients. Viscosity (dynamic) determines the "interlaminar stickiness" of the fluid, or how much a fluid particle is likely to be retarded by any lack of synchronized activity of adjacent particles. Density determines the "mass of a fluid particle," or its tendency to proceed as it has been regardless of the activities of its fellows. The ratio, or kinematic viscosity, as Batchelor (1967) puts it, "measures the ability of the molecular transport to eliminate the non-uniformities of fluid velocity."

Note in Table 2.1 that the kinematic viscosities of air and water are not especially different—only about 15-fold at 20 °C. Moreover, air is the *more* kinematically viscous fluid, demonstrating, if nothing else, that this property does violence to the raw intuition. In air, kinematic viscosity increases with temperature slightly more than does dynamic viscosity. In water, the kinematic viscosity shows the same dramatic decrease with increasing temperature as does the dynamic viscosity. So that's what's υ.

Values of density and dynamic and kinematic viscosities of gases, liquids, and aqueous solutions (including sea water) are most easily obtained from recent editions of the *Handbook of Chemistry and Physics.* More limited data are given in the appendices of most textbooks of fluid mechanics, hydraulics, and aerodynamics.

CHAPTER 2

measuring viscosity

If pure water, sea water, or some simple solution is in use, it is simpler to look up a value for either sort of viscosity than to make any measurements. Sometimes, though, a solution is used for which tabulated values are unavailable. It is a simple matter to measure the viscosity of most liquids; one just needs a viscometer. (The word "viscosimeter" is synonymous, but is a linguistic barbarism analogous to "orientate.") The simplest and cheapest commercially available device is the Ostwald viscometer (Figure 2.3). Improvised alternatives may be made of capillary tubing or fine pipettes, catheter tubes, and so forth.

FIGURE 2.3

(*a*) An Ostwald viscometer, slightly fatter than life. The time needed for a meniscus to drop from upper to lower mark as fluid passes through the capillary is proportional to the kinematic viscosity of the fluid. (*b*) A viscometer which measures dynamic viscosity. If the motor turns the outer cylinder at a constant rate, the torque on the inner cylinder extends the spring and rotates the dial.

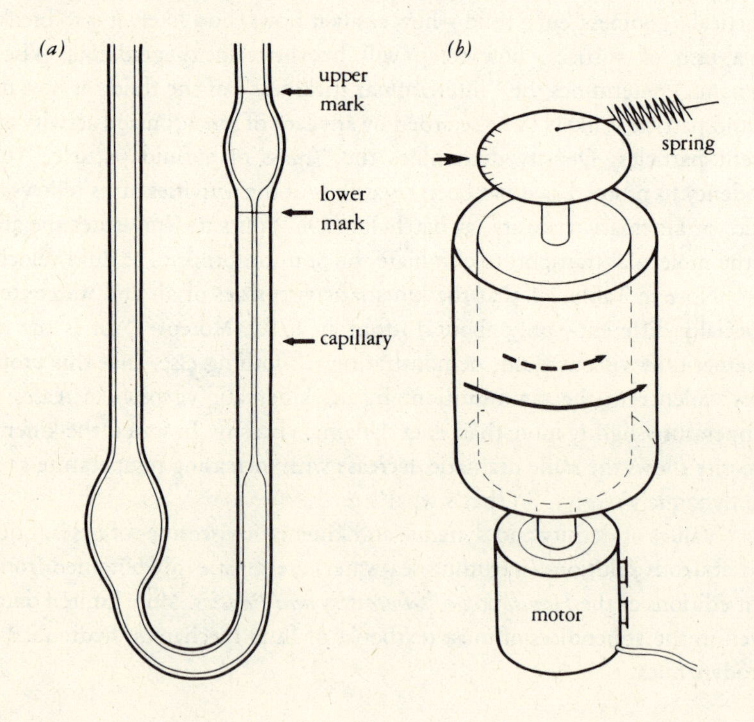

If an Ostwald viscometer is purchased from your favorite purveyor of scientific glassware, make sure it is appropriate for the expected kinematic viscosity; aqueous solutions run with dismaying rapidity through meters designed to test motor oils. The cheapest viscometers lack a glass connection across the top of the arms of the "U" and as a result are very fragile. One should provide some appropriate brace before use. Also, these glass devices should be kept scrupulously clean and should be rinsed with acetone or other solvent between uses.

For use, the viscometer is mounted in a constant temperature bath and a known (usually 5 cm^3) quantity of liquid introduced into the arm without the capillary. The liquid is then sucked up through the capillary tube until the top meniscus is above the upper reference line. The liquid is allowed to fall, and the time recorded for the top meniscus to drop from the upper to the lower reference line. Since the rate of flow is proportional to the density of the liquid and inversely proportional to the dynamic viscosity, the *time measurement is directly proportional to the kinematic viscosity*. The proportionality constant is obtained by timing the fall of a liquid of known kinematic viscosity. If dynamic viscosity is needed, a separate measurement of density needs to be made.

The biologist will rarely have occasion to measure the viscosity of a gas, and the Ostwald device won't work for gases. But an analogous arrangement can be set up; I'll leave the design as an exercise for the reader.

consequences of the inverse viscosity—temperature relationship

At 5 °C water is 2.5 times as kinematically viscous as it is at 35 °; there are organisms which experience both extremes of water temperatures. Does this variation in viscosity have any biological implications? I'd like to draw attention to the wide range of possibilities with the caveat that few cases can be considered well-established at this time.

The body temperatures of many terrestrial or amphibious animals vary seasonally, daily, or in response to changes in habitat or activity. At elevated temperatures less power should be required to maintain circulation if the viscosity of whole blood shows normal liquid behavior. Is this a fringe benefit of keeping body temperature elevated? The viscosity of human blood is nearly

50% higher at 20° than at 37 °C under some test conditions (Altman and Dittmer, 1971).[†] Are there compensating adjustments necessary in the circulatory systems of vertebrates which tolerate a wide range of temperatures? The red blood cells of cold-blooded vertebrates (and presumably their capillary diameters) are typically much larger than those of birds and mammals (Chien et al., 1971). The shear rate of blood is highest in the capillaries; must these be larger in order to permit circulation at adequate rates in a cold body? Or have some animals arranged for blood to behave like "multiviscosity" motor oils which resist excessive thickening when cold and thinning when hot? There is a marine iguana in the Galapagos which basks on warm rocks, heating rapidly, and then jumps into the cold Humboldt current to graze on algae, cooling only slowly. Circulatory adjustments as the animal takes the plunge have been postulated (Bartholomew and Lasiewski, 1965); is part of the circulatory reduction in cold water just a passive consequence of an increase in viscosity?

Guard and Murrish (1975) looked at the variation of blood viscosity with changes in temperature in Antarctic mammals and birds. They found that the apparent viscosity changes even more drastically in seals and penguins than in humans and ducks, and they suggest that the great increase in viscosity at low temperatures reduces heat loss by reducing blood flow in cold, poorly insulated extremities. Their results imply that the temperature coefficient of viscosity is a parameter over which some control may be exerted, either in an evolutionary or an immediate sense.

Do steeper velocity gradients at higher temperatures resulting from lower viscosity aid in gas exchange in fish gills and compensate in part for the drop in solubility of oxygen? Are the dimensions of filter-feeding devices adjusted intra- or inter-specifically to reflect the differences in kinematic viscosity of waters of differing temperatures? The latter seems reasonable according to the theoretical models of Rubenstein and Koehl (1977).

Planktonic organisms are quite commonly more dense than the water around them; unaided, they should sink more rapidly at higher temperatures according to Stokes' law (see Chapter 13), although as Hutchinson (1967) points out, passive changes in their own densities will partially offset the change. It is commonly reported (Sverdrup et al., 1942, for example) that

---

[†]Whole blood, as mentioned, is a rather non-Newtonian fluid. Its apparent kinematic viscosity drops with increasing shear rate (shear thinning), which aids its passage through small vessels. Thus the measured viscosity of blood depends on the conditions of measurement.

plankton from tropical waters are smaller and more angular and ramose than plankton from polar waters. Is this an adaptation to differences in viscosity?

From time to time the suggestion has been made that warm, swimming animals might release heat through the skin in such a way that kinematic viscosity is locally lowered and drag (see Chapter 5) is reduced by one of several possible mechanisms. Aleyev (1977) cites Parry (1949) to support a claim that the scheme is well-established for cetaceans; I find no support in Parry's paper except for some calculations that cetaceans produce a whale of a lot of heat and some evidence that a complex microcirculation in blubber can actively control the outward passage of heat. The latter is probably more reasonably viewed as a heat elimination system for use during and after high speed swimming (Palmer and Weddell, 1964). Aleyev also cites Walters (1962) who speculated on the possibility that tuna, whose muscle temperatures can reach 20 °C above ambient (Carey et al., 1971), locally release heat behind their so-called "corselet" to reduce drag. Walters is dubious, Webb (1975) is dubious, and so am I. Still, a study involving heated models in a flow tank might turn up something positive without excessive effort.

A long-standing suggestion is that the phenomenon of "cyclomorphosis" in the microcrustacean, *Daphnia*, is an adaptation to viscosity differences between warm and cool water. In warmer water, many species of *Daphnia* have larger heads and much longer and more curved crests or helmets (Figure 2.4). The difference is evident between generations raised at different

FIGURE 2.4

The cladoceran water flea, *Daphnia*, (*a*) in warm water form, and (*b*) in cold water form.

temperatures (''true'' cyclomorphosis), seasonally within a single species, as seasonal replacement of one group of species by another, and in comparisons of the fauna of climatically different ponds. The larger head and more flattened body should reduce sinking rate in less viscous water, but Hutchinson (1967), in a thorough discussion, finds substantial difficulties with this explanation. Other proposals, involving such factors as predation and nutrition, are likewise not fully satisfactory. Recently, Hebert (1978) has suggested a different hydrodynamic explanation. He points out that the propulsion of *Daphnia* will be less effective at high temperatures due to the reduction of kinematic viscosity, certainly a reasonable idea. In horizontal motion the poorer locomotion will be compensated by a lower drag, but in upward migration these denser-than-water animals will have to overcome an equal or greater sinking rate than they would experience in colder water. He suggests that the larger head should accommodate larger muscles moving the second antennae, the main propulsive organs, and permit a greater stroke amplitude. But again, the evidence is circumstantial and not conclusive.

CHAPTER 3

fluids can neither hide nor cross streamlines

WE ARE NOW READY TO EXAMINE what happens when fluids flow, beginning with the most universally applicable notions and then moving to increasingly specific phenomena. Expecially in this and the next few chapters, the reader should bear in mind the ultimate artificiality of a linear narrative. In particular, the examples may be nothing but the truth; but, for want of material to be developed further along, they are hardly the whole truth.

the principle of continuity

Consider a pipe open at both ends. If the pipe has rigid walls, it therefore must have a constant volume. With our assumption that fluids are incompressible, if a liter of fluid enters one end then a liter has to come out the other. This trivial idea, termed the ''principle of continuity'' turns out to be strangely potent in biological implications and has historically been less than obvious to practicing biologists.

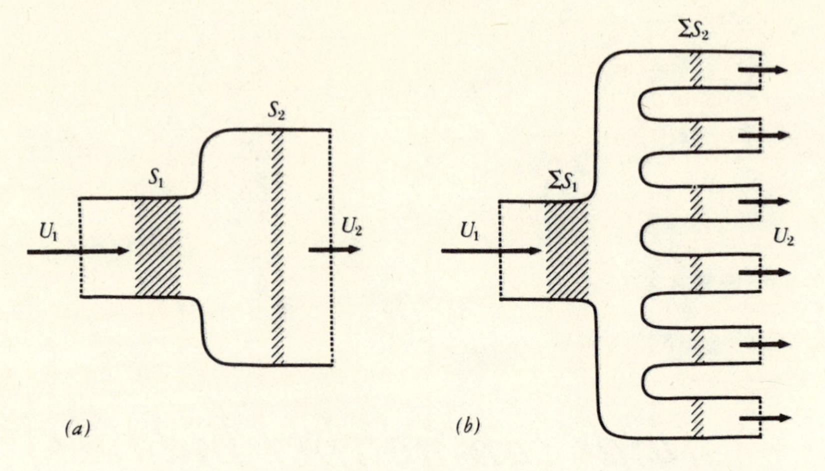

(*a*) An increase in the overall cross-sectional area of a pipe must be concomitant with a decrease in the speed of flow. Dividing the pipe into a parallel array (*b*) makes no difference. Local velocities are proportional to the length of the arrows and the width of the cross-hatched areas.

FIGURE 3.1

To put the matter more formally, consider a pipe whose cross-sectional area varies from one part to another (Figure 3.1a), with an area of S_1 near the entrance and S_2 near the exit. If a small volume of fluid, $S_1 dl_1$ enters the pipe in each incremental interval of time dt, then an equal volume, $S_2 dl_2$ must leave the pipe in these same increments of time. Thus

$$\frac{S_1 dl_1}{dt} = \frac{S_2 dl_2}{dt} \tag{3.1}$$

But any dl/dt is, of course a velocity, so

$$S_1 U_1 = S_2 U_2 \tag{3.2}$$

In short, then, the product of cross-sectional area and average velocity normal to the plane of that area is the same in either region of the pipe. No matter how the pipe expands, contracts, or changes shape, the product of cross-sectional area and velocity will remain constant: the volume flow rate or volume flux (called Q) does not change within a conduit.

Now consider a pipe which branches (Figure 3.1b). Again there is no hiding place for fluids inside, so for every volume which enters an equal volume must leave. We can, of course, look not just at entry and exit but any any cross-sectional plane; thus we can say that any one section of the array must have the same area–velocity product as any other section:

$$\Sigma S_1 U_1 = \Sigma S_2 U_2 \qquad (3.3)$$

Note that no assumptions were made about energy or about friction; the argument is purely geometric and is thus of very wide applicability. Indeed, were incompressibility not assumed, we would merely have to substitute ρS for S and mass flux for volume flux and the principle would still work. Continuity has the same role and the same generality for fluid mechanics that conservation of mass has for solids; it is, in fact, merely a special case of the idea of conservation of mass.

One garden-variety device based on continuity is the nozzle of the garden hose. By constricting an aperture, the fluid is made to speed up, and its increased momentum carries it a far greater distance than it would go without a nozzle. No energy need be put in at the nozzle; indeed it entails an additional loss of energy from the stream of water.

One way to illustrate the wide applicability of the principle of continuity is to describe a relatively unusual use. Consider some liquid gently flowing out from a downward-directed orifice and then falling freely (Figure 3.2). The

FIGURE 3.2

The column of fluid descending from a spigot contracts as it accelerates, as it must according to the principle of continuity.

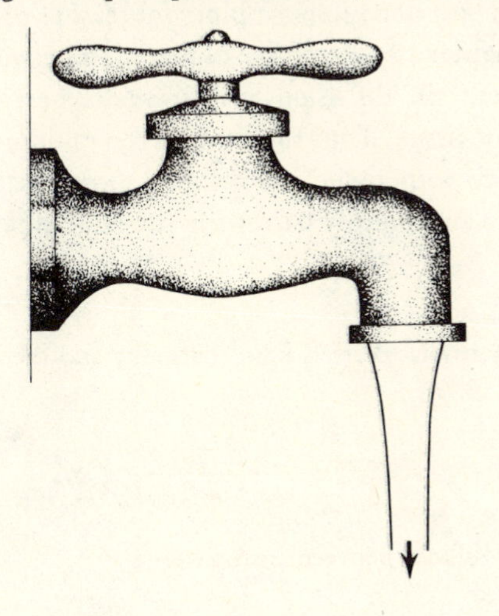

column of liquid contracts as it falls downward, the effect being most noticeable with a highly viscous liquid where the initial speed can be made low. It contracts from acceleration because its cross-sectional area must decrease in proportion to the increase in velocity. Knowing the acceleration of gravity, it is possible to figure the rate of discharge of the pipe with nothing more than two measurements of the width of the column and the vertical distance between them. Alternatively, given these measurements, a stopwatch, and container of known volume, one can make a fair estimation of the acceleration of gravity.†

Continuity has made large-scale mischief from time to time. The old London bridge (1209–1832) rested on a set of boat-shaped piers so wide that almost half of the river was blocked. As a result, the velocity between the piers should have doubled the already high tidal current. In fact, the situation was even worse because the cutwaters protecting the piers from scour reduced the waterway still further. Only small boats could pass, and these had to take careful aim and considerable risk in what was known as "shooting the bridge" (Gies, 1963).

moving fluids within organisms

Largish creatures whether plant or animal devote considerable anatomy to internal fluid transport systems. We give them various names—circulatory, respiratory, translocational—but they all work in much the same manner. Every such system must follow several partly conflicting physical imperatives. Long distance transport is best done in large pipes for reasons of energy economy (see Poiseuille's law, Chapter 10). Exchange of materials between fluid and tissue is ultimately dependent on diffusion, which works best over short distances within either fluid or tissue. Thus the fluid in the middle of a large pipe is not in diffusive intimacy with tissue, so without some small pipes its transport would be largely in vain. Simply narrowing the large pipes at sites of diffusive

---

†The relevant formula, derived from continuity and the basic equations of motion is:

$$S_2^2 = \frac{1}{1/S_1^2 + 2gh/Q^2}$$

where h is the distance between cross sections.

exchange would, according to continuity, result in very high speeds. Energy losses would be great, and little time would be available for diffusion to occur between tissue and any element of the fluid. In fact, all internal fluid transport systems, whatever their function, have both large (relatively) and small pipes. But fluid moves fastest in the large pipes and slowest in the smallest ones. There is no violation of the principle of continuity; rather it is just the general rule to make the total cross-sectional area of the small pipes very much larger than the cross-sectional area of the larger ones (Figure 3.3).

People. We provide perfectly ordinary examples. The output of the heart of a resting 70-kg human is about 6 liters per minute or 100 $cm^3\ s^{-1}$. Thus in an aorta of 3 cm^2 cross-sectional area the velocity is, on the average, 30 $cm\ s^{-1}$. A single capillary is about 6 μm in diameter or about $2.8 \times 10^{-7}\ cm^2$ in cross section. At rest, blood flows through at about 0.05 $cm\ s^{-1}$. The product of velocity and cross section for a capillary is some 6×10^9 (six billion) times less than the same product for the aorta. Ignoring shunts and other complexities, there must therefore be about six billion capillaries in parallel receiving the output of the aorta. Actually there are probably far more, since at rest only a minority of capillaries are open and operational. Even those open capillaries, though, represent a total cross-sectional area some 650 times greater than the area of the aorta (Burton, 1972; Altman and Dittmer, 1971).

FIGURE 3.3

An exceedingly diagrammatic view of the avian or mammalian circulation. Lower speeds in capillaries than in the heart are possible only if the total cross-sectional area of capillaries exceeds that of each half-a-heart.

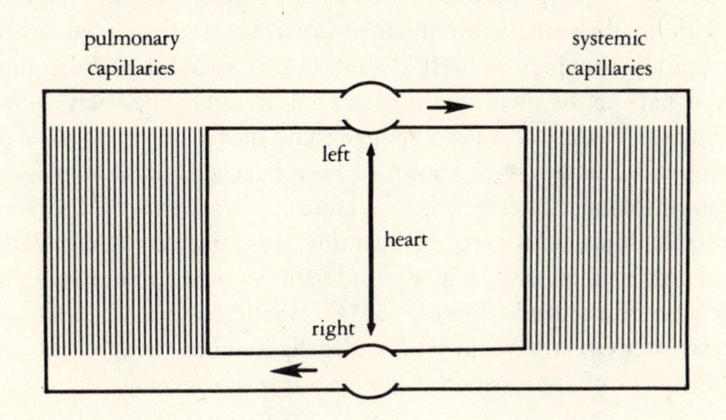

Trees. This kind of analysis may prove quite useful when one of the four variables in equation (3.3) cannot easily be measured. Such a case is provided by the system involved in conveying water from the roots to the leaves of trees. The problems of measurements in this system are formidable since even the largest conduits are only a fraction of a millimeter across and since negative pressures of several atmospheres are common. To illustrate how the principle of continuity can be used, I've taken some data from Lundegårdh (1966) on an inch-diameter oak tree (*Quercus robur*) with 300 to 400 leaves. I've combined this with stem sap velocities and transpiration rates (for *Q. rubra*) from Kramer and Koslowski (1960) to adress a question raised by Kramer (1959).

If the tree has 2 m^2 of leaf area (S_1) and transpires liquid water at 1.5×10^{-8} m^3 s^{-1} for each square meter of area, then the volumetric water loss (S_1U_1) is 3×10^{-8} m^3 s^{-1}. The trunk has a cross section of conductive xylem vessels of 1.5×10^{-4} m^2 (S_2); these 100-μm vessels make up about 7% of the total cross section of the trunk. Dividing S_1U_1 by S_2, we calculate a rate of ascent of sap in the vessels of 2×10^{-4} m s^{-1}, a fiftieth of a centimeter per second. But the rate of ascent of sap is not hard to measure; a pulse of heat is applied to the trunk, and the time is noted for the arrival of hot sap near a sensor a few centimeters higher. Such measurements give ascent rates of about 1.0 cm s^{-1}, fully 50 times higher than the calculated rate. Lundegårdh found such comparisons puzzling, although Kramer had earlier suggested that most of the xylem vessls might be filled with air and thus be non-functional. Application of continuity to some morphological and functional data permits one to estimate, non-destructively and under a variety of circumstances, the fraction of the xylem vessels conducting. The results of a systematic study might just bear upon the problem of repair of air embolisms in these vessels, a problem which has been troublesome ever since the negative pressures were recognized.

Sponges. Sponges are little more than highly elaborate manifolds of pipes, with small apertures covering their surfaces and one or a few large (commonly apical) openings as well. Grant (1825) established that flow was unidirectional—in the small apertures and out the larger ones—and he was impressed by the volume of water a sponge could move. But what was doing all the pumping? Sponges were known to have flagella, but spongologists persisted in invoking muscles which, unfortunately, persisted in remaining undetectable; it just wasn't credible that tiny, slow, flagella could be sufficient. The pumping *is* impressive; a sponge ordinarily pumps water equal to its own volume every five seconds (Reiswig, 1974). Finally Bowerbank (1864), almost forty years after Grant, recognized that flagella could do the job. The problem was really a lack of appreciation of the principle of continuity.

The situation is easily seen if it is put into quantitative terms. (I've mainly used morphological data from Reiswig, 1975a). A sponge of 100 cm^3 will ordinarily have an output opening 1 cm^2 in cross-sectional area and an output velocity of 20 cm s^{-1}. It is difficult to envision a flagellum only 25 μm long pumping water at more than 50 μm s^{-1}, a velocity 4000 times lower than the output velocity. But the total cross-sectional area of flagellated chambers proves to be nearly 6000 times greater than the area of the output opening (Figure 3.4). So one needn't make any heroic assumptions about the pumping speeds of flagella, much less make muscles mandatory. Interestingly, small sponges put a final constriction on their output apertures. Bidder (1923) recognized these as nozzles designed to increase the speed and coherence of the output jet and minimize the chance that the animal might refilter its own output. Table 3.1 summarizes some applications of continuity to living fluid transport systems.

Similar arrangements are evident in other filter-feeding animals such as bivalve mollusks and ascidians. In the latter, though, both intake and output apertures have low cross-sectional areas and consequently high velocities; they

FIGURE 3.4

Continuity and a typical sponge. (*a*) A flagellar pump is capable of generating high output flows by virtue of the large total cross-sectional area of the flagellated chambers. The numbers in the text should point out the difficulty of drawing the hydraulic system of a sponge with a consistent scale! A trio of the pumping and filtering cells, the choanocytes, are shown in (*b*).

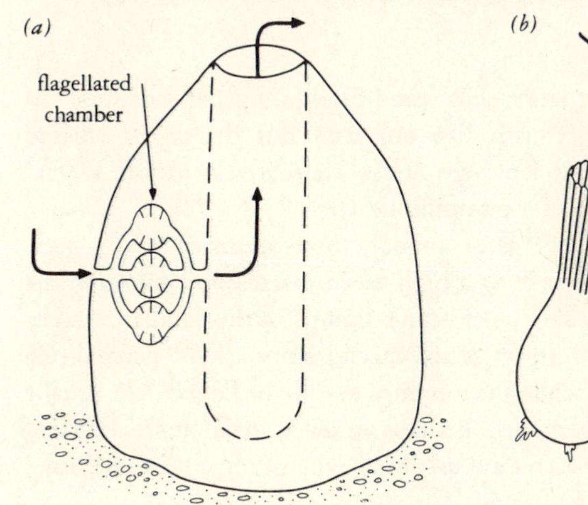

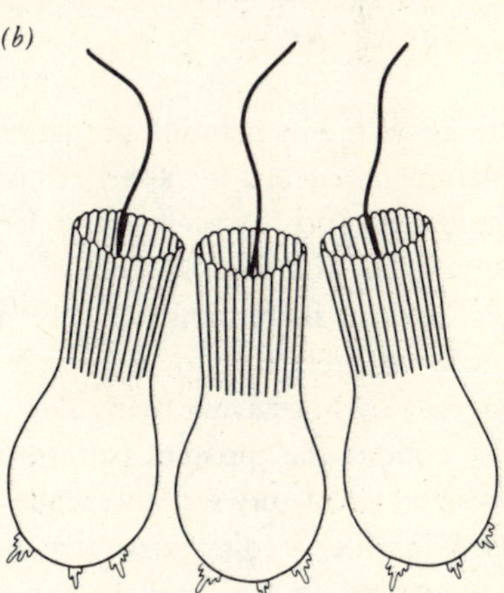

TABLE 3.1

Size and flow in fluid transport systems

	area (cm^2)	*number* –	*total area* (cm^2)	*velocity* ($cm\ s^{-1}$)
Small oak tree (1)				
Xylem vessels	0.000079	380	0.03	1.0
Evaporative surface	50.	(350 leaves)	20,000.	0.0000015
Dog circulation (2)				
Aorta	2.0	1	2.0	20.
Large artery	0.2	15	3.0	10.
Arteriole	0.00002	6,000,000	125.	0.75
Capillary	0.0000003	2,000,000,000	600.	0.07
Idealized human lung (3)				
Trachea	2.5	1	2.5	197.
5th bronch.	0.09	34	2.9	161.
10th bronch.	0.013	1000	13.3	38.
15th bronch.	0.0034	33,000	111.	4.4
20th bronch.	0.0016	1,000,000	1670.	0.3

Data derived from (1) Lundegårdh, 1966; (2) Caro et al., 1978; (3) Pedley, 1977.

are connected by manifolds to large, slow-speed filter pumps. Indeed, one can envision a circulatory system much like our own but driven by ciliated capillaries. Such a system could achieve quite substantial velocities in a great aorta in the complete absence of a pumping heart.

In the fluid transport systems of animals there seems to be no overwhelming preference for either having a high-speed macroscopic pump in the largest vessel or having many slow microscopic pumps in the smallest vessels. I've noticed one apparent pattern: muscular macropumps are the general rule in blood circulatory systems, while micropumps of cilia or flagella are usually used in filter-feeding arrangements. Perhaps muscles make more efficient pumps than do cilia, and the latter are used only when they serve some other role (such as filtration) as well. Some inveterate invertebrate sort might find the situation worth investigating.

Streams. At times the rate at which water is carried by a stream or river (the "discharge") increases substantially. A small stream may metamorphose into a raging torrent in which the depth and width obviously increase and the velocity seems far greater than normal. How much does the velocity really increase? If it were to increase in proportion to discharge, then, by the principle of continuity, the river shouldn't rise! Clearly velocity doesn't increase as fast as discharge; in fact, the increase in velocity turns out to be quite modest. From the graphs presented by Dury (1969) or Leopold and Maddock (1953), it seems that mean current speed at a particular station along a river only doubles for every tenfold increase in discharge. Since peak velocity in a larger cross section will occur further from bottom and sides of a channel, the actual velocities near the interfaces should increase even less. Being swept downstream by high water speeds of floods may just not be as worrisome to small attached or bottom-living organisms as we might otherwise think! As the discussion in Hynes (1970, pp. 224–229) implies, abrasion and alternations of the form of the bottom are more important than velocity *per se*.

streamlines

So far the discussion of the principle of continuity has been in the context of either flow through pipes or flow in discrete, homogeneous streams. Can we apply the concept to open fields of flow or, more generally, to cases in which the velocity of flow varies *across* the field of flow? It turns out that the principle is, if anything, even more potent in such situations. We're less likely to use it intuitively, and the results of its application are less self-evident. One additional device is needed, a curiously abstract yet practical one. *Streamlines* have a very special meaning in fluid mechanics, a meaning which bears only an indirect connection with the common use and wide abuse of the term.

A streamline is a line to which the local direction of fluid movement is everywhere tangent. Loosely, then, a streamline traces a path through a field of flow along which some particles of fluid travel. Let us assume we know the direction of flow at all points in a flow field at some instant in time. To create a streamline, we start at some upstream point, draw a very short line in the direction of flow, consider the direction of flow at the far end of that line, and extend the line a short distance in the new direction. We continue the process until the line wends its way across the entire field (Figure 3.5). The procedure, however trivial, is rather awkward. But what have we gotten? The line follows the stream. If the direction of flow is along the line, then the velocity component normal to the line must be zero; in short, fluid does not cross the line.

It is no more difficult to create a second streamline, running in tandem with the first. For a simple, two-dimensional flow (no convergence or divergence outside the plane of the paper), *the principle of continuity must apply between the pair of streamlines.* If, in three-dimensional flow, we make a set of streamlines which surround some fluid, we have a so-called "stream tube" within which continuity is again applicable. Streamlines, then, provide a conceptual device for dividing a complex field of flow into an array of pipes with non-material walls. Viscous and other effects are transmitted across the walls just as heat passes through the walls of material pipes, but fluid (i.e., matter) does not pass. Where a pair of streamlines diverges or a stream tube becomes wider we know immediately that the fluid is traveling more slowly. If streamlines converge, that is a definitive indication that velocity is increasing.

pathlines and streaklines

But how to make streamlines in the real world? The direction of flow at all points is rarely a matter of public record. Two fundamentally different schemes can be used. In the first, a visible marker or particle of some sort is released near the upstream end of the flow in question. Ideally, the marker is neutrally buoyant and very small, so it always travels in the local flow direction. A time exposure or repetitive photograph of its travel gives a solid or dotted line recording the history of the marker; it's called a *pathline* or particle path (Figure 3.6). Alternatively, a continuous stream of particles, dye, or smoke may be introduced at a fixed point; some time after the start of injection of the marker an instantaneous photograph is taken. It gives the present position of fluid which has, over a period of time, passed by the injection point and is called a *streakline* or filament line. Most often, markers are introduced from an array of points normal to the flow direction, so a set of streaklines is simultaneously produced. (Figure 3.7).

If the flow is steady, that is, if velocity does not vary with time, then pathlines and streaklines coincide, and both mark the streamlines. If flow is unsteady, neither, strictly speaking, marks streamlines, and everything

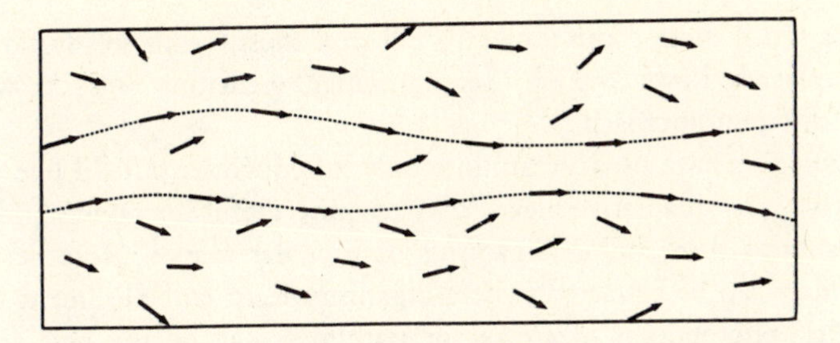

Streamlines (dotted) drawn by following the local direction of flow across a flow field.

FIGURE 3.5

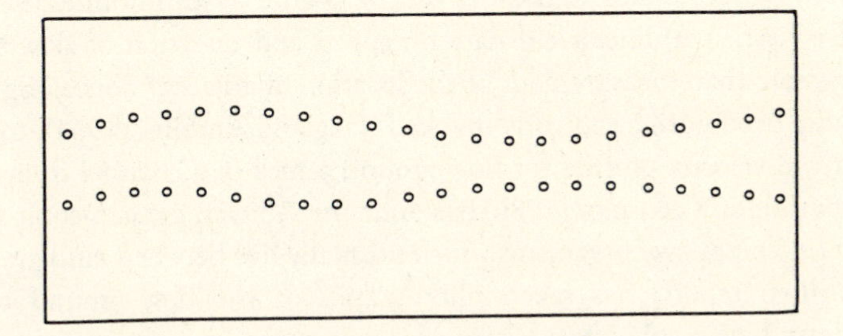

Pathlines appear as a result of a repetitive photograph of a pair of particles moving across the field of view.

FIGURE 3.6

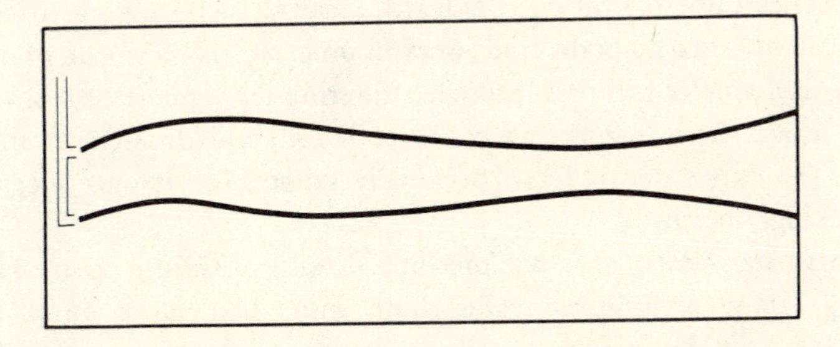

Streaklines result from an instantaneous photograph of a stream of dye or smoke which has been released over a period of time in the immediate past.

FIGURE 3.7

becomes much more complicated. We'll deal mostly with steady flows and, where unsteady flows crop up (beating wings, oscillating tails) we will take quite a different approach.

Biologists have used streamlines only infrequently, but I'd like to stress their utility. As qualitative views, they're quite useful for understanding the complex flows surrounding organisms of irregular shapes. Moreover, quantitative data can be extracted; where organisms resist embellishment with instruments, photographs of streak- or pathlines may be the best and least abusive approach. And streaklines or pathlines provide a vast quantity of data for a small investment of time by an often unwilling subject. Thus Kokshaysky (1979) has made nice photographs of the vortex rings behind a flying bird by inducing the bird to fly through a cloud of wood particles while these were photographed with a repetitive stroboscope.

Both pathlines and streaklines can be useful. With illumination from a repetitive flash, pathlines yield data on speed and direction of flow through nothing more than measurement of the location of adjacent dots along a path. Continuity need not be explicitly invoked. Jaag and Ambühl (1964), for example, derived velocity profiles for flow around a rock in a channel from a set of dotted pathlines. Crenshaw (1980) has done the same on a microscopic scale for flows in fine pipes over organisms which normally live between sand grains. Lee (unpublished report) has taken photographs of the flow around tethered echinoderm larvae only a few tenths of a millimeter long; from these, she derived local velocities. There are many possibilities.

Streaklines permit one to look at flows through filtration devices, and a single streakline can be manipulated as the investigator watches, without waiting for a *post facto* photo. Very fine streaklines are possible if dye is injected from a micrometer-driven syringe through a piece of drawn out polyethylene catheter tubing. Lidgard, using encrusting bryozoa, and LaBarbera, with brachiopods, find (work in progress) that a stream of dye abut 0.1 mm in diameter will pass through a filtering lophophore of tentacles and emerge intact. It thus shows an entire path for fluid through an animal or colony. The experiments, I can personally report, are at once elegant and breathtakingly beautiful.

Small size clearly does not preclude making pathlines or streaklines; if anything, these techniques really come into their glory on a coarsely microscopic scale. Little else is usually available, and the absence of turbulence actually simplifies the technical problems. Figure 3.8 gives a view of pathlines around a model of a fruit-fly wing immersed in a moving liquid. Appendix 2 suggests techniques for flow visualization.

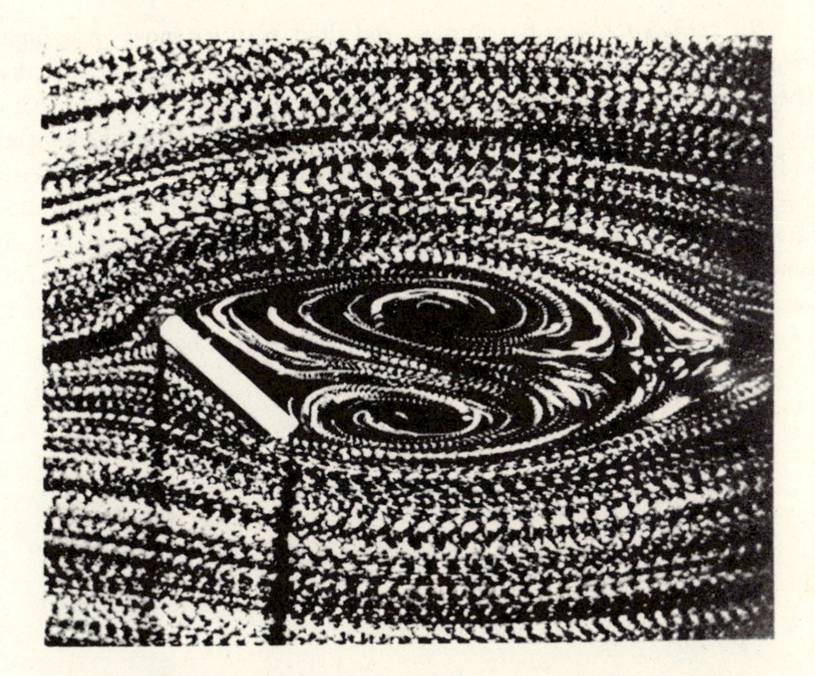

Pathlines around an inclined flat plate immersed in a rotating bowl of water in which particles have been suspended (Vogel and Feder, 1966). The plate is 8.5 mm across, and the rate of flashing of the stroboscope is 12 s^{-1}.

FIGURE 3.8

laminar and turbulent flow

Having mentioned turbulence, I should no longer avoid the matter of the distinction between laminar and turbulent flow. Another of the strange complications in the behavior of fluids is the existence of these two radically different regimes of flow. In introducing viscosity, another tacit assumption was made in addition to that of the no-slip condition. We assumed that the "layers" of fluid slipped smoothly across one another, that the particles of fluid moved in an orderly, unidirectional fashion differing only in their speeds. This situation, in which all fluid particles move more or less parallel to each other in smooth paths, is called *laminar flow*. In it, the large- and small-scale movements of the fluid are the same.

In *turbulent flow,* by contrast, the fluid particles move in a highly irregular manner even if the fluid as a whole is traveling in a single direction. There are intense small-scale motions present in directions other than that of the main large-scale flow. Turbulence is essentially a statistical phenomenon and descriptions of overall motion in turbulent flow must not be taken as describing the paths of individual particles. The easiest analogy is to diffusion, in which Fick's law works admirably for the overall phenomenon but says almost nothing abut what any molecule is likely to be doing at any particular instant. The distinction between these two regimes of flow has been known for a long time, as has the abrupt character of the transition between them. One

FIGURE 3.9

The behavior of a stream of dye in laminar and turbulent flow. The apparatus is not entirely impractical. Flow in the upper pipe will be slower than that in the lower pipe, and the rates can be adjusted by changing the height of water in the tank so that the upper stream is laminar and the bottom one turbulent. Under some conditions, though (see Chapter 10), flow through the upper pipe will, paradoxically, be more rapid.

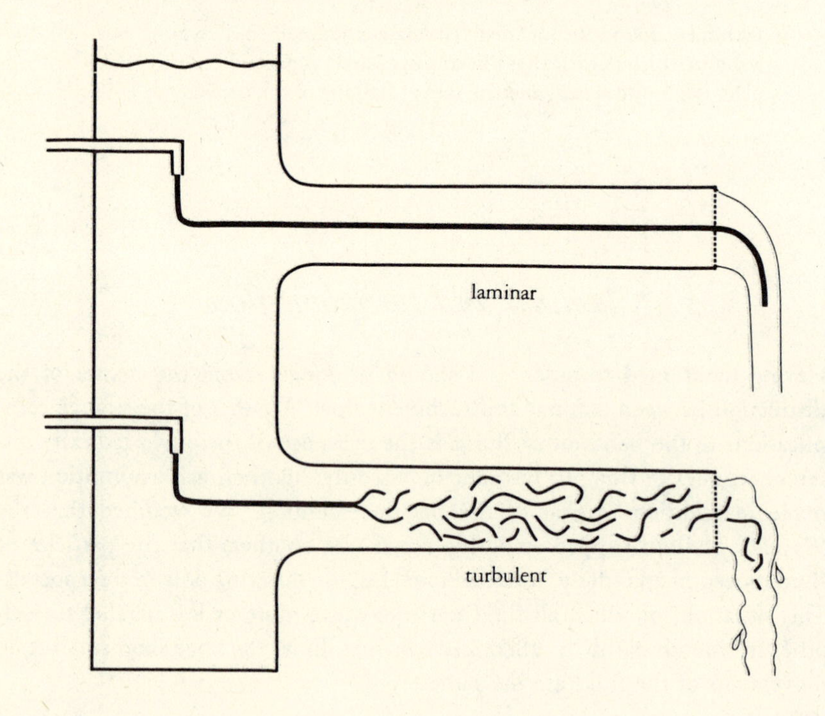

can do no better than to quote the description of Osborne Reynolds (1883), who introduced a filament of dye into a tube of flowing water and watched how it behaved as the current was altered (Figure 3.9):

> When the velocities were sufficiently low, the streak of colour extended in a beautiful straight line across the tube. If the water in the tank had not quite settled to rest, at sufficiently low velocities, the streak would shift about the tube, but there was no appearance of sinuosity. As the velocity was increased by small stages, at some point in the tube, always at a considerable distance from the trumpet or intake, the colour band would all at once mix up with the surrounding water. Any increase in the velocity caused the point of break-down to approach the trumpet, but with no velocities that were tried did it reach this. On viewing the tube by the light of an electric spark, the mass of colour resolved itself into a mass of more or less distinct curls showing eddies.

While engineers are almost exclusively concerned with turbulent flows, both sorts of flows are of biological interest. In general, small, slow organisms or tiny pipes experience laminar flow and large, fast organisms or large pipes, turbulent flow. Reynolds is best known for having deduced the basic rule governing the transition point (see Chapter 5). At this point, I'm mainly concerned that the reader recognize (1) that there are two such regimes, (2) that the transition may be abrupt, (3) that the practical formulas for dealing with the two are quite different, and (4) that much of biological consequence occurs near the transition point where our *a priori* uncertainties are maximal and where nature has unusually rich opportunities to surprise us.

CHAPTER 4

energy and momentum: more conservation laws

NOTHING HAS BEEN SAID ABOUT energy, and someone may have wondered when we would get around to it. Energy is, of course, a very peculiar stuff. It's usually defined only by what it can do—"the ability to do work"—really quite a monstrous evasion. So let us honestly admit to treating energy as an imaginary device to simplify the task of describing the operation of the real world. It permits us to express a conservation law, and, of the latter, the more the better.

Bernoulli's principle

We demonstrated the principle of continuity with the use of geometry alone and obtained a concept at once simple, general, and underappreciated, with no exceptions lurking about to trap the unwary. I now want to introduce a principle of quite opposite proclivity, one involving such a disquieting assumption that you might well wonder if it is ever useful. Oddly enough, this is the only principle of fluid dynamics that just about everyone has met—the principle (or equation) of Bernoulli. (Daniel Bernoulli, 1700–1782, was one of several eminences of the same surname and disputatious family.)

Bernoulli's principle is very nearly just a simple translation of the idea of conservation of energy into terms useful for moving fluids. The basic trouble is that the principle works only for so-called "ideal" fluids. The ideal fluid is a really bizarre notion: it is a fluid of zero viscosity, the property we used to recognize fluids in the first place. Lacking viscosity, it slips easily along surfaces and doesn't care about shear rate, but it consequently flows without frictional losses of energy. No energy is lost, either, to the walls of the container, to any immersed object, or to internal heating of the fluid.

For the energy balance sheet of a moving, isothermal, incompressible, continuous, inviscid fluid, what are the relevant terms? Three ought to be considered. First, kinetic energy, since the fluid is moving, $mU^2/2$. Second, potential energy in the form of internal pressure, pressure times volume or pV. And often we need to include gravitational potential energy, mass times height times gravitational acceleration, mgh. As an element of fluid flows along, say, in a magically frictionless pipe, the sum of these terms should, by conservation of energy, remain constant, so,

$$\tfrac{1}{2}mU^2 + pV + mgh = C \qquad (4.1)$$

Mass is awkward when dealing with continua, and incompressibility has been assumed, so let us divide terms by m and multiply by ρ:

$$\tfrac{1}{2}\rho U^2 + p + \rho gh = C' \qquad (4.2)$$

Every term now has dimensions of pressure, an easily measured quantity, which gives rise to the usual textbook drawing (Figure 4.1). I've omitted the customary manometer tubes since they end up only confusing matters; I intend the pressure gauges just as black boxes. At each of the three stations along the pipe we're measuring not C' but the second term, p. An increase in either the first term (here achieved with the aid of continuity) or the third (hold the book upright) must be reflected in a decrease in the second.

The constants in the previous equations are a minor nuisance. One way to circumvent them is to consider the changes in the relative magnitude of each term as the fluid moves from an upstream point (subscript 1) to a downstream point (subscript 2). Bernoulli's equation then becomes

$$\tfrac{1}{2}\rho(U_2^2 - U_1^2) + (p_2 - p_1) + \rho g(h_2 - h_1) = 0 \qquad (4.3)$$

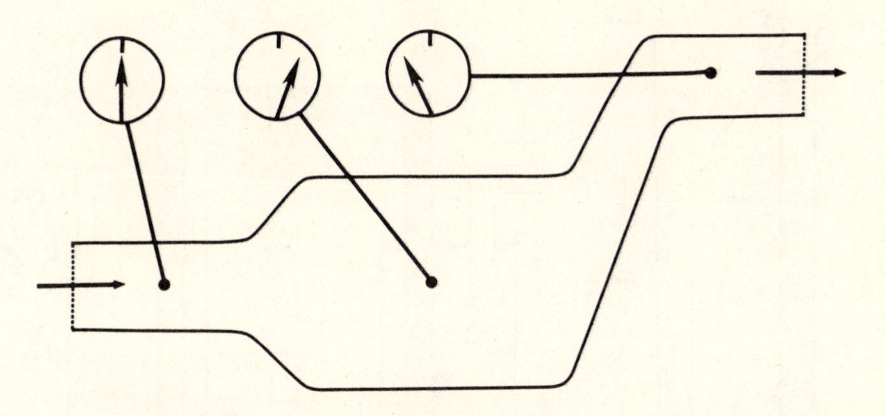

Bernoulli's principle: An increase in the cross-sectional area of a pipe causes a decrease in velocity and therefore an increase in pressure (center). An increase in elevation of the pipe (right) causes a decrease in pressure.

FIGURE 4.1

Frequently the gravitational term can be omitted: even if it's relevant one can often concoct some simplified situation in which the fluid has no net upward or downward motion, so,

$$\tfrac{1}{2}\rho(U_2^2 - U_1^2) + (p_2 - p_1) = 0 \tag{4.4}$$

manometry

To illustrate how Bernoulli's principle can operate, let's consider some useful devices based upon it, machines which can be used, among other things, to measure the speed of flow of a fluid. As a start, assume a fluid at rest. Therefore the first term in equation (4.3) is zero, and

$$\Delta p = \rho g \Delta h \tag{4.5}$$

This equation turns out to be the basic formula for manometry: a difference in height is used as a measure of difference in pressure, as in Figure 4.2a. The equation is also useful to get "real" units of pressure (SI) from quaint quantities such as inches of water or millimeters of mercury.

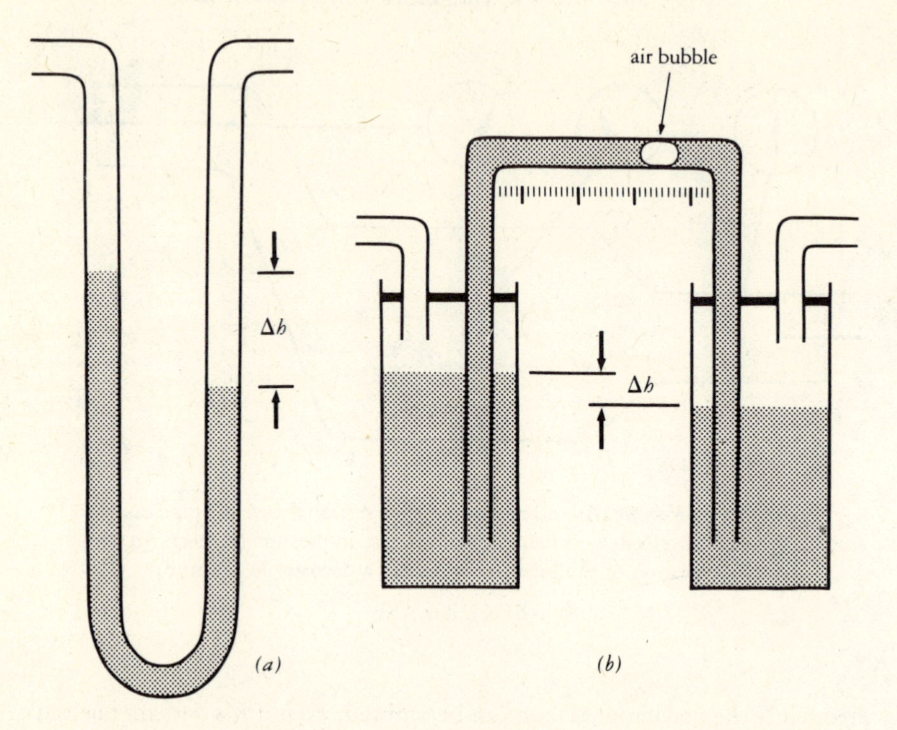

(*a*) A U-tube manometer: the liquid in the lower arm must make a sharp interface with the ambient fluid in the upper portion and elsewhere. (*b*) A multiplier manometer, in which a small change in the heights of liquid in the two jars causes a very large movement of the air bubble.

FIGURE 4.2

As we'll see, the pressure differences involved in low-speed flows are miniscule compared to very ordinary hydrostatic pressures (which is how we escape difficulties with the assumption of incompressibility). The ordinary mercury manometer appropriate to warn of hurricane or high blood pressure proves hopelessly inadequate for the pressure differences resulting from small velocity differences. Equation (4.5), though, suggests two ways of achieving high sensitivity. A liquid of low density may be used in the manometer, such as water, xylene, isooctane, or acetone. Or some way of reading very small height differences may be devised. Both have been used in fluid mechanics, and the technology behind inclined-tube manometers, Chattock gauges, and their ilk is discussed in older sources such as Prandtl and Tietjens (1934) and Pankhurst and Holder (1952).

Direct manometry shouldn't be dismissed as anachronistic. For an investment of ten dollars, the device shown in Figure 4.2b can provide a differential sensitivity of a millionth of an atmosphere (0.1 Pa) under field conditions. With acetone as a low viscosity manometer fluid, equilibration takes only five to ten seconds. Using half-pint vacuum bottles and glass tubing of about 3 mm internal diameter, motion of the air–acetone interfaces is multiplied about 200 times. Thus a 1-cm travel of the top bubble marks a 50-μm change in the height difference between the two bottles of acetone. From equation (4.5), that change corresponds to a pressure difference of 0.386 Pa. The arrangement, though, works only for air, not for water.

For measuring pressures in water, one can take advantage of the fact that water-immiscible liquids can be made with densities quite close to that of water. In manometry, it isn't really the density that matters (equation 4.5 is slightly misleading) but rather the difference in density between ambient fluid and manometer fluid. It's just that when the ambient fluid is a gas and the manometer is filled with liquid we can safely ignore the density of the former. In practice, though, with simple liquid manometers, rather wide-bore glass tubes must be used to keep interfacial effects within tolerable limits, so again high sensitivity to pressure changes is traded against a requirement for large volume changes. Incidentally, these two-liquid manometers may be inverted if it proves convenient to use a manometer fluid of lower density than that of the ambient water.

anemometry

Just as measurement of a change in height can give a measure of change in pressure, in the same way a measurement of pressure can provide a measure of velocity. And velocity is something we very much need a way to measure. If the difference in height between upstream and downstream points is negligible, then equation (4.4) becomes the following:

$$\Delta p = \frac{\rho}{2}(U_1^2 - U_2^2) \qquad (4.6)$$

Two practical problems should be immediately evident. We're not measuring velocity directly but the *difference* between velocities. And the pressure change is proportional to the *squares* of velocities. The latter will hurt sensitivity at low speeds, which is the main reason for carrying on at length about pressure-measuring devices of very high sensitivity.

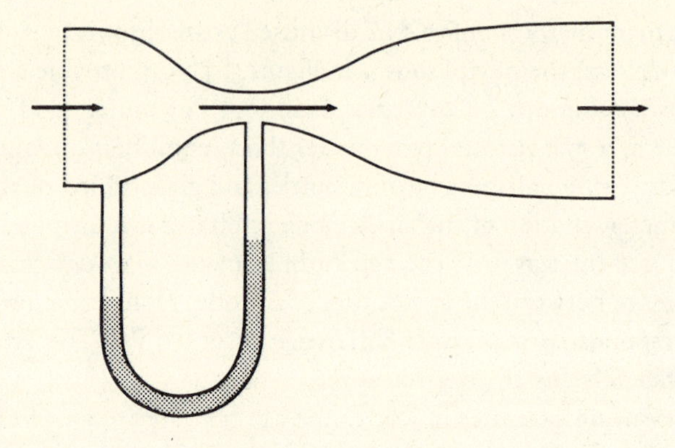

A Venturi meter, in which a contraction speeds the flow, lowers the pressure, and permits the velocity or volume flow rate to be determined.

FIGURE 4.3

Venturi meters. In one very useful scheme, the principle of continuity is used to circumvent (partly) the limitation of equation (4.6) to a differential measurement. It's called a "Venturi meter" and consists of a predetermined contraction in a pipe which locally speeds the fluid (Figure 4.3). If U_1 is the item of interest, and S_1 and S_2 are known, then equations (4.3) and (4.6) may be combined to eliminate U_2 and give

$$\Delta p = \frac{\rho U_1^2}{2}\left(1 - \frac{S_1^2}{S_2^2}\right) \tag{4.7}$$

A Venturi meter is worth remembering as a cheap and dirty substitute for an expensive in-line volume flow meter. There are, you'll note, no arbitrary constants in equation (4.7) to complicate calibration. As we'll see, some biological schemes are rather closely analogous to Venturi meters. Laboratory aspirators and automobile carburetors work on the same principle as well.

Pitot tubes. Most often the biologist is interested in measuring the current speed at a single point rather than across some closed conduit; for this latter purpose the Venturi meter is nearly useless. Yet in practical terms, the Bernoulli equation requires some reference or comparison between the pressures at two points. The first and crucial term in equation (4.2) may have dimensions of pressure, but some trick is needed to convert it to a physical pressure. That trick consists of suddenly reducing the current speed to zero in such a way that the energy released is all immediately reinvested as pressure. The pressure that results from stopping a fluid is called the "dynamic pressure" to distinguish it from the pressure of the undisturbed fluid, or "static pressure." Only the former need be considered at a particular point; ordinarily the static pressure varies little in a direction normal to that of the overall flow. The device for suddenly bringing a moving fluid to a halt is called a "Pitot tube" (or Pitot-static tube); to it applies a minor varient of the Bernoulli equation,

$$\frac{\rho U^2}{2} + p = H \qquad (4.8)$$

where H has the amusing name of "total head" and represents the sum of the dynamic and static pressures, the two terms on the left side of the equation.

How does a Pitot tube (Figure 4.4) work? The aperture facing upstream is carefully designed so that it samples fluid suddenly brought to rest and is exposed to both dynamic and static pressures, thus to the total head. The "static hole" (or holes), whether adjacent or remote, is exposed only to the static

FIGURE 4.4

A Pitot-static tube, in longitudinal section and somewhat fatter than life. One aperture faces upstream while a ring of small "static" holes are parallel to the flow. If, as here, the static holes are on the probe itself, the device is called a Pitot-static tube; if a remote static hole is used, the device is simply termed a Pitot tube.

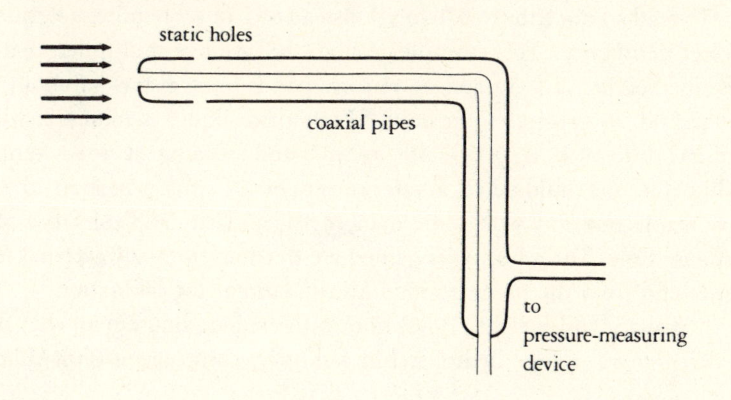

pressure. A differential pressure-indicating device such as a manometer between the upstream-facing and static holes therefore measures $(H - p)$, or the dynamic pressure. Recall our general formula for manometry, equation (4.5), now written with densities for both ambient (ρ_a) and manometric (ρ_m) fluids,

$$\Delta p = g\Delta h \, (\rho_m - \rho_a) \tag{4.9}$$

This may be combined with a rearrangement of equation (4.8),

$$\Delta p = H - p = \frac{\rho U^2}{2} \tag{4.10}$$

and we get a practical formula for using Pitot tube with an attached manometer:

$$U^2 = \frac{2g\Delta h(\rho_m - \rho_a)}{\rho_a} \tag{4.11}$$

When used in air, the manometer fluid is much more dense than the ambient fluid, so the ρ_a in the numerator may be omitted.

The practical version, the Pitot-static tube, is inexpensive and rugged. Much effort has gone into the design of the upstream end to make it minimally sensitive to minor misalignments with respect to the current direction and to determine the best size and location for the static holes. Nevertheless, for the biologist interested in low-speed flows, it has a serious problem of insensitivity; again the *square* of velocity in the Bernoulli equation rears its anterior extremity. To read a Pitot tube to 0.1 m s^{-1} instead of 1 m s^{-1} requires a hundredfold-greater sensitivity. For example, using the acetone-multiplier manometer described earlier, a 1-cm deflection corresponds to an airflow of about a meter per second, or two miles per hour. Using a two-liquid manometer where the density difference is 0.1×10^3 kg m^{-3} and reading it to a tenth of a millimeter, one could detect a water current of 1.4 cm s^{-1}. In short, the scheme is workable but only with some manometric audacity and tolerance of a slow response time. The advantages gained are freedom from any external source of power and from the necessity for a known current for calibration.

Wallace and Merritt (1980) note with evident amusement that the larva of *Macronema,* a lotic hydropsychid caddisfly, constructs a Pitot tube in the middle of which it spins a catch-net. One opening faces upstream and is exposed to dynamic pressure less some loss due to flow through the structure. The other opening is normal to the current and must experience static pressure. Many bivalve mollusks and some ascidians appear to do likewise.

limitations

At the start of the chapter I made some derogatory remarks about Bernoulli's principle; the reader may be wondering whether they were gratuitous. Let's return to that assumption of zero viscosity. Energy must always be conserved in the cosmic view, and in fluids viscous effects provide a major route of energy transduction; their neglect poses perils. Bernoulli's principle should be applied only along a streamline or within a streamtube, and even there only when times are short or shear rates low. It is distinctly unsafe to use the principle for points along some line normal to the direction of flow, particularly in the velocity gradients near walls. As fluid velocity drops off near a wall due to the no-slip condition, static pressure, not total head, remains constant. Speed may drop, but energy is converted to heat in the shear rather than being converted to pressure potential.

I make this point about the non-constancy of total head with some passion, since I once narrowly escaped misattributing a set of interfacial phenomena entirely to Bernoulli's principle. By the way, the heat from shear is quite real. Several years ago I encountered an annoying problem of drift of the null point of a flowmeter in a flow tank. It turned out (after false leads were chased) that about one horsepower was being inserted into the water by a pump and was warming the water by several degrees per hour.

As we'll see in the next chapter, the direct application of Bernoulli's principle (or its three-dimensional equivalent, the equations of Euler) gives dismally unreal results for the way fluids flow around the rear of almost any object at any speed. Pitot tubes, fortunately, face upstream. In general, the more important the influence of viscosity, the less strictly applicable the principle. Bernoulli's principle does have its uses, as the examples which follow should point out. But in each of these, you should be aware that the explanation is either quantitatively incomplete or else requires some degree of qualification.

CHAPTER 4

applications

Where the fish eye is at. A swimming fish bears a distinct resemblance to a Pitot tube, especially if its mouth is open. On the front of the fish, fluid is rapidly brought to rest, so the fish experiences a pressure greater than ambient. At its point of maximum width, the fish experiences a pressure below ambient; in getting around the body, streamlines are crowded, fluid velocity must therefore increase, and, by Bernoulli's principle, pressure must decrease. Pressures at these locations have been measured by implanting flexible tubing within fish swimming in a flume (DuBois et al., 1974). The pressures turn out to be about what one would calculate, around 9000 Pa or 0.1 atmosphere for a 0.6 m bluefish at its top speed of 5 m s^{-1}. For a fish which swims near the surface this is of the same magnitude as the hydrostatic forces to which it is subjected whether swimming or stationary. These pressures, then, might subject a fish to some distortion of shape merely as a result of its swimming. Gill ventilation will be augmented (the so-called "ram ventilation" of fish gills is well known) because water enters at the high pressure location and leaves near the low. More interesting, though, is the suggestion that the eye of an ordinary bony fish is located at the point where the pressure changes caused by swimming are least (DuBois et al., 1975). This arrangement should keep the focus of the eye minimally sensitive to speed-dependent pressure changes.

How the plaice stays in place. Plaice (*Pleuronectes platessa*) are bottom-living flatfish similar to flounder. At rest they constitute low, rounded humps on smooth sandy bottoms. As elevations, they speed the flow over themselves and, again according to Bernoulli's principle, they experience lift. Despite its fine low-drag shape, a quiescent plaice has some tendency to slip downstream. This tendency is offset by its friction with the bottom and its submerged weight; it is an especially dense fish. But the weight of a plaice is reduced by its lift when currents are substantial, and the lift force is ten to twenty times the drag force. So the plaice slips only in part because of drag but as much because the lift reduces effective purchase on the bottom (Figure 4.5a). In practice it remains quiescent in currents up to a "slip speed" and beats it posterior marginal fins in stronger currents up to a "lift-off speed" at which its net weight is zero, and it must either dig in or take off. All this and more is described, modelled, and calculated in an engaging paper by Arnold and Weihs (1978).

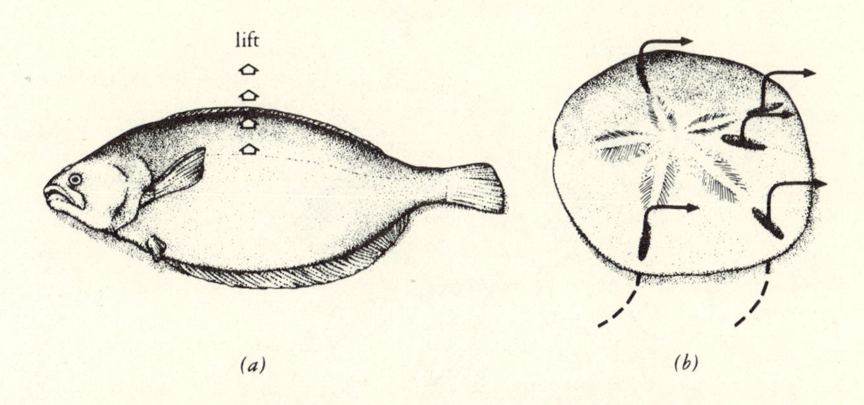

(*a*) A plaice, lying on sand and exposed to flow, develops lift as a result of the operation of Bernoulli's principle. (*b*) The same phenomenon draws water up through the holes in a perforate sand dollar.

FIGURE 4.5

A slotted sand dollar. A sand dollar resting on the substratum presents a hump similar to that of a plaice. As a result (although no measurements seem to exist) it should also experience lift and have critical speeds for slip and lift-off. At least one species, *Mellita quinquiesperforata,* has a set of five radially-oriented slots which extend from lower (oral) to upper (aboral) surface. In a current, water is drawn up through the slots by the reduced pressure on the top (Figure 4.5b). This upward water current may reduce the lift; it certainly augments the feeding activities of the animals. If slots are plugged, stomachs remain empty. In a sand dollar, the anus, curiously enough, is on the underside. But in this species, one slot is adjacent to the anus; it appears that Bernoulli does his part at that end of the digestive system as well (Alexander and Ghiold, 1980).

The beetle's bubble. Quite a different application of the principle was reported by Stride (1955) for an elmid beetle (*Potomodytes tuberosus*), the adults of which inhabit rapidly-flowing streams in Ghana. These beetles congregate on submerged rock surfaces within a few centimeters of the surface of the water. Most of each beetle is enclosed in a large air bubble supported by the thorax and forelimbs and extending a centimeter or two beyond the end of the

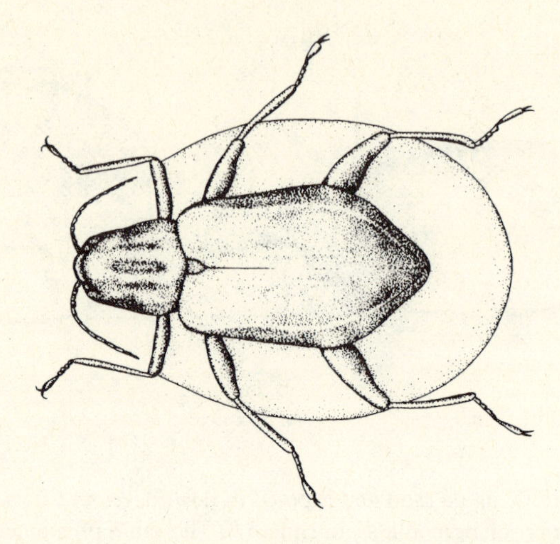

A permanent bubble surrounds much of a submerged elmid beetle; its permanence requires flow as well as a low hydrostatic pressure and well-aerated water.

FIGURE 4.6

abdomen (Figure 4.6). The bubble appears to act as a physical gill, but it is permanent only in moving, air-saturated water, to which these beetles are restricted. In the laboratory, a beetle can maintain a bubble of normal size only when it is near the surface in flowing, well-aerated water; otherwise the bubble gradually shrinks, and the beetle dies. What happens is that the pressure reduction caused by the flow around the bubble is sufficient to reduce its internal air pressure to or below that of the atmosphere, so gas doesn't diffuse out into the water. Stride showed that air pressures in these bubbles were subatmospheric at depths of one centimeter and currents over about $0.8\ \mathrm{m\ s^{-1}}$. This curious adaptation is possible, of course, only near the surface: the greater the depth, the greater must be the current in order to counterbalance the hydrostatic pressure.

Potomodytes may not be the only insect which extracts air from moving water with the aid of Daniel Bernoulli. Some pyralid lepidoptera pupate in air-filled cocoons in swift streams; water can pass beneath the air within the cocoon (*Aulacodes,* Nielsen, 1950; *Paragyractis,* Lange, 1956). Also, some aquatic beetles pupate in gas-filled cocoons (Leech and Chandler, 1956).

Aspirating pits. The movement of sap in trees has already been mentioned. In conifers, the main conduits in the trunk are not wide vessels but shorter and narrower tracheids, and the velocities are much lower than those in oaks and other hardwoods. Tracheids are interconnected via pits, of which one common type is the so-called "bordered pit". In this structure the normally fused cell walls of a pair of tracheids arch apart and provide an opening between the tracheids through an atrium (Figure 4.7). Traversing the center of the atrium is a thick disk, the "torus," surrounded by radial fibers arranged like the spokes of a bicycle wheel. The resemblance of the structure to a valve has been noted in just about every description of a bordered pit.

Chapman et al. (1977) have developed a theoretical model for the operation of a bordered pit. Ignoring viscosity and invoking Bernoulli, they observe that streamlines will be symmetrical on either side of the torus at very low speeds. At higher speeds the torus will be displaced downstream slightly. (Strictly speaking, viscosity is needed to get the initial displacement.) As a result, by continuity, the streamlines near the downstream margin will crowd and the pressure will drop. This will displace the torus still further, eventually closing ("aspirating") the passage between the tracheids. So for slow flows the pit should remain open in either direction while for more rapid flows it should close. What happens next is unclear. The pit might remain shut by a hydrostatic pressure difference or (as Gregory and Petty, 1973, pointed out) by the pressure generated by the surface tension at a sap–air interface. The latter would function as part of a scheme to prevent spread of air embolisms. Or the valve might "chatter," opening and closing repeatedly.

FIGURE 4.7

(*a*) Diagrammatic transverse section through a bordered pit, with the torus dotted and the radial fibers shown as thin lines. In its extreme positions the torus closes the pit to flow. (*b*) The torus and fibers in a facing view.

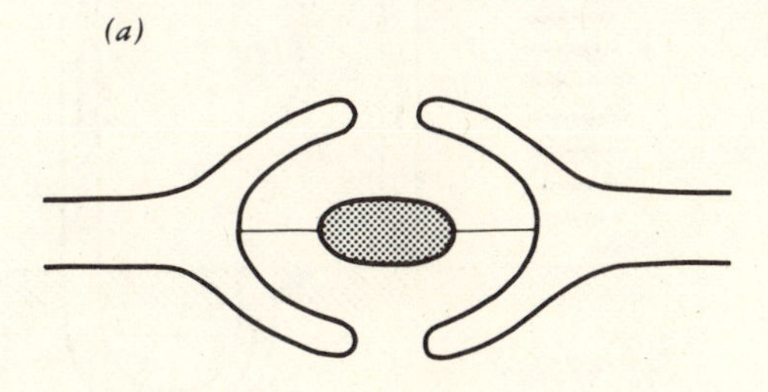

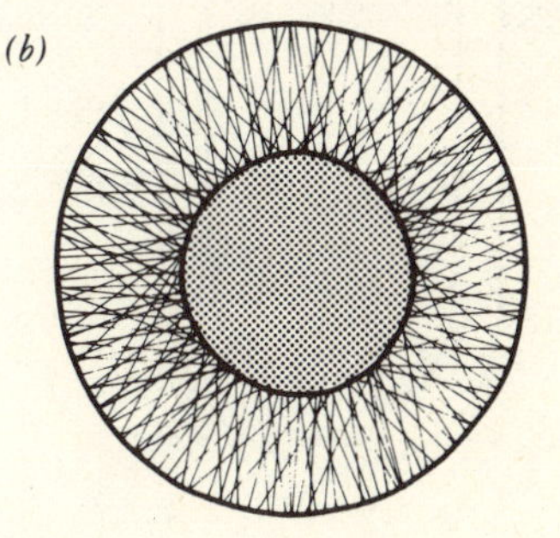

Forcing further flows. If the manometer fluid (pictured in Figure 4.3) were drained, then the main flow could, by Bernoulli's principle, induce a secondary flow through the manometer tube. This will still be the case if the upper wall of the main pipe is removed; that is, if we replace the pipe with an open channel in which the flow must go up and over some obstruction (Figure 4.8a). With this scheme, an organism living at a solid-fluid interface could use the energy of the flowing external medium to drive a flow through itself or its domicile. The direction of flow of the air or water is not crucial; it merely must change its speed by going over some barrier. Surprisingly, the wide range of biological examples of this simple arrangement escaped comment until very recently (Vogel and Bretz, 1972). It now looks as if this scheme helps prairie dogs ventilate their burrows (Vogel et al., 1973) and some termites ventilate their mounds (Weir, 1973). It reduces the cost of filtering water in sponges (Vogel, 1977); it aids flow through keyhole limpets (Murdock and Vogel, 1978); it augments airflow through the thoracic tracheae of some large beetles in flight (Miller, 1966); and it does lots of other biologically interesting and useful things (Vogel, 1978a).

FIGURE 4.8

(*a*) If a primary flow speeds up in crossing an obstruction, then the reduction in pressure can drive a secondary flow from *A* to *B* according to Bernoulli's principle. (*b*) If a small pipe is mounted transversely in the velocity gradient near a surface, viscous entrainment at *B* will draw fluid from *A* to *B*. (*c*) If the lower opening of the small pipe is looped under the substratum, the arrangement begins to resemble that of (*a*), and the two bits of physics get a bit tangled.

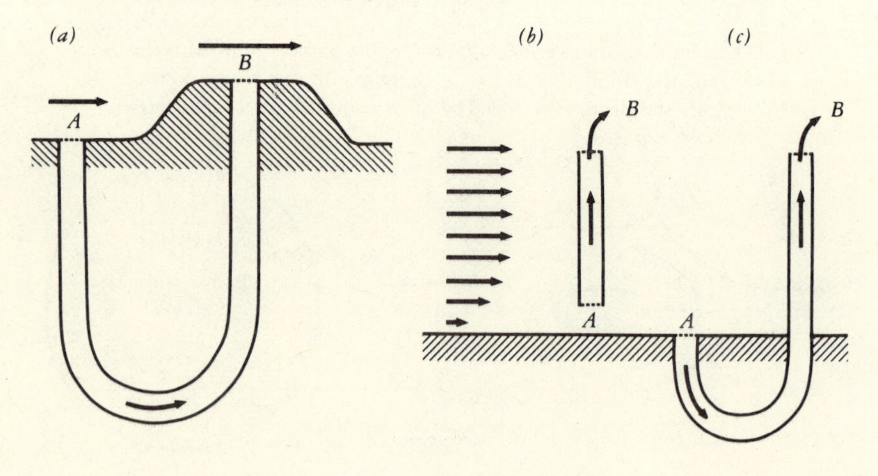

In all of these cases, though, Bernoulli's principle is not the complete extent of the relevant physics. The no-slip condition remains, and velocities near surfaces are consequently lower than elsewhere. As mentioned earlier, pressure rather than total head remains constant in the velocity gradients near surfaces. If a tube is placed transversely in a channel, with one end near a wall (Figure 4.8b), then fluid will move from wall to free stream by a mechanism of quite a different origin. Viscosity, you may recall, is the resistance to rapid shear rates. At the end of the tube away from the wall, shear rate will be greater. To reduce the effective shear, fluid will move out of the tube and into the channel. At an equal rate, other fluid will move into the tube at the wall end. The phenomenon, rarely of interest to fluid mechanists, is termed "viscous entrainment" (Prandtl and Tietjens, 1934). It's one reason why the static holes in Pitot tubes must be small.

The difficulty of dissecting the physical mechanism of a given induced flow is evident if the tube in Figure 4.8b is looped around as in 4.8c, the result looks all too similar to 4.8a! In actuality, some mix of the two mechanisms probably always operates, and the nature of the mix is not of overwhelming concern to the biologist (Vogel, 1976).

solids and fluids: a look back

As large organisms whose density far exceeds that of the surrounding medium, we are raised to regard solids as more palpably real than fluids; our training in physics begins with the solidest of solids, and we forget the fascination of the two-year-old who pours water repeatedly from one container to another. Having left that familiar world of solids some pages back, perhaps it may be useful at this point to summarize with a list of some loose analogies between the physics of solids and that of fluids. Thus...

solids	*fluids*
mass	density
elasticity	viscosity
interfaces	streamlines
friction	drag
conservation of mass	continuity
conservation of energy	Bernoulli's principle

CHAPTER 4

momentum

Newton recognized quite clearly something basic about a concept of "quantity of motion," and he saw that it was proportional to neither mass nor velocity, but to their product. To this product, mU, we give the name *momentum*. Newton's second law is commonly stated as the assertion of equality between force and the product of mass and acceleration; it may be a better practice to think of the law as defining force as the rate of change of momentum. The usual statement just simplifies things by assuming that mass is constant:

$$F = \frac{d(mU)}{dt} = \frac{mdU}{dt} \tag{4.12}$$

The assumption of constant mass, incidentally, may sometimes be inappropriate. My student, Thomas Daniel, reminds me that a pulsating squid or jellyfish, ejecting water to move, is best treated as a variable mass system.

Every elementary physics book notes that in all collisions between bodies, momentum is conserved. Energy *may* be conserved, but only in perfectly elastic collisions can we neglect the conversion of mechanical to thermal energy. Thus the conservation of momentum is by far the more potent generalization for mechanical problems. It must be emphasized that velocity is properly a vector quantity, so direction must be taken into account in applying the principle of conservation of momentum.

In fluid mechanics, collisions are not generally of much consequence on a macroscopic scale. Momentum, though, is still conserved; the total momentum of an isolated system is constant in magnitude and direcion. But a moving fluid is rarely an isolated system—it passes walls, obstacles, pumps, and so forth. Equation (4.12) gives a useful measure of this lack of isolation. The momentum of a system can only be changed by some external force acting on the system, and the rate of change of momentum is the force, just as with solids. The trouble with the equation is that it involves mass and time, neither of which are handy quantities when dealing with steady flows of fluids. We can do better by considering what must go on within a streamtube. How fast does mass move through the tube? If we consider uniform flow by taking a small enough tube so transverse velocity differences are trivial, we conclude that the mass flux (m/t) will equal the product of the density of the fluid, the cross-sectional area of the tube, and the velocity of flow. Which is the same as the product of density and volume flux:

$$\frac{m}{t} = \rho SU = \rho Q \tag{4.13}$$

Similarly, the momentum flux is obtained by multiplying both sides by velocity:

$$\frac{mU}{t} = \rho SU^2 \tag{4.14}$$

Momentum flux is, of course, force by equation (4.12) so

$$F = \rho SU^2 \tag{4.15}$$

This combination of variables, ρSU^2, will occur repeatedly in chapters to come; for now let us just note it as a sort of dimensional definition of limited direct applicability.

We can use Newton's second law to define some forces. In these terms, *drag* becomes the rate of removal of momentum from a flowing fluid. *Thrust* is then the rate of addition of momentum to a moving fluid. And *lift* is the rate of creation of a component of momentum normal to the flow of the undisturbed stream. These definitions, of course, say absolutely nothing about the physical origin of forces such as drag. This latter and most awkward subject will be deferred to the next chapter.

indirect force measurements

We ought to be more interested than we usually are in the forces exerted on organisms by air and water currents. Drag is likely to be a parameter of considerable adaptive significance; and, due to the irregular shapes and common flexibility of organisms, its behavior will rarely be self-evident.

Drag, however, may not always be an easy thing to measure under anything approaching normal conditions. Not all creatures can be detached with impunity and remounted with instruments in flow tank or wind tunnel. When tolerable, the procedure is attractively direct. But when impossible, there is an alternative almost universally overlooked by biologists. It emerges from our formal definition of drag, combined with the ideas of continuity and streamlines, and is justified by Newton's *third* law.

By the third law, if a body exerts a force on another, then the second body exerts a force on the first of equal magnitude and opposite direction. *If the fluid exerts a drag on the body, then the body must remove momentum from the fluid at a rate which balances its drag.* If this rate of removal of momentum is measurable, then it is possible to obtain the drag of an organism without touching it, much less detaching it! The procedure, though, can become a bit laborious. To understand how the scheme works and when it can be legitimately employed, it's necessary to derive an appropriate procedural equation.

Consider an attached object subjected to a flow (Figure 4.9), an object which (for simplicity) has a bulk concentrated well away from the substratum so it is not in a velocity gradient. Consider, as well, two imaginary parallel planes, one upstream (S_0) and one downstream (S_1) of the object, each sufficiently far from the object that the flow is normal to its surface. Consider, finally, a stream tube which leads fluid from part of the upstream plane (dS_0) to a corresponding part of the downstream plane (dS_1). On the upstream plane the velocity (U_0) is uniform, while on the downstream plane the velocity (U_1) varies, being minimal in the center of the wake of the object. The momentum flux across dS_0 will be, from equation (4.15),

$$\rho U_0^2 dS_0$$

FIGURE 4.9

A stream tube, a pipe with nonmaterial walls, passes an attached object; the presence of the object removes momentum from fluid passing through the stream tube. The rate at which the object removes momentum from all such stream tubes is its drag.

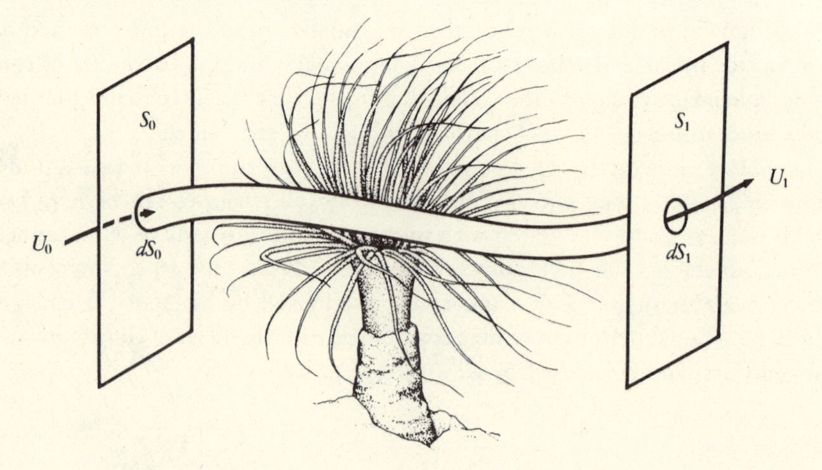

and the momentum flux across dS_1 will be

$$\rho U_1^2 dS_1$$

The difference between these will be the part of the drag acting on that stream tube, so,

$$dD = \rho U_0^2 dS_0 - \rho U_1^2 dS_1 \tag{4.16}$$

By continuity, though,

$$dS_0 U_0 = dS_1 U_1 \tag{4.17}$$

Combining equations (4.16) and (4.17) to get rid of dS_0 gives

$$dD = \rho U_0 U_1 dS_1 - \rho U_1^2 dS_1 \tag{4.18}$$

All that remains is integrating over the whole plane, S_1. The double integral sign just means that we have two dimensions to worry about, that $dS_1 = dy_1 dz_1$.

$$D = \rho \iint U_1(U_0 - U_1) dS_1 \tag{4.19}$$

This last equation may look complex, but the complexity is more apparent than real. It is rarely going to be possible to solve it explicitly; instead it constitutes the procedural guide we seek. What has to be done to measure drag indirectly should now be evident. First, one has to measure the velocity, U_1, at a set of points on plane S_1. Naturally it is only necessary to do so in the wake of the object—where U_0 and U_1 are appreciably different—so the limits of the planes of integration present no problems. Second, U_0 must be measured somewhere on the upstream plane; it makes no difference where, since we're assuming that U_0 is constant across S_0. Then all that must be done is for each $U_1(U_0 - U_1)$ to be multiplied by its corresponding area, dS_1 (or ΔS_1, to be realistic about it), the products added up, and the result multiplied by the density of the fluid. Voilá, drag!

A number of suggestions and cautions are in order. First, the location of the downstream plane may be a problem. If you're too close to the object, the assumption that the flow direction is uniform and normal to the plane may be violated; the plane should be at least several times the maximum diameter of the object behind it. If you're too far away, viscosity will have begun to eliminate the wake altogether, to reaccelerate it to free stream-velocity. A few trial runs are needed. Second, there is the decision as to how many velocity measurements to take in S_1. One hundred is a reasonable number; that's an array of ten by ten measurements. Rather than mechanically advancing your measuring instrument between each determination, it may be easier to attach it to a motor-driven manipulator and make it go back and forth over S_1, with a recorder taking the data. It's even possible to get the data directly in computer-readable form so the investigator need never deal with the hundred or so data behind each figure for drag.

It is possible to modify the procedure to work within the velocity gradients near surfaces. One must return to equation (4.16) and apply continuity to the data, increment by increment, up from a known streamline—a path on the surface. And, of course, the distribution of velocities across S_0 as well as S_1 must be considered, along with the drag of the surface itself.

The only use of this technique in a biological investigation of which I am aware was a study of variations in the drag of the fuselage of fruit flies, a part of my own doctoral thesis. Even with objects as small as fruit flies it was possible to get the requisite resolution to measure drag with about 5% accuracy. A more valuable application should be in measurements of the drag of attached organisms *in situ* with flowmeters such as those described in Appendix 2. If available and applicable, direct force-measuring technology will almost always be less cumbersome. Further information on indirect force measurements can be found in Prandtl and Tietjens (1934, pp. 123-130), Goldstein (1938, pp. 257-263) and Maull and Bearman (1964).

Thrust can be measured in much the same manner; only the sign of the result is reversed. Lift requires some way of determining the deflection of the wake from the free-stream direction. I once measured the lift of a fixed fly wing by first measuring the airspeed at each of a series of points behind it; then for each point I centered the wake of a tiny wire on the airspeed transducer to get the local wind direction. From these data it was a simple matter to calculate the downward component of momentum.

CHAPTER 5

drag, scale, and Reynolds number

SO FAR, FLUID MECHANICS HAS PROBABLY struck the reader as a decent, law-abiding branch of physics. We've touched on viscosity, continuity, momentum, and have even mentioned drag and how it might be measured—all quite ordinary topics. I now want to pursue the business of drag somewhat further, asking in particular about its actual physical basis. With this innocent question, Pandora's box springs a leak, and the true mire of fluid flow is shamelessly exposed. So queer is this aspect of the physical world that we'll have to defer the relevant biology to the next two chapters in order to spend this one on the physics.

from whence drag?

It's easy, of course, to define drag as the rate of removal of momentum from a moving fluid by an immersed body. Similarly, it's no trick at all to get a dimensionally-correct formula for the drag force from that definition, as done in equation (4.15), $D = \rho S U^2$.

This is, in fact, the formula Newton suggested from just such dimensional considerations. In physical terms, though, it implies a rather curious situation. Fluid meets object, the latter having a projected area normal to the flow of S. Particles of fluid make inelastic collisions with the object, which implies that they stick to it and are carried along with it, not with the flow. It is about as useful a steady-state mechanism as a badly designed snow plow which accumulates all that it encounters—not as much impossible as unwieldy. A bird in flight which opens its beak and inflates its air-sacs will experience a slight drag very briefly in just this manner.

Aristotle, by the way, was even further from the mark. He didn't believe in drag at all, figuring that an object needed air to give it thrust; in a vacuum an arrow would tumble earthward upon leaving the bow. Newton's first law, a counter-intuitive idea if there ever was one, remained almost 2000 years in the future.

Alternatively, we can try Bernoulli, Euler, and the ideal fluid theorists of the 19th century. Theoretical streamlines can be calculated for smooth flow around, say, a circular cylinder with its axis normal to the flow (Figure 5.1). At the upstream and downstream extremities are so-called "stagnation points" where the fluid is locally stationary with respect to the cylinder. Laterally, the fluid reaches maximum speed. As a result, at the front and back the pressure on the surface will reach H, the total head. At the sides, where streamlines are closest together, the pressure will be less, $H - 2\rho U^2$, to be exact. However, the

FIGURE 5.1

Theoretical streamlines for flow perpendicular to the long axis of a circular cylinder, as shown in perspective in the inset at the right.

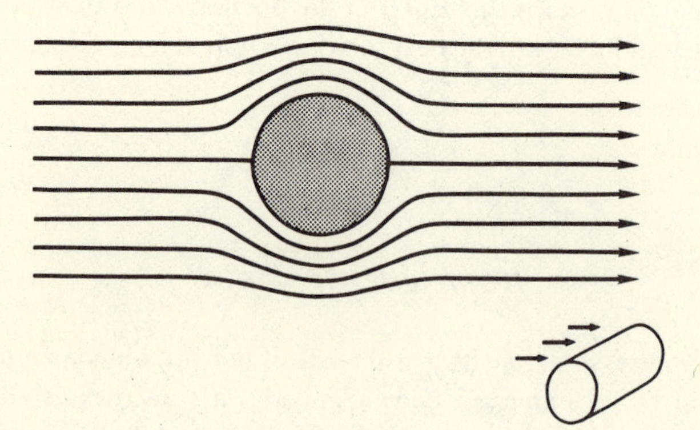

whole diagram is symmetrical about a plane normal to the flow and along the axis of the cylinder. Not only do the pressures on each side cancel each other, but those of front and back do as well. As a result, there is no net pressure tending to carry the cylinder with the flow, and thus there is no drag at all. This amazing result can be generalized to cover bodies of any shape; it's called d'Alembert's paradox, the ultimate pursuit of a will-o'-the-wisp into a cul de sac. Lord Rayleigh (1842–1919) pointed out that according to such theory a ship's propeller wouldn't work but, on the other hand, it wouldn't be needed anyhow! Further,

> There is no part of hydrodynamics more perplexing to the student than that which treats of the resistance of fluids. According to one school of writers, a body exposed to a stream of perfect fluid would experience no resultant force at all, any augmentation of pressure on its face due to the stream being compensated by equal and opposite pressures on its rear....On the other hand it is well known that in practice an obstacle does experience a force tending to carry it downstream.

As a first step toward reconciling theory and reality, it is quite a simple matter to measure the pressure distribution around a circular cylinder in a wind tunnel or flow tank. One just drills a tiny hole in the cylinder and hooks a manometer between this hole and a static orifice elsewhere. Rotating the cylinder allows one hole to serve all orientatations. The results of such an experiment are shown in Figure 5.2. For these particular data, the conditions are equivalent to a sapling in a light breeze or a human foreleg wading in a very gentle stream. In the figure, net pressure measurements (total head less static pressure) have been divided by the dynamic pressure ($\frac{1}{2}\rho U^2$, where U is the free-stream velocity) to express them as values of a dimensionless pressure coefficient.

At the upstream center, the pressure is, as expected, H, so the pressure coefficient comes out to be unity. Between 30° and 35° around from the upstream center, the pressure becomes less than the static pressure—quite as we'd expect from Bernoulli's principle and the obvious convergence of streamlines in Figure 5.1. So far, theory and measurement agree reasonably well. But proceeding further along the cylinder, the rest of the decrease in pressure does not materialize entirely; and, at 70°, the pressure actually begins to rise somewhat. From here to the rear, ideal fluid theory bears no resemblance to reality. About the only point on which theory and practice agree is that, for the cylinder as a whole, the net overall force is negative, so the cylinder will

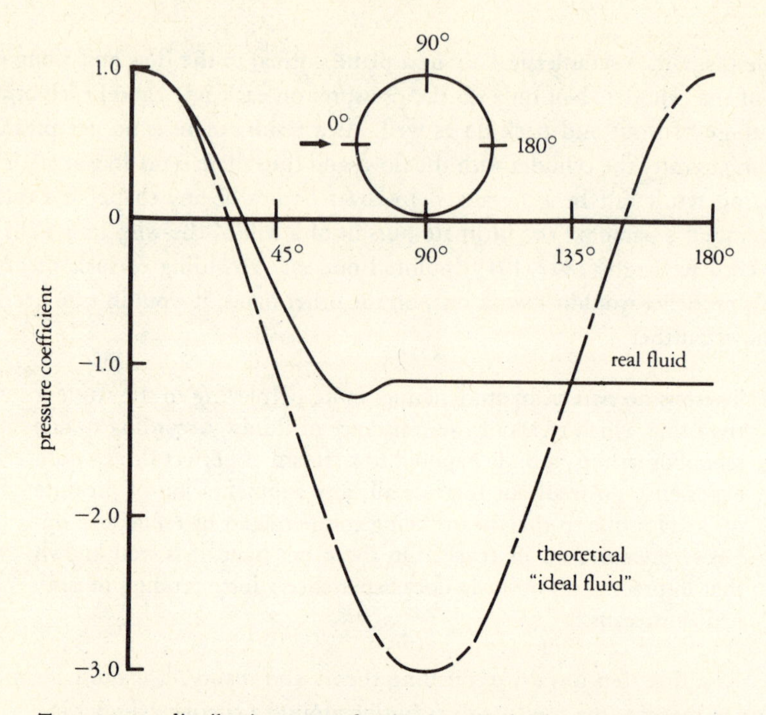

Two pressure distributions around a circular cylinder from upstream (0 °) to downstream (180 ° extremities). If flow followed the streamlines of Fig. 5.1, the dashed curve would result. In fact, the solid line represents reality for one particular set of conditions; theory and practice are most divergent at the sides and rear.

FIGURE 5.2

tend to expand rather than collapse in the current. I mention this point since we invoked Bernoulli's principle in the last chapter to explain things like lift in flatfish and the maintenance of the beetle's bubble.

It is this difference between the dynamic pressure on the front and the *absence* of the predicted counteractive pressure at the rear which we subjectively feel as drag. Although, in a sense, we've not solved anything, we have at least put a finger on the location of the problem. The flow of fluid around the cylinder is not symmetrical, front to back. And it is in the rear where something strange is happening. We'll return to the problem of the pressure distribution around a cylinder after developing the needed analytic tools; for now just bear in mind that for limiting drag in fluids, discriminating design of the derriere is de rigueur!

Unfortunately, these pressure measurements don't indicate some simple component of drag (perhaps a direct function of viscosity) heretofore overlooked. Both the pressure distribution and the drag itself often vary with speed, for instance, in queer and irregular ways. Even for a case as overtly simple as a cylinder normal to flow, drag varies peculiarly. At low speeds it is proportional to the first power of velocity, then it gradually approaches the second power for a while, drops abruptly, and resumes its second-power behavior but with a lower constant of proportionality at very high speeds. Perhaps it is just this sort of lawlessness which has left the fluid state largely to engineers even though the solid state is a common concern of physicists.

on what might drag depend?

Perhaps at this point we should remind ourselves that one objective in science is to obtain the minimum number of the simplest rules which account for the largest variety of phenomena. We don't have and may never have a universal and practical rule for predicting the drag of even as simple a shape as a cylinder. But it is certainly still worth trying to simplify the situation as much as possible; anything is better than bludgeoning the problem with massive empirical tables of how drag varies with shape, speed, size, and other relevant variables. For each additional independent variable the number of data will rise by one or two orders of magnitude. And such tables would have little use for the shapes a biologist is prone to meet. At least we should consider the possibility that some of the variables have the same effects as others, and each need not be regarded as behaving in a completely unprecedented fashion.

This latter possibility turns out to be both real and useful. Its pursuit leads to the peculiarly powerful *Reynolds number,* the centerpiece of biological fluid mechanics. The utility of the Reynolds number extends far beyond mere problems of drag; it's the nearest thing to a completely general guide to what is likely to happen when solid and fluid encounter each other. For a biologist, dealing with systems spanning an enormous size range, the Reynolds number is the central scaling parameter which makes order of a diverse set of physical phenomena. Its role is comparable to that of a surface-to-volume ratio in physiology.

This almost magical parameter can be most easily introduced as it was originally through a series of empirical experiments done by Osborne Reynolds (1842–1912). We've already mentioned these in connection with laminar and turbulent flow (Chapter 3). Reynolds, as you may recall, introduced a dye stream into a pipe of flowing liquid. Sometimes the resulting straight streak indicated laminar flow, and sometimes the dispersal of the streak signalled turbulent flow. The transition was fairly sudden, both in location in the pipe and as the parameters of the flow were altered. He found that the fluid could be persuaded to shift from laminar to turbulent flow in several ways: by *increasing speed; increasing the diameter* of the pipe; *increasing the density* of the liquid; or *decreasing its viscosity*. Each change was as effective, quantitatively, as the others; and they worked in combination as well as individually. The rule appeared to be that when a certain combination of these variables exceeded 2000, the flow became turbulent. The particular combination is now know as the Reynolds number, *Re*†, with the length factor, *l,* representing the diameter.

$$Re = \frac{\rho l U}{\mu} \qquad (5.1a)$$

It is one of the marvelous gifts of nature that this index proves to be so simple—a combination of four variables, each with an exponent of unity. It has, however, a few features worthy of comment. The Reynolds number is dimensionless (as you can verify through reference to Table 1.1 (page 8), so its value is independent of the system of units in which the variables are expressed. It marks the promised reappearance of the kinematic viscosity (after a three-chapter absence), here located in the denominator, so:

$$Re = \frac{lU}{\nu} \qquad (5.1b)$$

It is not the dynamic viscosity and the density which matter so much as their ratio: the higher the kinematic viscosity, the lower the Reynolds number. Notice, also, the use of "*l*", a measure called the "characteristic length." For a circular pipe, the diameter is used as this measure; the radius might have been but for a mere convention. For a solid imersed in a fluid, *l* is typically taken as the greatest length of the solid in the direction of flow. But it is a very crude measure of size, and it emphasizes the crude nature of the Reynolds number as a yardstick when objects of different shapes are being compared. Indeed, the

†It's a small point, but while Newton's law, Bernoulli's principle, and Poiseuille's equation are possessives, Reynolds and the other named dimensionless numbers are not. Also, it's *Re,* not R_e.

value of the Reynolds number is rarely worth worrying about to better than one significant figure. Still, that's not trivial with biologically-interesting flows which span at least fourteen orders of magnitude.

Of greatest importance in the Reynolds number is the product of size and speed, for the two work in concert, not in counteraction. For living systems, ''small'' almost always means slow and ''large'' often means fast. Thus, those fourteen orders of magnitude far exceed the range of lengths of the organisms themselves. At this point it might be worthwhile just to calculate a few Reynolds numbers. Since we're only interested in order of magnitude, I've taken the liberty of using educated guesses for speeds and sizes.

	Re
A large whale swimming at 10 m s^{-1}	300,000,000.
A tuna swimming at the same speed	30,000,000
A duck flying at 20 m s^{-1}	300,000
A large dragonfly going 7 m s^{-1}	30,000
A copepod in a pulse of 20 cm s^{-1}	300
Flight of the smallest flying insects	30
An invertebrate larva, 0.3 mm long, moving at 1 mm s^{-1}	0.3
A sea urchin sperm advancing the species at 0.2 mm s^{-1}	0.03

a physical view of the Reynolds number

It's possible to come up with this same dimensionless index in another way. Consider what must have gone wrong with the attempt to explain drag by ideal fluid theory; failure must be traced to the neglect of viscous forces. It would be reasonable to suppose that not one but two sorts of forces are at work when a moving fluid crosses an immersed body. There are, first, inertial forces, those derived directly from Newton's second law which essentially defines inertia, and expressed in that equation which we obtained so easily and found so nearly useless in practice:

$$F_I = \rho S U^2 \qquad (4.15)$$

Under the imperatives of its inertia a particle of fluid keeps on doing its usual thing unless aggressed upon by external authority. And then there are the previously ignored viscous forces. Fluids don't like to be sheared. If a flow involves shear, then viscous forces will oppose the persistence of the motion just as friction opposes the movement of one solid across another. Viscous force has already been defined in the process of defining viscosity:

$$F_V = \mu SU/l \qquad (2.3)$$

Now let us make an intuitive leap and suggest that what distinguishes different regimes of flow is the *relative importance of inertial and viscous forces.* The former keeps things going; the latter makes them stop. High inertial forces favor turbulence, with the substantial internal shearing which that implies. High viscous forces should prevent sustained turbulence by damping incipient eddies and other such approaches to discontinuity and should thus ensure laminar flow. To get an indication of the relative magnitude of the two, one need only divide equation (4.15) by equation (2.3) to get (of all things) the Reynolds number.

$$\frac{F_I}{F_V} = \frac{\rho SU^2}{\mu SU/l} = \frac{\rho lU}{\mu} = Re \qquad (5.1c)$$

This time its lack of dimensions is seen as the natural outcome of dividing one force by another; it is the ratio of inertial to viscous forces.

One point should be made emphatically. If, for example, the Reynolds number is low, one faces a highly viscous situation. The flow will be dominated by viscous forces, vortices will be either nonexistent or non-sustained, and velocity gradients will be very gentle unless large forces are exerted. But it is the value of the Reynolds number that indicates the character of the flow and not the value of the viscosity *per se.* A tenfold reduction in size (length) will increase relative viscous effects with precisely the same efficacy as a tenfold increase in viscosity itself. If, in nature, small means slow and large means fast, then small creatures will live in a world dominated by viscous phenomena and large ones by inertial phenomena: this, even though the protist swims in the same water as the whale.

Consider an object of a given shape and orientation immersed alternately in two flows. *Equality of the Reynolds number for the two flows guarantees that the physical character of the flows will be the same.* A moment's thought should persuade you that this last statement is powerful but still quite reasonable, given the origin of the Reynolds number. As it turns out, equality of the Reynolds number does not necessarily mean that forces are unchanged, but the patterns of flow as might be revealed with a set of streamlines will be the same even if one flow is of a gas and the other of a liquid.

We come, parenthetically, to a mild curiosity and convenience. It was mentioned earlier that, at normal temperatures, air is about fifteen times *more* kinematically viscous than is water. As a result, for an object of a given size, the same Reynolds number will be achieved at a velocity fifteen times higher in air than in water. I doubt whether it is more than coincidental, but in nature, air flows are normally about fifteen times as rapid as flows of water. Thirty meters per second (68 mph) is a hurricane of air; two meters per second is a torrent of water. Thirty centimeters per second in air we find barely detectable; two centimeters per second in water appears to be a roughly similar threshold. In short, organisms (sessile ones at least) of any given size operate at approximately the same Reynolds numbers whether they live in air or water, quite a convenience for any biologist who is trying to understand the relationship between flow patterns and morphological adaptations.

A recent experience of mine gives some idea of the potency of this idea that equal Reynolds numbers mean geometric similarity in flow patterns. Malcolm Brown, of the University of North Carolina, has been making freeze-fracture photographs of the insides of the cell walls of plants. The orientation of cellulose fibers in the pictures suggests flow, but nothing is so fixed as a microscopist's specimen. We noticed, though, that the pattern of fibers near "fixed" points bore a strong resemblance to pictures of glaciers surrounding small mountain peaks. A few exceedingly rough calculations demonstrated that these two phenomena, vastly different in scale, operate at similar, very low, Reynolds numbers. If, indeed, the cell wall is flowing, it *ought* to look like a glacier in motion!

Reynolds number and the drag coefficient

Back to drag: Newton figured that it was proportional to three quantities: first, to the density of the medium; second, to the projected area of the body; and third, to the square of the velocity. And, although his theory was wrong, drag often does behave in this manner, at least crudely. One can do somewhat better, however, with a technique called *dimensional analysis*. (Rouse, 1938, has a particularly illuminating discussion of the approach.) Dimensional analysis leads to an empirical formula for drag with all of the peculiarities lumped into one composite variable; I'll not go through the derivation but merely proclaim proof by intimidation.

The analysis begins by identifying the parameters upon which drag ought to depend—the *size* of the object, its *speed* relative to the fluid, and the *viscosity* and *density* of the fluid. It concludes by viewing drag as equal to the product of the two composite variables. The first is the dynamic pressure times the area of the object; it accounts for the ordinary variations Newton identified. The second variable is dimensionless, and it accounts *in toto* for all of the peculiarities in the variation of drag:

$$D = \tfrac{1}{2}\rho S U^2 \left(\frac{\rho l U}{\mu}\right)^a \tag{5.2}$$

The dimensional analysis tells us nothing about how this second term behaves, merely that it is the *only* other term which should matter. The second term (less its exponent) happens to be the Reynolds number, so the latter has appeared in a third context, this time as a dimensionless combination of the four variables of interest.

In short, the behavior of drag is describable as the product of two variables—the dynamic pressure times the area of the object—and some function of the Reynolds number. In practice, that awkward exponent, a, about which we know nothing *a priori,* is absorbed by the so-called "drag coefficient," C_D:

$$C_D = (Re)^a = f(Re) \tag{5.3}$$

Thus the drag coefficient is a function only of the Reynolds number and accounts for the peculiarities in the behavior of drag. The drag coefficient is most easily envisioned as a dimensionless form of the drag—the drag per unit area divided by the dynamic pressure. Or, as usually given,

$$D = \tfrac{1}{2} C_D \rho S U^2 \tag{5.4}$$

This last equation is most emphatically not a simple formula for drag, no matter what one sees in the biological literature. We have no grounds for declaring C_D a constant, and it does not usually turn out in practice to be a constant. Equation (5.4), though, represents a huge simplification, the one we have been working toward. Instead of having to record how drag varies separately with speed, size, viscosity, and density, we merely need to know how the coefficient of drag varies with the Reynolds number. The fact that both drag coefficient and Reynolds number are dimensionless just takes a little getting used to. From equation (5.4) we can always get the drag from its coefficient.

The inclusion of a factor for area, S, does present a minor difficulty, one similar to that of choosing a length in the Reynolds number. The commonest area for reference is the frontal or projecting area of an object—its maximum projection onto a plane normal to the direction of flow. Hereafter, S will refer to frontal area unless otherwise specified. But three other areas are used, and it is crucial to know just which is meant or to specify just which you mean. "Wetted area" is the total surface exposed to the flow; it's commonly used for streamlined bodies. For airfoils, the "plan" or "profile" area is often used; it's the maximum projected area of the airfoil, the area which one would see if the airfoil were laid on a table and viewed from above. Finally, the two-thirds power of volume has been used as a reference area, originally for airships, where volume was proportional to lift. $V^{2/3}$ is probably the most biologically appropriate, what with guts and gonads within, but it is also the least common. So we have four possible areas, and a "best shape" which minimizes the drag coefficient with respect to one may not minimize it with respect to another.

With only two variables for a given shape—coefficient of drag and Reynolds number—graphs of the two are of great convenience. We have, remember, lumped all the peculiarities into the relationship between these variables; if drag behaved as Newton (and some biologists) believed it should, then all such graphs should take the form of horizontal, straight lines. In its deviation from such a line, the graph shamelessly dissects and exposes the queerness. Figures 5.3 and 5.4 give C_D versus Re plots for cylinders and spheres, respectively. Neither of these variables is handy at first for intuitive reckoning, but one does after a while acquire a properly contemptuous familiarity with them. If your intuition needs a crutch, think of these graphs as representing, for a given object and fluid, the drag divided by the square of speed (ordinate) plotted against the speed (abscissa).

It may come as a surprise to biologists to realize that a simple physical case such as the drag of a circular cylinder can generate the sort of graph we think of as the special curse of our more complex systems. It is my impression, however, that the abruptness of the transitions seen in these graphs of C_D vs. Re for cylinders and spheres are much more drastic than are equivalent transitions for "biological" objects. The very symmetry and regularity of these simple shapes means that transitions from one flow regime to another tend to take

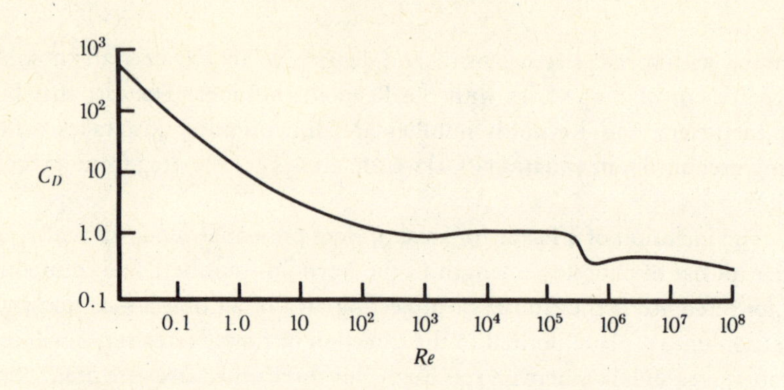

Drag coefficient versus Reynolds number for flow normal to the long axis of a long circular cylinder. The drag coefficient has been calculated from the drag using the frontal area of the cylinder as reference area. The irregularities of the plot get a bit suppressed in this double logarithmic presentation.

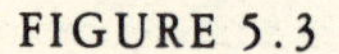

FIGURE 5.3

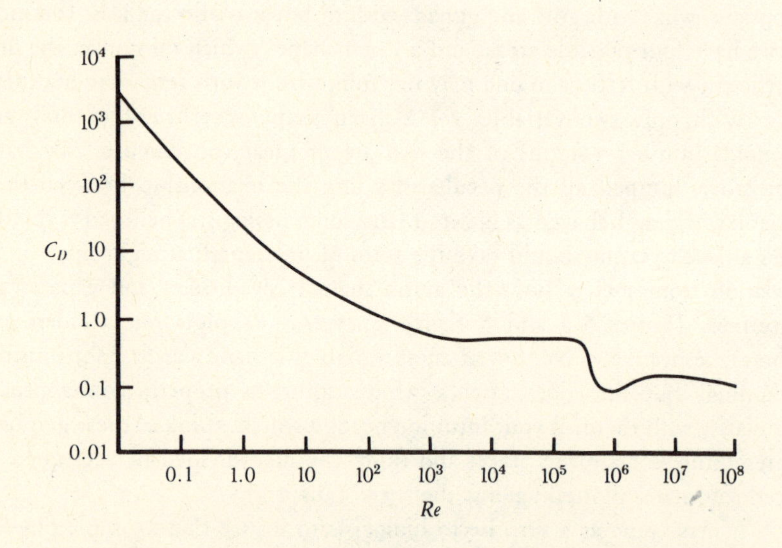

Drag coefficient (based on frontal area) versus Reynolds number for flow around a sphere.

FIGURE 5.4

place synchronously for the whole object. One should expect the more irregular biological objects (leaves and branches, for example) to give less irregular graphs of C_D vs. *Re*.

The drag coefficient is not just a convenience for graphing. Drag, by itself, depends on the product of the two terms of equation (5.2). The coefficient of drag depends only on the Reynolds number as in equation (5.3) for an object of a given shape and orientation. Put another way, *equality of Reynolds number implies equality of drag coefficient.* This last statement is sometimes called the "law of dynamic similitude;" it is, we'll see, the basis of almost all testing of model systems in flow tanks and wind tunnels.

flow around a cylinder revisited

Woven in and out of this chapter has been the question of the flow around and the drag of a circular cylinder. We've seen that the pressure distribution diverges markedly from that expected for an ideal fluid and that the plot of C_D vs. *Re* is, to say the least, bizarre. Let us tie up one loose end by considering the actual patterns of flow which are observed as the Reynolds number is slowly raised, combining a few paragraphs of verbal description with the illustrations in Figure 5.5. You might refer, as well, to Figure 5.3 as the description proceeds.

At Reynolds numbers well below unity, smooth and vortex-free flow surrounds the cylinder. Vortices are rare in the real world of low Reynolds number flows of any sort. The influence of the cylinder in lowering relative velocities around itself is felt for quite a distance: the cylinder is surrounded by a cloud of retarded fluid which effectively obscures its shape. Measurements of drag may give exaggerated results if there is a solid wall even a large number of cylinder-diameters away. For example, White (1946) found that at $Re = 10^{-4}$, the presence of walls five hundred diameters away doubled the apparent drag of a cylinder. As the Reynolds number approaches and then exceeds unity, the volume of the disturbed fluid gradually diminishes and we enter a regime in which the drag curves for different shapes diverge.

At Reynolds numbers between about 10 and 40, the cylinder bears a pair of attached eddies on its rear. With flow from left to right, the upper eddy rotates clockwise and the lower one counterclockwise. Above about 40, the pattern is no longer stable, and the vortices alternately detach, producing a wake of vortices with each rotating in a direction opposite that of its predecessor further downstream. This pattern of alternating vortices is known as a "von

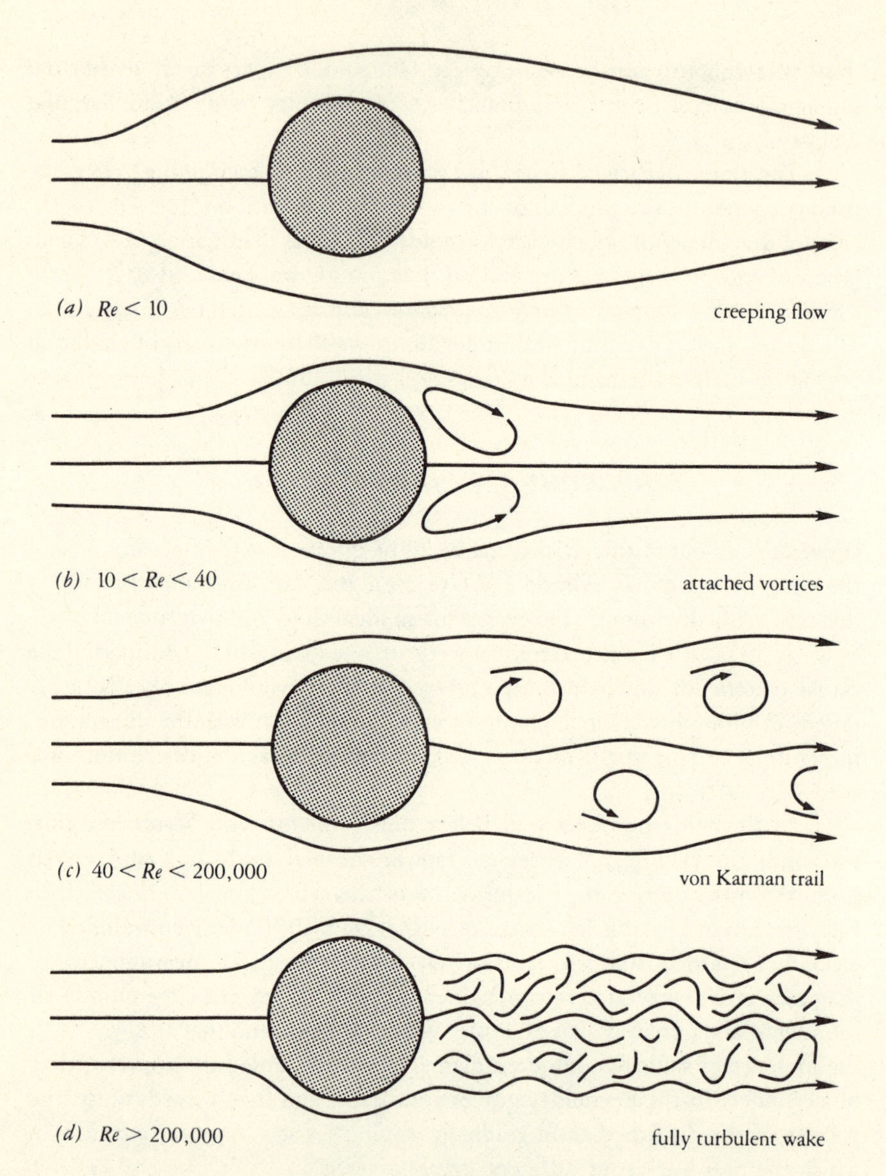

Real patterns of flow around a circular cylinder. Note, in particular, the absence of any vortices at very low Reynolds numbers (*a*) and the constriction of the wake between (*c*) and (*d*). This last change is concomitant with the sudden drop in drag coefficient—the great "drag crisis"—at Reynolds numbers in the low hundreds of thousands.

FIGURE 5.5

Karman trail'' (see Chapter 14). Periodic shedding of vortices continues up to a Reynolds number of about 100,000, but with an increasingly turbulent wake behind the cylinder.

Somewhere between 100,000 and 250,000 another transition occurs. The wide wake of turbulent eddies narrows rather abruptly. Concomitantly, the drag coefficient drops by about two-thirds. The ''somewhere'' is a deliberate evasion, for results seem to depend upon circumstances. The less turbulent the basic flow to which the cylinder is exposed, the higher the Reynolds number for the transition. Not only does the drag coefficient drop, but the drop is so sudden that the drag itself briefly undergoes a paradoxical reduction with an increase in speed. Then, with further increase in Reynolds number, there is little change in the pattern of flow up to numbers so high that compressibility is no longer negligible and so high as to be biologically irrelevant.

The abrupt drop in drag and narrowing of the wake is worth some attention, for it reflects a phenomenon which is of importance at all Reynolds numbers reasonably above unity, the phenomenon called ''separation of flow.'' Recall from Figure 5.2 that pressure increases from the widest part of a cylinder toward the rear. Just where and how much it increases depends, of course, on the Reynolds number, but there is always some region of increase in going from the front to the rear. In moving around the cylinder from the vicinity of the front stagnation point, fluid goes from high to low pressure, an easy and natural journey. But beyond the point of minimum pressure, fluid must progress ''uphill'' to make any more headway toward the center of the rear. Such motion against a pressure gradient is possible only at the expense of preexisting momentum, which never seems sufficient for the flow to get to the rear stagnation point. What happens is that fluid is robbed of its momentum by viscous effects—shear at the surface and adjacent to it. It is just like a sled which coasts down one hill but, having experienced friction, can't make it up another hill of equal height. At some point, then, the fluid stops following the surface of the cylinder and heads off more or less straight downstream; the point at which this occurs is called the ''separation point.'' It corresponds very nearly to the point at which the ideal and real pressure curves diverge (Figure 5.2). Downstream from the separation point are eddies, turbulence, and usually, quite near the surface, some flow from the rear forward towards the separation point (Figure 5.5).

The noise you hear when facing directly into a moderate wind is the turbulence around your ears, which are just downstream from the separation point at about the position of the cheekbones (Kristiansen and Petterson, 1978).

The location of the separation point on the surface of the cylinder depends on the Reynolds number. Above about 100,000 or so, turbulence (which has already been present) suddenly invades the region of fluid very near the surface of the cylinder. The result is an increase in the momentum of the fluid very near the surface; the result of that is that fluid flows further around the cylinder before separating, thus narrowing the wake; the further result is an abrupt reduction in the cylinder's drag and drag coefficient.

Similar phenomena occur in the case of a sphere. In fact, the Reynolds number at which this turbulent transition occurs has been used as a measure of the turbulence level for comparisons among different wind tunnels. In short, separation, operating under the aegis of viscosity, is the physical culprit we were looking for at the beginning of the chapter.

shape and the two kinds of drag

The other loose end, the amount of drag, was formally tied up with Figure 5.3. However, some additional complications remain. The total drag of a circular cylinder can be dissected into two components, the "skin friction" and the "pressure drag." Ultimately both come from viscous effects, but they do so in quite different ways.

Skin friction is the direct result of the interlamellar stickiness of fluids, the mechanism by which one plate exerts a force on another (as shown in Figure 2.2). It always exerts its force parallel to the surface and in the local direction of flow. As you might expect, it's more significant in more viscous situations; that is, at low Reynolds numbers. But no Reynolds number is so high that skin friction disappears. As long as the no-slip condition holds, there has to be a shear region in a fluid near a surface. The general rule is that the more skin (surface) a body exposes to the flow, the more skin friction it will experience.

Pressure drag is a less direct result of viscosity. It occurs because of separation and the consequent fact that the dynamic pressure on the front is not counterbalanced by an equal and opposite pressure on the rear. In effect, energy is being invested to accelerate fluid to get it around the object, but the energy is not being returned to the object in the deceleration near its rear. Instead, the energy is dissipated (to heat, eventually) in the wake. While it is commonly regarded as a high-Reynolds number phenomenon, pressure drag is significant at all Reynolds numbers; only its magnitude relative to skin friction is greater when the Reynolds number is high. For a circular cylinder (again), pressure drag constitutes 57% of the total drag at $Re = 10$, 71% at $Re = 100$, 87% at $Re = 1000$, and 97% at $Re = 10{,}000$.

Although pressure drag ultimately requires viscosity to occur, it mainly reflects inertial forces. As such it is roughly proportional to surface area (or to the square of a linear dimension) and to the square of velocity. The drag coefficient is inversely proportional to these factors, so where pressure drag dominates at high Reynolds numbers, the drag coefficient will not vary widely. By contrast, skin friction reflects mainly viscous forces and follows the first powers of linear dimensions and velocity as in equation (2.3). Thus at low Reynolds numbers, where skin friction dominates, the drag coefficient will be inversely proportional to the Reynolds number. This, broadly, is what we see in Figures 5.3 and 5.4. In between (*Re*'s of 1 to 100,000) are the peculiar transitions just discussed. An awesome range of biological situations occur in this latter range of Reynolds numbers!

An example may sharpen the contrast between pressure drag and skin friction as well as between high and low Reynolds numbers by putting the matter in a less abstract context. Consider two objects, the first with negligible pressure drag at any Reynolds number, the second with very high pressure drag at all Reynolds numbers. The objects are, first, a long, flat plate of negligible thickness parallel to the flow, and second, the same flat plate now oriented perpendicular to the flow. Table 5.1 gives the variation of the drag of these plates over six orders of magnitude of the Reynolds number together with the ratio of the perpendicular to parallel drag coefficients. The data come from a variety of sources, including some measurements of my own, so they may not all be precisely comparable in detail.

The first thing which emerges from Table 5.1 is the rise in drag coefficient at low Reynolds number, just as we saw for the sphere and cylinder. One then notices the differences: the absence of the sudden drop in the drag coefficient of the perpendicular plate at high *Re*. This indicates only that the separation point is fixed at the outer edge of the plate, so it can't move downstream. More important, though, is that ratio.

In general, at low Reynolds numbers, skin friction is of great importance, and it depends only a little on orientation. At moderate and high Reynolds numbers, pressure drag can be overwhelming, but whether or not it is depends on the orientation of the plate. Thus orientation has a far more potent effect on total drag at high and moderate Reynolds numbers.

Which brings us to the question of how drag might be reduced, clearly a vital matter for nature as well as for human designers. At moderate and high Reynolds numbers, any object from which flow separates will experience a relatively high total drag; the problem, as mentioned, is the energy loss in the rear. There's a trick, however, which as been recognized at least empirically for

TABLE 5.1

The drag of a long, flat plate

	C_D*		
Reynolds number	*parallel to flow*	*perpendicular to flow*	*ratio*
1,000,000	0.005	1.1	220
100,000	0.004	1.1	275
10,000	0.013	1.1	85
1,000	0.042	1.1	26
100	0.02	1.5	7.5
10**	1.5	2.2	1.5
1**	12.5	18.3	1.4

*C_D based on projected, or plan form, area; for a flat plate the latter is one-half the total surface area.

**Thin airfoils, actually. From Thom and Swart (1940).

a very long time. If the object is endowed with a long and tapering tail, fluid gradually decelerates in the rear, little or no separation occurs, and the object is literally pushed forward by the wedgelike closure of the fluid behind it. In short, the energy, instead of being lost in the wake, re-emerges as a forward-directed pressure from the rear which nearly counterbalances the dynamic pressure on the front. The trick is called "streamlining" (Figure 5.6) because, loosely speaking, the object is shaped to follow an advantageous set of streamlines. At least at high Reynolds numbers, streamlining can be immensely effective: a well-designed airship has only about one or two percent of the drag of a sphere of the same frontal area. The sphere, by the way, is called a "bluff body," antonymous to a streamlined profile. The trick, for any designer, is mainly in the design of the rear. As Hamlet put it, "There's a divinity that shapes our ends, Rough-hew them how we will."

But as the Reynolds number gets lower, streamlining isn't quite so overwhelmingly beneficial. The trouble is that the reshaping involved exposes more surface relative to either projected area or to volume contained. More surface inevitably means more skin friction. At the Reynolds numbers encountered by whales and large fish, skin friction is a minor matter. At the moderate Reynolds

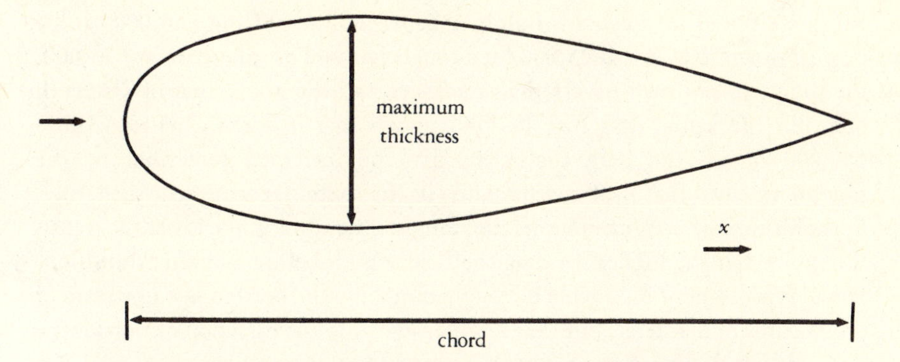

A streamlined shape, with the familiar blunt front and sharp rear. It may be viewed either as a body of revolution, symmetrical about its long axis in the plane of the page, or as a typical cross section of an elongate body extending far above and below the page. This particular shape has a "fineness ratio," or chord over maximum thickness, of about 3.5; its maximum thickness occurs 30% of the way from front to rear, that is, at $x/c = 0.3$. These are the two most often quoted descriptive data for streamlined bodies.

FIGURE 5.6

numbers of large flying and swimming insects, streamlining is still a good thing. At low Reynolds numbers the extra surface exposed may actually outweigh any reduction in pressure drag, which is undoubtedly why very small swimming creatures do not look obviously streamlined. Perhaps the main disadvantage of streamlining, from a biological point of view, is that it requires a particular direction of flow. A minor change in wind direction would more than wipe out the benefits of a streamlined tree trunk. If the direction of flow can be neither predicted nor controlled, then a bluff body is the unavoidable recourse.

What, if anything, can be done about the high drag of bluff bodies such as spheres and cylinders? Curiously enough, the situation can sometimes be improved by roughening the surface in one way or another. Generally, roughness is either neutral or bad. At low Reynolds numbers small bumps are within the slowly moving fluid near the surface and thus are of little consequence. At high Reynolds numbers, roughness increases the drag of streamlined objects. But there is a range in which roughness can help: a rough surface promotes turbulence close to a surface and can have the effect of postponing separation as the fluid travels around a bluff body. The transition to a lower drag coefficient which we noted for Reynolds numbers between 100,000 and 250,000 can be pushed down to as low as 25,000 for a rough cylinder or sphere. Below 25,000, the rough version will have a drag coefficient a little above that of a smooth one, but the actual drag may be low enough that it doesn't matter. Above 150,000 or so, the drag coefficient may be increased somewhat by the roughness, but that will matter only if the system encounters such high Reynolds numbers. Within limits, the rougher the surface, the lower the transition point but the higher the drag coefficient both before and after transition. Some improvement is possible by roughening only the portion just upstream of the maximum diameter, but that requires knowing the direction of flow; if the direction can be counted on, then streamlining is preferable.

The most common example of controlled roughness for drag reduction is a golf ball with its set of dimples. This originated with the observation that old, rough balls travelled farther than smooth, new balls. At the urging of a driver, a golf ball moves at a Reynolds number of about 50,000 to 150,000, so the scheme is reasonable. One wonders whether the rough bark of trees is a similar adaptation: slender branches are smooth and a trunk or large branches are much rougher. I think it is unlikely: for a trunk or branch 20 cm in diameter, a Reynolds number of 100,000 will be reached at a speed of about 8 m s^{-1}, less than 20 mph. Speeds high enough to be troublesome would automatically exceed the turbulent transition, and the roughness should not help. But nature is subtle, and trees do blow over with some presumed loss of fitness, so the matter might bear investigating. Another possible application of purposeful roughening might be in the design of large attached reef organisms which are exposed at times to fairly violent water currents.

Where practical, streamlining is the best approach to drag reduction. Quite a number of human cultures have developed streamlined throwing sticks (''boomerangs''), mostly of a nonreturning genre. These are most impressive weapons against small prey and might have really caught on if standardization and mass production had been feasible; not only do they cut a wide swath with lots of rotational inertia, but they travel far greater distances than any hand-thrown, spherical projectiles. I'm told that a boomerang could easily be thrown from home plate out of any baseball stadium although it is quite difficult to return a ball from the outfield wall. The difficulty with the streamlined stick, again, is the necessity for knowning or controlling the direction of flow—it loses all advantage if it tumbles.

These matters of drag, Reynolds number, separation, and so forth are particularly well-developed by Shapiro (1960), intuitively and anecdotally, and by Goldstein (1938), more thoroughly and formally.

modelling and the Reynolds number

Let us return to two comments made earlier. Consider situations involving objects in a flow or some flow through pipes. If the objects or pipes are geometrically similar, equality of the Reynolds number implies both geometric similarity of the pattern of flow and equality of the drag coefficient. You might reread that last sentence. Physical science has rarely given the biologist a better deal or one which is so commonly ignored. What is means is this: Say you have a situation involving fluid movement which you find experimentally awkward by virtue of inconvenient size, speed, or fluid medium. You are completely free to model that situation with another situation of different speed, size, or medium provided that you maintain the original Reynolds number and geometry. The restrictions on application of this rule are few. Compressibility must be negligible, no fluid-fluid interface may be involved, the situation must be isothermal, and the object or model being tested must not be abnormally distorted by the force of the new flow. Beyond that, the rule is both precise and general.

Put in the usual jargon, if the Reynolds numbers for two situations are the same, the situations are therefore "dynamically similar." Drag may be a complex phenomenon, theory may be inadequate and recourse to measurements and models necessary, but dynamic similarity greatly facilitates the business. If the original situation and the model system both involve the same fluid medium, equality of the Reynolds number is ensured merely by maintaining constant the product of length and velocity. If the medium is changed, the variables are three—length, speed, and kinematic viscosity (equation 5.1b)—and the juggling is only slightly more complicated Often one can take the same object and test it in a new medium, adjusting the speed to compensate for the change in kinematic viscosity. Thus flow visualization may be facilitated by a shift from air to water. To work on very small organisms, large models can be made and then tested in a highly viscous medium such as corn syrup, silicon oil, or glycerin. For inconveniently large systems, a smaller model may be tested at a higher velocity. A few examples of investigations which used such model systems follow.

Jefferies and Minton (1965) were interested in the use of drag-increasing devices to reduce sinking speeds in small, extinct bivalve mollusks. They found it convenient to test larger models with adjustable densities to control sinking speed, at speeds appropriately lower than those estimated to occur in nature. They also made what is, I think, the first mention of the general applicability of modelling at constant Reynolds number to biological problems.

As mentioned earlier, I once investigated the induced ventilation of the burrows of prairie dogs (Vogel et al., 1973). A prairie-dog burrow was an unwieldy subject, about 0.1 m in diameter, 15 m long, and located more than 2000 km from my laboratory. Just down the hall, however, was a wind tunnel originally designed by my colleague, Vance Tucker, for his work on bird flight. So I built a model burrow ten times smaller than life-size to fit in the tunnel. To see what would happen with an external wind of 1 m s^{-1}, the tunnel was run at ten times that speed. Perhaps a lot of stove pipe could have been buried in an open field, but the scale model was far simpler.

In a related project, I wanted to look at how flow induction varied with the Reynolds number. By combining a range of sizes of models with a range of speeds in a flow tank, I was able to measure the rate of induced flow over three orders of magnitude of the Reynolds number. The scaling rule worked perfectly. A single curve emerged, not a family of curves: and the upper end of the range tested with one model neatly coincided with the lower end of the range tested with the next larger one.

In a study as yet unpublished, Robert Zaret investigated the stability and preferred orientation of tiny oblate spheroids in free fall, asserting its relevance to the hydrodynamics of plankton. He used a brand of edible oblate spheroids, each abut 1 cm in diameter (M&M's) in a very viscous liquid. And Frederick Vosburgh (1977) covered a large chunk of reef coral skeleton with strain gauges and tested its mechanical properties in a wind tunnel at about fifteen times the speed of the water currents expected in the ocean.

William Balsam and I (1973) built a life-size model of an archaeocyathid (an extinct, vase-shaped, perforated, sessile beast) to see if secondary flows similar to those in sponges could be induced. At that time I lacked a suitable flow tank, so this aquatic organism got its test in a wind tunnel, with the requisite fifteenfold speed increase.

Sometime earlier, I was interested in whether fruit-fly wings ''stalled'' in a manner similar to airfoils at higher Reynolds numbers. These wings are only about 1¼ mm from leading to trailing edges, and air flows over them at 2 m s^{-1}. Thus a particle of fluid passes from leading to trailing edges in less than a millisecond; between the gaseous medium, the high speed, and the small size, the prospect of flow visualization was daunting. So I increased the wing size eightfold by using a 1-cm model and shifted to an aqueous system. In compensation, the speed was reduced 8 × 15, or 120 times, and photography with a repetitive flash and suspended plastic particles (Vogel and Feder, 1966) was a simple matter (see Figure 3.8). The results were informative as well as pretty (Vogel, 1967b).

I've also investigated (with less definitive results) whether an external wind can induce bulk airflow (to be distinguished from diffusion) within air passages in the spongy mesophyll of leaves. Not only are the passages tiny, but diffusion would be inconveniently rapid if a tracer gas were used. However, diffusion in water is about 10,000 times slower than in air (which is probably why lung capillaries are so much smaller than alveoli). I found that it is possible to replace the air within leaves with water by simply submerging them and applying and releasing a vacuum several times, for plant cells are sturdy creatures. The leaf can then be submerged in a flow tank of a non-staining dye such as Evans blue; the progress of the dye through the leaf is observed by comparing various end points. Dye movement is certainly easier to watch than the movement of any gas. In addition, it is possible to extract dye quantitatively and to follow the rate at which fluid entered the leaf with *post hoc* colorimetry.

So the possibilities are vast. Why am I one of the few people making use of the scheme? Someone might look into the effects of surface sculpture on sinking rate in pollen. The Reynolds number is exceedingly low, and surface detail should not, by the usual rules, matter much. Still, Zeleny and McKeehan (1909) found that some spores fall 30% more slowly than Stokes' law predicts, so nature may have some tricks to be discovered. Large models, variously ballasted and falling in highly viscous liquids should be easy to arrange. Small, passively sinking, biological objects come in a wide range of shapes: think of all the pollen, wind-dispersed seeds, ballooning spiders, and phytoplankton. Yet we know almost nothing about the relationship between all that morphology and drag.

One additional topic ought to be considered while we're on the subject of modelling. Recall that equality of Reynolds number means equality of drag coefficient. It does not necessarily mean equality of drag itself unless the original object and the model are both exposed to the same fluid medium. But if the medium is changed, the drag will typically change. For a shift from air to water (both at 20 °C) the drag will rise, although not enormously—3.5 times if the size of the object remains unchanged. The low factor of change is, of course, a result of the attendant velocity decrease.

It is entirely possible to pick a liquid of appropriate viscosity and density so that an object which normally experiences an air flow can be shifted to a liquid system without an alteration in drag. The utility of such a procedure should be great when dealing, as we often do, with organisms that deflect or change shape when air moves across them. We can, without difficulty, derive an expression which permits such a liquid to be chosen. Let subscript 1 indicate the original situation (air) and subscript 2 indicate the unknown liquid medium. Constant Reynolds number means that

$$\frac{\rho_1 l_1 U_1}{\mu_1} = \frac{\rho_2 l_2 U_2}{\mu_2} \tag{5.5}$$

And constant drag coefficient means that

$$\frac{D_1}{\rho_1 S_1 U_1^2} = \frac{D_2}{\rho_2 S_2 U_2^2} \tag{5.6}$$

Using the same object implies that $l_1 = l_2$ and $S_1 = S_2$, so that the S's and l's cancel in equations (5.5) and (5.6). We can then combine the two equations to eliminate U_1 and U_2 and obtain a most useful result:

$$\frac{D_2}{D_1} = \frac{\mu_2^2 \rho_1}{\mu_1^2 \rho_2} \tag{5.7}$$

In the shift from one medium to the other, drag will be unchanged if $D_2/D_1 = 1$. Table 5.2 gives D_2/D_1 values for the transition from air to a variety of liquids for an object of a given size. Water at 52 °C would seem a reasonable choice if the specimen can stand the heat. Alternatively, mixtures of acetone and water or ethanol or of carbon disulfide and methanol might be useful. As far as I know, no biological investigation has ever made use of a shift from air to a liquid which was picked so drag would remain unchanged.

TABLE 5.2

Factor by which the force on an object is increased if it is shifted from air to a liquid. Liquids are at 20 °C unless otherwise indicated.

liquid	D_2/D_1	*liquid*	D_2/D_1
Acetone	0.421	O-xylene	3.052
Carbon disulfide	0.379	Propanol	21.957
Carbon tetrachloride	2.069	Toluene	1.369
Ethanol	6.331	Water	3.467
Methanol	1.556	Water, 52 °C	0.994
Octane	1.441		

CHAPTER 6

the drag of simple shapes and sessile organisms

THE SUBJECT OF DRAG HAS NOW BEEN introduced, aspersions cast upon its origins, and defamatory remarks leveled at its character. In this chapter and the next we turn to the matter of what organisms do about drag: how much drag they encounter and with what adaptations they increase or decrease it. That drag depends very much on shape is no recent observation. It does so in ways so subtle that no single numerical index has been devised to relate the shape of an object to its drag. Drag coefficient and Reynolds number, one must remember, are just measures we apply; and most of their virtue lies in organizing our empirical observations.

the biology of drag

Nature long ago discovered that certain shapes experience particularly low levels of drag, and one of the earlier intentionally-streamlined bodies had a distinctly biological source. Sir George Cayley (1773-1857) recognized that in "spindle shaped bodies," "the shape of the hinder part of the spindle is of as

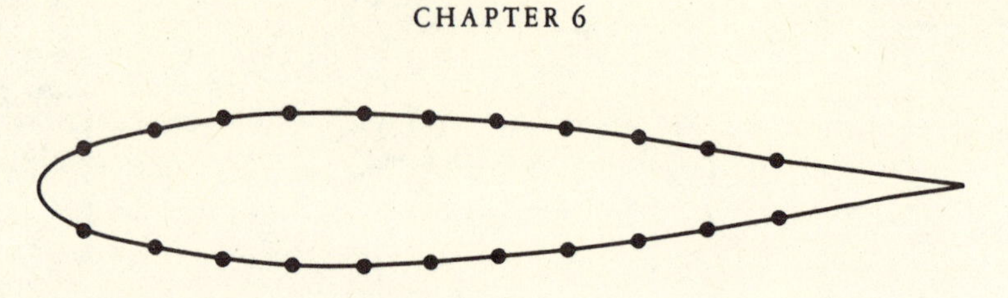

The profile of a modern, low-drag airfoil superimposed on Cayley's data from girth measurements of a trout, the coincidence pointed out by von Karman (1954).

FIGURE 6.1

much importance as that of the front in diminishing resistance.'' Since the subject seemed so mysterious as to defy reasoning, he quite frankly copied nature. Cayley measured the girths of various cross sections of a trout and then divided each girth by three. From these latter figures he constructed the profile shown in Figure 6.1. Von Karman (1954) has pointed out that Cayley's profile corresponds almost precisely to that of a modern, low-drag airfoil. The coincidence may seem somewhat surprising since the body of a trout is a thrust-producing as well as drag-experiencing object. Still, fish do glide or coast, and (as Weihs, 1974, has shown) they may improve their energy economy by alternately accelerating and gliding.

The biological relevance of drag needs little belaboring. The coincidence in shape between natural swimmers and modern submarines, between the fuselages of small aircraft and the bodies of large birds (Pennycuick, 1972), and between the hull shapes of ducks and boats (Prange and Schmidt-Nielsen, 1970); the fact that trees sometimes blow down and corals incur damage in storms—all these need only the assumption of natural selection to argue the point. Indeed, the difficulty we face is that we see the solutions rather than the basic problems, and the latter must of necessity be inferred. Measuring the drag of some natural object yields only a datum. Only through controlled alterations of the object and comparisons with models of varying degrees of abstraction can much be said about the design. And any such statements must be based, as well, on information about the characteristics of the flow the living object would encounter in nature.

The particular relevance of drag depends very much on the circumstances. For a fish or bird, propelling itself away from a possible contribution to the next trophic level, drag is clearly a Bad Thing. But for the same fish, faced with an urgent need to reverse course, or the bird, about to land on a fixed branch rather than a runway, drag is undoubtedly quite welcome. For the sessile inhabitants of streams, reefs, and forests, drag must be a nuisance endured for the sake of the carbon atoms brought in one form or another by the flow. Thus sea fans are exposed to strong, bidirectional waves on reefs. They begin attached life quite randomly-oriented but gradually all grow to the orientation with the highest possible drag—perpendicular to the currents. It appears that this "bad" orientation is important in maximizing the feeding efficacy of their thousands of tiny polyps (Wainwright and Dillion, 1969). For a moderate-sized animal freely falling earthward, drag may have value above all else, with nothing fixing its fitness better than parachute or patagium. I once dropped two mice onto concrete from about 15 meters. They were not obviously damaged by the experience; but as interesting as the lack of injury was their adoption of a nicely parachute-like posture, with all appendages spread wide and no tumbling at all. Drag maximization appears to be a built-in feature of their behaviorial repertoire.

We'll be mainly concerned here with the minimization of drag, and we'll consider the phenomenon of drag at Reynolds numbers of roughly 100 and above. Most of the interesting cases of maximimization of drag occur at lower Reynolds numbers and are thus in a different physical world; their consideration will be deferred to Chapter 13. There is, by the way, human relevance to these questions about the drag of organisms. Jobin and Ippen (1964) have emphasized the crucial role of drag on snails in irrigation canals. If the water velocity is everywhere sufficient, the snails are dislodged. Otherwise the local inhabitants may be afflicted with schistosomiasis. And, it has been claimed (by a blowhard, no doubt) that tornadoes will pluck chickens.

shape and drag

The distinction between streamlined and bluff bodies has already been mentioned. For organisms, the functional distinction is not merely a matter of whether drag is to be minimized or maximized. It depends, in addition, on whether the direction of flow is constant or controllable and the extent to which behavioral or morphogenetic reorientation is possible. Streamlining, in short, works only for a rather narrow range of relative orientations of object and flow. And streamlining isn't a simple and definitive "cure" for drag, with one standard shape to replace a cylinder and another to replace a sphere. Things are more complicated. First, the optimal shapes depend on the Reynolds number.

At low Reynolds numbers (below 100 or so) minimization of pressure drag entails a significant rise in skin friction, so one actually minimizes total drag with a somewhat stubbier shape than that which would give the lowest pressure drag. And second, at all Reynolds numbers, alternatives are available, trading off, as it were, a convexity here for a slight concavity there.

Moreover, the function of the streamlined object must be considered. If an organism with guts and gonads is to be put inside, then what really matters may not be a minimal drag coefficient as usually given but perhaps minimization of drag per unit of body volume. If the object is, say, a compression-resisting strut, then it is important to have a high "second amount of area" to resist "Euler buckling" without excessive weight (Gordon, 1978). In either of these cases, a wider cross section should be functionally best even if its drag is not the lowest possible. Then, with an aero- or hydrodynamically suboptimal shape, separation reasserts itself, and the turbulent transition may not lead to any increase in drag coefficient. Hoerner (1965) provides a compendium of relevant data and references.

One reasonably safe generalization is that at moderate and high Reynolds numbers, the higher the thickness-to-chord ratio (the fatter the airfoil), the higher the drag coefficient based on wetted area, as shown in Figure 6.2. Still,

FIGURE 6.2

The variation of drag coefficient with fineness ratio (Fig. 5.6) for an airfoil section at a Reynolds number of 600,000. The drag coefficient is based on plan area (top view). As an exercise, the reader might try to visualize such a graph with the drag coefficient based on frontal area.

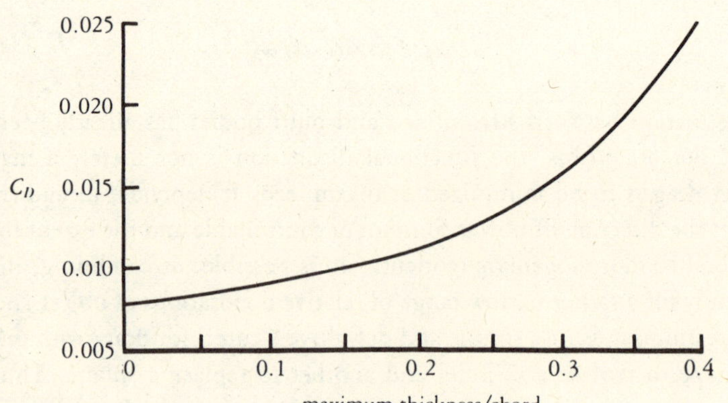

tricks are possible. Laminar and unseparated flow may be maintained by sucking fluid in through the porous skin of an object. Alternatively, separation may sometimes be postponed by ejecting high velocity fluid in a downstream direction at the surface near the normal separation point.

Airfoils with especially low drag have been developed in which the drag is actually a little less than that of a flat plate parallel to the flow. They're called "laminar flow" airfoils, and are distinguished by having their maximum thickness unusually far behind the leading edge. Also, they work only when very smooth and very closely oriented to the oncoming flow; otherwise, turbulence develops.

There is an apparent physical difference between the behavior of streamlined and bluff bodies. It looks as if the transition to turbulent flow occurs at different Reynolds numbers, between 1×10^5 and 2.5×10^5 for bluff bodies and between about 5×10^5 and 1×10^7 for streamlined bodies. But that's mostly an accident of the conventions for picking the characteristic length. If maximum length in the direction of flow is used, then the relatively more elongate shape of streamlined bodies will automatically result in higher Reynolds numbers. The real difference is in what happens at transition. For the bluff body the onset of turbulence leads to a narrower wake, less pressure drag, and hence less total drag than prior to the transition. For the streamlined body, separation is absent or minor and pressure drag is quite small. Turbulence, if it does anything, increases drag. Thus roughening a bluff body may (by giving premature turbulence and delayed separation) decrease drag, while roughening a streamlined body usually (by increasing skin friction) increases drag (see Figure 6.3).

Streamlined and bluff bodies are best regarded as the extremities in a continuum of shapes. Again, Hoerner (1965) provides data and examples. For relatively bluff bodies, sucking fluid through a porous surface works in the same manner as it does for relatively streamlined bodies; at a Reynolds number of 100,000, it can reduce the drag coefficient from 1.2 to 0.15. "Splitter plates" (Figure 6.4) behind bluff bodies greatly reduce the rate at which vortices are shed, and the drag may be reduced at Reynolds numbers between 10,000 and 100,000.

Nor is the shape of the front of a bluff body inconsequential. Rounding or fairing the edges in a downstream direction reduces the width of the wake and the drag coefficient. For a flat plate normal to flow, the momentum of the fluid moving outward past the edge carries it well beyond the edge, so the wake is substantially wider than the plate itself. Rounding or deflecting the edges leads to a wake of width more nearly equal to that of the plate (Figure 6.5). This, by the way, is why whirling cup anemometers turn. The cup facing the wind has a higher drag not due to any air it contains but because it has a wider wake than that of the cup facing downward. Hoerner gives drag coefficients for Reynolds numbers between 10,000 and 100,000 for open hemispheres (Figure 6.5); note that an open hemisphere concave upstream has more drag than a sphere. (Figure 5.5, page 74). This difference occurs with long half-cylinders as well, which have C_D's of 1.17 and 0.42 when facing upstream and downstream, respectively. This difference is the basis for the operation of the "Savonius rotor," a pair of half-cylinders on opposite sides of a vertical shaft: it makes an

FIGURE 6.3

The effects of surface roughness on the drag coefficient of streamlined and unstreamlined shapes. Notice that at low Reynolds numbers roughness is inconsequential in either case. Also, note that for the cylinder there is a range in which a rougher surface confers a lower drag; no equivalent range exists for the streamlined shape. In (*a*), C_D refers to plan area; in (*b*) it refers to either frontal or plan area.

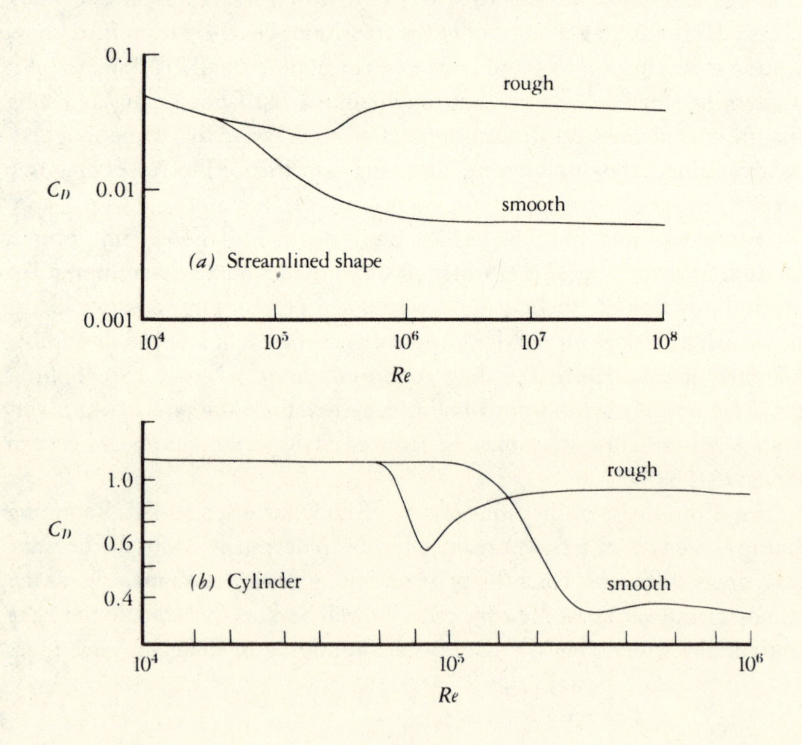

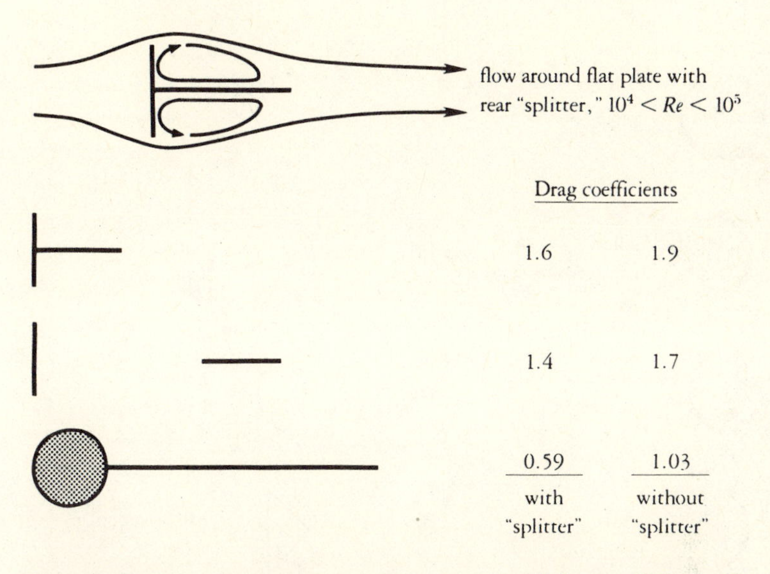

The influence of "splitter plates" on flow patterns and drag coefficients of cylinders and long flat plates. Reynolds numbers are based on maximum dimensions normal to flow and drag coefficients on frontal areas.

FIGURE 6.4

inefficient windmill which enjoys some countercultural appeal since the open half-cylinders can be just the halves of an ordinary oil drum which has been cut lengthwise.

For slender cylindrical bodies oriented parallel to the flow, a concave face is the worst; a flat face slightly better, a sharply pointed front still better; a convex hemispherical front better yet; and a parabolic nose the lowest drag arrangement of all. The same ordering holds for a "rectangular section," a long beam of uniform thickness with a chord (length in the direction of flow) much larger than the thickness (length normal to flow). Rounding the upstream edge, much as the exposed edge of a stair tread is rounded, can reduce the drag coefficient by 40%. By rounding the front, the drag coefficient of the original Volkswagen "Microbus" was reduced from 0.73 to 0.44—not the order of magnitude improvement possible with full (and impractical) streamlining, but not an inconsequential difference. Air deflectors atop the cabs of articulated trucks function in a similar manner (Figure 6.6).

A streamlined airfoil over which flow goes from trailing to leading edge is a kind of bluff body. But it is not particularly bad in terms of drag. For an airfoil for which the maximum thickness is 12% of the chord, the drag coefficient

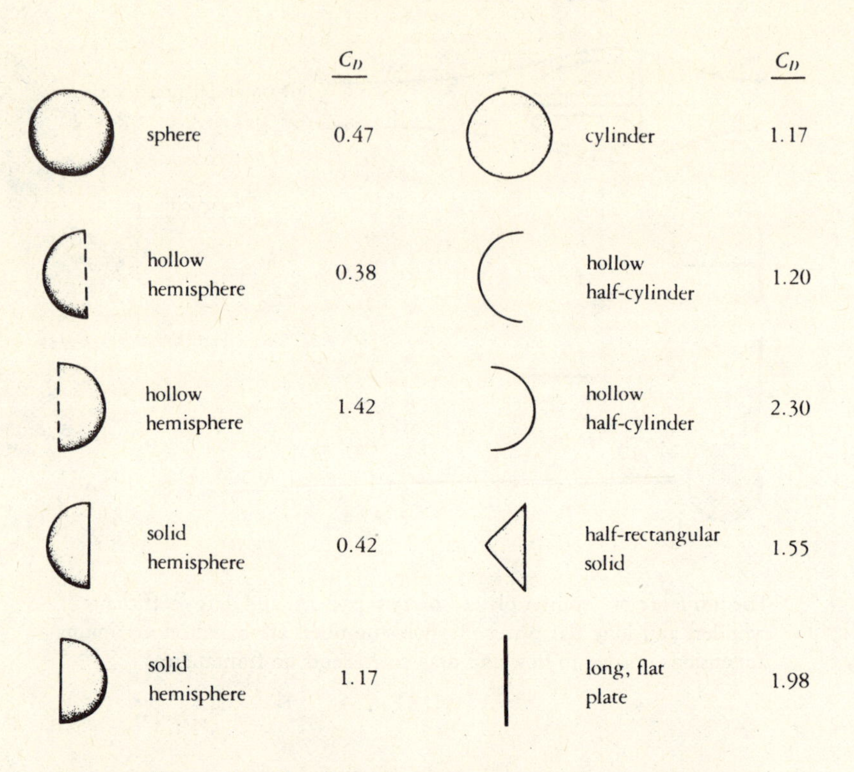

Drag coefficients, based on frontal area, for a variety of shapes and orientations at Reynolds numbers between 10^4 and 10^6. The data may not be precisely comparable due to variations in the experimental conditions.

FIGURE 6.5

(based on frontal area) is 0.06 for normal flow and 0.15 for reversed flow. The drag coefficient may be more than twice as great when flow is reversed, but it's still only an eighth of that of a cylinder at the same Reynolds number of 10^6. It is probably possible to design more-or-less streamlined shapes which have relatively low but equal drags for opposite flow directions, although I haven't determined whether the ends should be pointed or rounded. While they may be irrelevant to aircraft, which are rarely called upon to fly backwards, they may be important in understanding the design of organisms exposed to alternating wave or tidal flows.

Objects which flutter in the wind or current ordinarily incur more drag than ones which don't. The drag of a waving flag, for example, may be several times greater than the skin friction of a rigid, flat plate of the same shape. According to Hoerner (1965), local separation causes flutter, which causes further separation; the result is substantial pressure drag. The oscillations of a flag become more violent further downstream from the flagpole; and, as expected, long, narrow flags have higher drag coefficients (based on surface area) than do short, tall flags. Thus the flexibility of leaves and algal lamellae is unlikely to function as a drag-reducing arrangement, at least in strong, steady flows.

Most of the data discussed in the past few pages are strictly applicable only at Reynolds numbers considerably higher than those commonly encountered by ordinary organisms; that is, in the range of roughly 10^4 to 10^6. There seems to be a great dearth of comparable data for the range of 10^2 to 10^4 and almost nothing for 10^0 to 10^2. Furthermore, extrapolation is hazardous. It's perhaps reasonable to extrapolate to lower Reynolds numbers to generate hypotheses; it's probably unsafe to use the extrapolations as hard evidence. I ran across one cautionary example recently. For flow around a sphere above the turbulent transition, separation occurs about 140° around from the upstream stagnation point (Re = 440,000; Hoerner, 1965); just below transition, separation occurs only 75° around from the front (Re = 170,000). At much lower Reynolds numbers does separation occur even earlier? It actually occurs later again. At high speeds the greater momentum of turbulent flow postponed separation; at lows speeds the greater relative viscosity limits the occurrence of the high shear rates concomitant with separation. According to Seeley et al. (1975), separation from a sphere occurs at 96° at Re = 3000; 102.5° at Re = 1300; 111.5° at Re = 750; and 121° at Re = 300.

FIGURE 6.6

A truck equipped with an air deflector designed to reduce its drag.

CHAPTER 6

drag in the forest

Most often, organisms which live attached to a substratum and protruding into air or water show some sort of radial symmetry. The advantages are obvious: facing all directions at once is handy for grabbing prey or intercepting sunlight. Among the disadvantages are the high drag forces encountered by bluff bodies. One evasion, crouching down against a surface, we'll explore in Chapters 8 and 9; here we'll consider the magnitude of the drag forces and some other minimizations and circumventions.

Pre-eminent among large organisms for whom drag is important are trees. They often exceed 50 m in height, have enormous surface area, withstand major windstorms, and constitute a major component of terrestrial biomass. I am astonished to discover whole books on the functioning of trees which make no mention of their splendid mechanical and aerodynamic performance, but the fact that we know very little about the relationship between tree design and wind resistance does not diminish the interest or importance of the problem. In the hope that someone will take up an investigation, I'll present an introduction to what is known in an admittedly speculative context.

Horizontal winds should act mainly on the crown of a tree, both because wind speed increases with height above the ground and because the surface area of a tree mainly resides in the crown. Acting high up in the tree, the wind will produce a torque at the base of the tree which is the product of the drag and the height at which the drag force acts, as shown in Figure 6.7. Now, healthy trees commonly uproot rather than break when stressed by extreme winds; the phenomenon is called "wind-throw" (see, for example, Brewer and Merritt, 1978). If the tree trunk is relatively stiff, the uprooting torque will be opposed by a torque composed of the weight of the tree times the distance between the line of action of the weight and a rotation point somewhere lateral to the center of the tree's base. Under these circumstances, a tree might withstand considerable wind with no more attachment to the ground than enough friction to keep it from sliding downwind! Perhaps that is part of the reason why large trunks are so heavy, and why trees don't resorb their heartwood and become hollow cylinders like bamboo stems or flagpoles. Put another way, their weight may help them resist overturning, since the first motion in overturning will require lifting of the trunk. If, on the other hand, a tree trunk is less stiff and given to significant swaying, then the gravitational torque may actually act in concert with the drag torque (Figure 6.7). In that case the roots must literally grab the ground or penetrate deep enough to resist major turning forces. As Alexander (1971) points out, the local forces in the trunk generated by winds

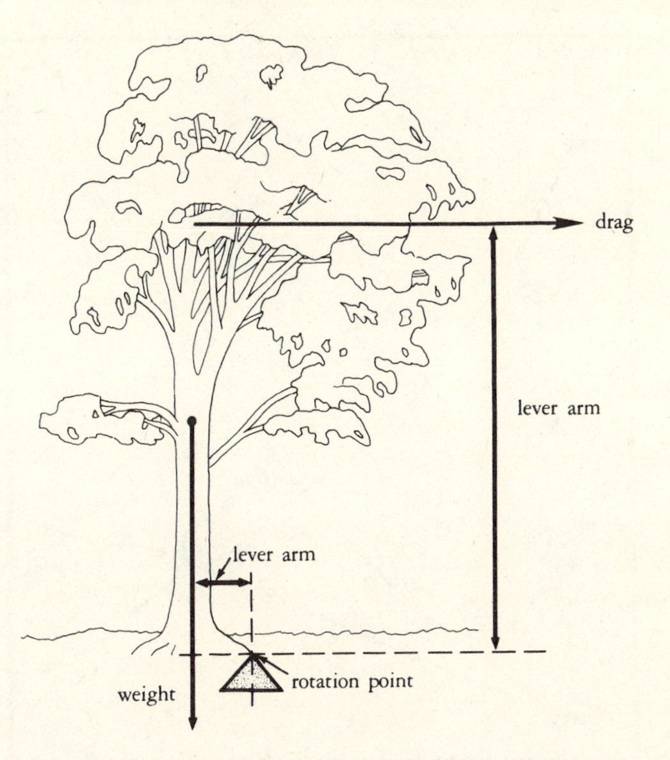

The principal forces and their lines of action when a wind blows on a stiff tree.

FIGURE 6.7

may far exceed those due to the weight of the tree: a tree may fall under wind stress, not under its weight alone.

So what are the drag forces on a tree? The practical problem is formidable. You can't just measure the drag on a leaf or branch and blithely multiply. On one occasion a series of trees were tested in a large wind tunnel (Fraser, 1962) and drag measured at a series of speeds between 9 and 26 m s^{-1}. Some time later, the original data from these experiments, including previously unpublished items, were further analyzed (Mayhead, 1973a). As speed increased, the drag increased in all cases, with no discontinuities anywhere in the data. There were, as we'd expect, no sharp transitions with such irregular objects. But as the drag increased, the coefficients of drag dramatically decreased. What was happening was that the drag coefficients were calculated based on frontal area in still air. However, as the wind increased, the trees steadily

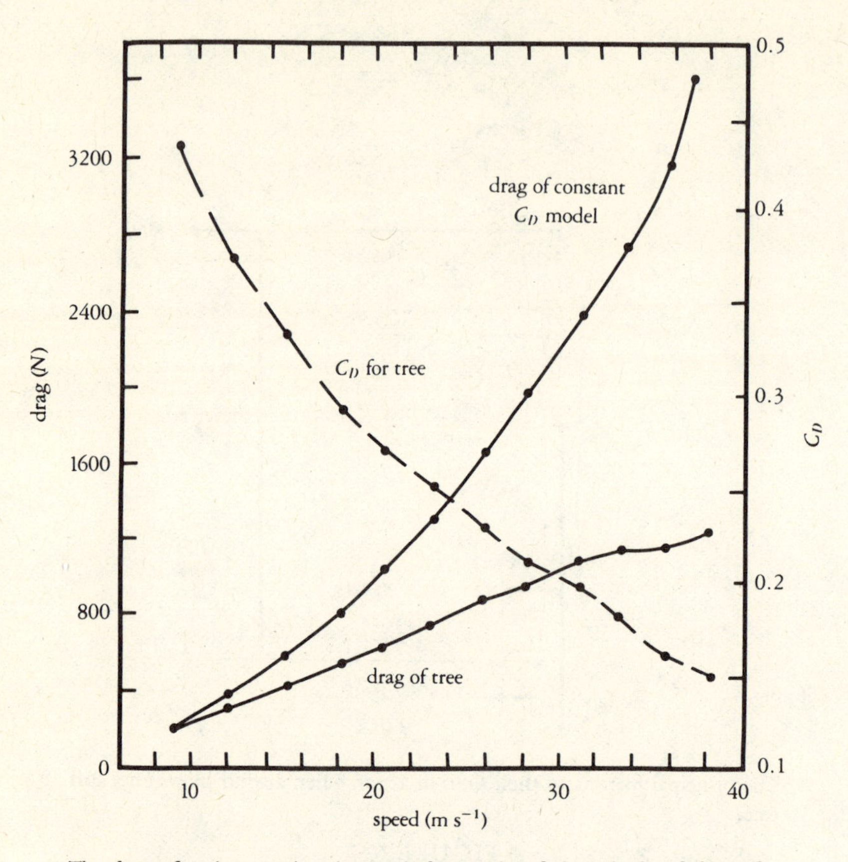

The drag of a pine tree in a large wind tunnel and the calculated drag of a model with the frontal area and drag coefficient of the tree in a gentle breeze (solid lines). The dashed line shows the decreasing drag coefficient which the pine would need in order to produce its nearly linear plot of drag versus speed.

FIGURE 6.8

decreased their exposed area, so drag, overall, was more nearly proportional to the first than to the usual second power of velocity. Probably the drag coefficients at very low speeds would have been even higher than those calculated for the lowest available tunnel speed. Figure 6.8 gives the drag and drag coefficients for one tree which was tested to speeds beyond 26 m s^{-1}, in fact to the actual breaking point following irreversible distortion. For comparison, I've included the drag of an object for which C_D is constant at a value equal to that of the tree at the lowest speed tested.

These trees, then—firs, pines, spruces, and hemlocks—have a trick. At low speeds they have, by any reasonable comparison, very high drag coefficients, from even a conservative extrapolation of the curve in Figure 6.8. At low speeds, though, wind is of litle mechanical consequence to trees, and there are more important considerations in their design such as sufficient interception of light (Horn, 1971); efficient water use (Parkhurst and Loucks, 1972); and adequate convective heat dissipation (Vogel, 1970). During the small fraction of their existence when mechanical stress is serious, they put aside, to some extent, these other considerations and assume their progressively more drastic reorientation to reduce their drag.

The story has many more pieces. For one thing, the trees tested by Fraser (1962) were all conifers. I'd expect the drag of broad-leafed trees such as oaks to be greater. It would not be surprising if greater drag of the crown underlies the broader, stiffer, and denser trunks of temperate hardwoods; with greater drag, they might have to be stiff to keep that gravitational torque working for rather than against them. And the hardwoods commonly have very wide and stiff bases. Many square meters of ground rotate upward when a large oak blows over, at least here in central North Carolina. Broad leaves have special opportunities for reorientation, yet I know of no drag measurements on single leaves or groups of leaves of deciduous trees at high speeds. The results of such measurements are likely to be quite different from those of Thom (1968) on rigid models of single leaves at low speeds. Our local oaks, when they are windthrown, are fully leaved. I am puzzled that they make no attempts to save themselves with a bit of pre-season leaf abscission.

Also, the forces necessary to pull trees down with a winch correspond to higher speeds than those which blow them over (Fraser, 1962). Trees don't slowly deflect to some point of no return; instead they sway back and forth in response to gusts and their own natural periods (Mayhead, 1973b; Holbo et al., 1980) and, according to Oliver and Mayhead (1974), gradually get their roots loosened. Furthermore, the natural period is highly size-dependent (McMahon, 1975) which raises complex questions about scaling.

Consider also the situation in the great forests of the tropics. Tropical trees are reported to be more slender for their heights (Hartshorn, 1978) and more elaborately buttressed (Smith, 1972). To what extent do these characteristics reflect differences in extreme wind speeds, mechanical properties of soils, or the interconnection of trees by strong vines?

And then there are the drag problems of alpine trees and shrubs which, between the effects of altitude, slopes, and lack of shelter, are clearly exposed to both strong and relatively continuous winds. Their deformations are so dependable that they may serve as excellent climatic indicators (see, for example, Yoshino, 1973; D. Thomas, 1958; and T.M. Thomas, 1973). The standard explanation is that winds deform these plants, stunting them and causing their branches to point downward. I find no suggestion that this causation might be ultimate as well as proximate. Perhaps the deformed shapes are adaptive, producing a design which exposes a reasonable leaf area to the sun even in moderate winds and which experiences reduced drag in the face of winds from the prevailing direction. Some of these plants would fit into wind tunnels of modest proportions and power, and a few deliberate experiments might be more revealing than a huge amount of field work.

and drag at the seashore

The role of drag in determining the structure and ecology of sessile organisms may be even more important in water than in air. For inhabitants of streams and rivers, current is at least relatively constant in both speed and direction, but for communities on reefs or rocky coasts current can be exceedingly irregular. Under moving (but not breaking) waves, individual fluid particles move in circular orbits, completing one revolution during each wave period. In shallow water, these circular orbits become increasingly elliptical at greater depths until they reduce to purely back and forth motions just above the bottom (Figure 6.9). Flexible organisms attached to the bottom wave back and forth under even very mild surface waves; they are exposed to what is mainly an alternating current in contrast to the pulsating direct current impinging on terrestrial vegetation. A simple and engaging account of wave-induced water movements is given by Bascom (1980), and a more detailed discussion can be found in Ippen (1966) and other standard works on coastal oceanography. Where waves encounter near-vertical obstructions, the situation is far less regular or predictable.

There exist quite a number of studies on the relationship between currents and the distribution of attached marine organisms; the reader might find interesting Lewis (1964), Riedl (1971a), Schwenke (1971) and Mathieson et al. (1977). Mostly, though, these are field investigations looking for correlations and are of limited utility, at least in the present context. It's no easy matter for a correlative study to rise above correlation to causation. So, while the distribution of organisms in the rocky intertidal may correlate with current levels, it also may correlate with relative exposure to air, light, and temperature

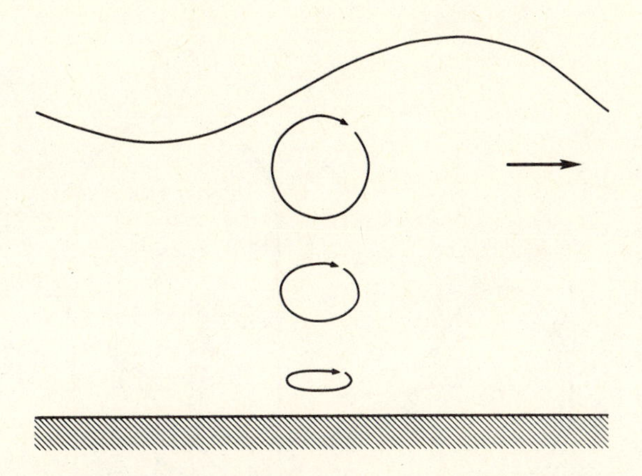

The orbits of three particles of fluid as a wave passes in shallow water.

FIGURE 6.9

fluctuations as well as with changes in biological factors such as predation and food supply.

To make matters worse, water movement may be viewed as three almost distinct agencies. There is, first, velocity *per se,* which generates drag. Velocity must be of great significance under the rare but crucial circumstances when it achieves extreme values. In addition there is something which has been called "exposure" and is operationally defined by the rate at which a standard hunk of plaster of Paris is eroded by water movement and wave action (Muus, 1968; Doty, 1971). Exposure entails drag, but only in extremely viscoelastic organisms such as anemones is chronic low-level drag likely to approach acute high-level drag in importance (Gosline, 1971). Finally, there is shock or impact pressure, treated well by Carstens (1968). A wave of a dense and nearly incompressible substance like water may generate a very brief, local, and intense pressure when suddenly stopped by a rock or sea wall. Carstens cites a wave shock of 2×10^6 Pa (20 atmospheres) hitting a ship. But, as with sound waves, the displacement is low; and only a little "give," or flexibility, is presumably sufficient to render these pressures harmless. It is possible, of course, that some organism has devised an appropriate impedence-matching device (such as that we use in our middle ear) to increase displacement and do something interesting with shock pressures. Shock pressure might shear off an attached body which wasn't centered in the pressure zone or extend cracks in brittle objects, but its biological role seems as yet undocumented.

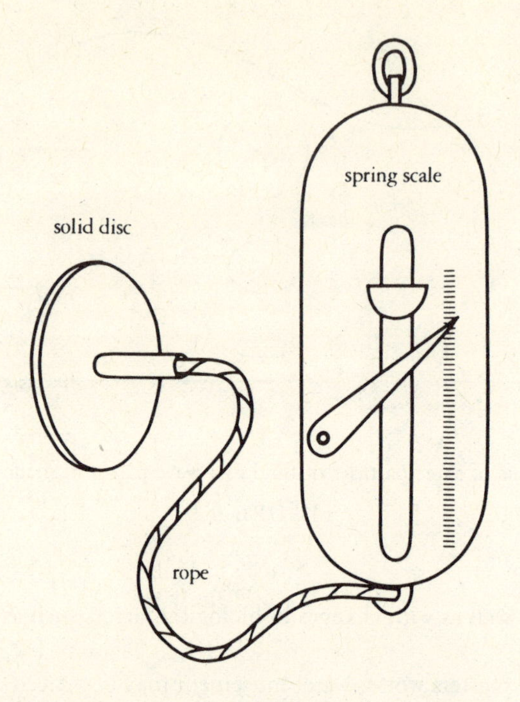

The device of Jones and Demetropoulos (1968) for measuring maximum currents near the edge of the ocean. The spring scale is modified by the addition of a lever which remains at the extreme position to which it is pushed as the spring extends.

FIGURE 6.10

Of these several mechanical aspects of water movement, velocity and its concomitant drag has been regarded as the most important (Jones and Demetropoulos, 1968), a judgment with which I concur. These investigators have devised a simple device to record the force of the maximum current in near-shore habitats. They use a disc normal to the flow trailing behind a spring scale that is equipped with an arm which remains in whatever extreme position it reaches (Figure 6.10); the whole device is free to rotate downstream from its attachment. The meter is inexpensive and weathers storms unattended, so if peak velocities are what matter, it is a most appropriate kind of measuring tool. The actual data of Jones and Demetropoulos are slightly misleading on two accounts. First, they measured drag rather than (as claimed) dynamic pressure. To get speed from drag a drag coefficient must be assumed, and Hoerner (1965) gives a C_D of 1.2 for a circular disc normal to flow at Reynolds numbers of between 10^3 and 10^7. (The constancy of C_D makes a disc a particularly good test object for the present purpose.) Using this drag coefficient, one must multiply their readings (in pounds) by 5.6 and take the square root of the product to get velocity (in meters per second). The results, though, will still be misleading, since the disc (4 cm in diameter) was only 17.5 cm behind the unstreamlined scale and was thus in the middle of a substantial wake. The real velocities must have even been higher than the estimates quoted from their data by Charters et al. (1969) and Wainwright et al. (1976). (If, by the way, you build one of these J-D "dynamometers," I suggest getting a proper figure for the drag coefficient in a fast flow tank or by towing alongside a boat.)

The fastest flows recorded by Jones and Demetropoulos (1968), calculated as above, were about 14 m s^{-1} and resulted from the action of waves of about 6.5 m in height. My own guess is that these flows were really about 16 m s^{-1}, which would correspond to drag forces about 30% higher still. Either way, these maximum speeds are far above the 5 m s^{-1} quoted by Riedl (1971a) for "shallow or narrow coastal areas." But that doesn't make the figures at all unreasonable: a steep bank can be very effective in converting even small waves into locally-rapid currents. This difference between 5 and 16 m s^{-1}, incidentally, corresponds to a difference of an order of magnitude in drag, assuming constant C_D.

A current of 16 m s^{-1} should not be taken lightly. In terms of flow pattern, it corresponds to a wind of 240 m s^{-1} (540 mph). In terms of drag, it corresponds to a wind of no less than 460 m s^{-1} (1000 mph), were the latter subsonic and the drag coefficient unchanged. Certainly it's nothing you would swim in and enough to thoroughly disembowel any ship. Still, life abounds on rocky coasts, perhaps in part living on the premasticated food breaking waves provide. In fact, the drag on an organism needn't prove intolerable. Consider a limpet, a very wide and low, elliptically-conical beast, attached to a rock. If it has a frontal area of 1 cm^2 and a drag coefficient of 0.3 as a semi-streamlined protrusion (Hoerner, 1965), then its drag comes out to less than 4 N (about 0.9 lb.). Clearly this is not devastatingly high for an organism which is nearly all foot. In my experience, it takes more than a pound of force to pull a limpet off a rock or to push enough to dislodge it. And we've considered a worst case in which the limpet derives no protection from the irregularities on the surface of its rock. Barnacles seem a little less streamlined, but they're much more thoroughly cemented to the substratum and probably mutually sheltered. A snail, though, might well seek shelter in such a storm: its foot is smaller than the limpet's, and its drag coefficient must be much greater (Moore, 1964).

Algae. On surf-pounded, rocky coast, algae often are large in size and comprise a major portion of the local biomass. They have somewhat the same problem as forest trees: exposure to sunlight is crucial for photosynthesis, yet exposing a large surface almost inevitably involves a large drag coefficient. A common arrangement for large algae is a "holdfast" which grabs a rock; and a "stipe," often cylindrical in cross section, connecting the holdfast to one or more "laminae." The dominant force on these algae is drag. While drag has been measured only at modest velocities, it is clear from extrapolation to higher speeds, from measurements of the strength of the stipe (Delf, 1932), or from observation of storm damage to algal beds (Schwenke, 1971) that drag forces may reach very high values. These algae experience major drag forces as almost purely tensile loads: at low speeds the stipe, with or without the help of floats, may hold the laminae up near the surface, but at high currents the stipe's flexibility is sufficient so that it certainly reorients parallel to flow. The laminae are inevitably very flexible.

Two studies provide some view of the response of an alga to drag. Charters et al. (1969) measured reductions in drag coefficients with increases in speed for *Eisenia arborea;* their results were quite analogous to what we noted for trees. At 0.26 m s^{-1} the drag coefficient (based on cross-sectional area) was 2.41, while at 0.49 m s^{-1} it had dropped to 1.41. Their data, however, are very limited and somewhat confused as a result of an attempt to separate the total drag into pressure drag (on cross section) and skin friction (on wetted area); I think this separation is hard to justify either theoretically or for any possible utility. It would be nice to have figures for drag coefficient over a wider range of speeds.

More recently, Koehl and Wainwright (1977) investigated the way in which *Nereocystis luetkeana* resists tensile forces. This remarkable creature (Figure 6.11a) may attain a length of 40 m in its single growing season. Nature seems to make them shoddily as well as rapidly, and late in the summer enormous numbers break loose and entangle just about everything on parts of the northeast Pacific coast. At 1 m s^{-1} the stress from drag is about a tenth of the breaking strength of the stipe; this means that at constant C_D they should

FIGURE 6.11

(*Facing page*) Various organisms which live attached to rocks and exposed to rapid and irregular currents. (*a*) *Nereocystis,* a very large alga with laminar blades and a tubular stipe; (*b*) a laminate alga with markedly undulate margins; (*c*) a highly dissected alga with cylindrical branches; (*d*) a cluster of sponges; (*e*) *Leptogorgia,* the "sea whip," a flexible, branched coral; (*f*) *Gorgonia,* the "sea fan," a planar, fenestrated, flexible coral.

(a)
(b)
(c)
(d)
(e)
(f)

withstand currents of no more than 3.2 m s^{-1}. But we know nothing about the actual drag coefficients nor about the role of reorientation with increasing speed. The data hint that non-steady-state behavior of the algal stipe may be critical: stipes aren't particularly strong (stress to break) but are sufficiently stretchy so the *work* needed to break one is as high as that for wood or bone.

One common feature of large laminate algae is a so-called "undulate," or ruffled margin (see, for example, Bold and Wynne, 1978, and Figure 6.11b). This is the character which results in the folds one sees in the edges of laminae when they are pressed flat for museum collections. Ruffling is rare in terrestrial leaves but occurs also in some submerged leaves of stream vegetation (Sculthorpe, 1967). Regarding an undulate margin as an adaptation to wave action is attractive both as pun and possibility. I'd guess that the feature doesn't reduce drag, but may help the leaves or laminae resist currents and abrasion by minimizing stress concentrations and making tears more difficult to initiate.

Another frequent characteristic is a high degree of "dissection" of foliage: elaborate branching combined with individual elements having very small cross-sectional areas (Figure 6.11c). The arrangement occurs not only in marine algae but in many fresh-water plants and in many colonial hydroids and bryozoans. With somewhat flexible branches, the scheme seems ideal for gradually reducing exposure as current speed increases. One might guess that it would be more useful where velocity is variable than, say, in reasonably steadily-flowing streams: the branches could collapse against each other in periods of rapid flow. Again, no one seems to have investigated the matter.

Sponges. These are less common on rocky coasts, but they resist large currents on coral reefs where they attain heights of a meter or more. Typically, sponges are much stiffer than algae but much less stiff than corals: they appear to deflect but not collapse in high currents, and their volume and orientation are essentially fixed. A common shape is a cylinder, or barrel form (Figure 6.11d). Many sponges are rather bumpy on the outside (rugose, conulose, or verrucose, to be pedantic); it is possible that the bumps promote the turbulent transition, minimize wake diameter, and keep drag down. Riedl (1971a) suggests that sponges reduce drag by sucking in water around them, a possibility we raised earlier in the chapter. Sponges, of course, do suck in water at a high rate: they're active suspension feeders in the sense of Jørgensen (1975). A typical sponge passes its own body volume of water about every five seconds, which should, I calculate, be enough for drag reduction. Nevertheless, I'm dubious about the idea. Sponges shut down their pumps during storms

(Reiswig, 1971a), presumably to prevent clogging with sediment. And if, as I think (Vogel, 1978b), they have dermal valves, then water enters the sponge too far upstream to do much good in drag reduction. However they manage, large sponges can remain erect in high currents with a remarkably small investment (less than 5% of wet weight) in skeletal material. As a rule, though, sponges grow taller in more protected habitats (DeLaubenfels, 1947).

Cnidaria. Conspicuous among what are occasionally called the "rheophile fauna" are the sea fans, sea whips, sea anemones, corals, and other attached cnidaria. They are structurally and dimensionally diverse but most of them live by separating small (and sometimes not so small) organisms from the water which passes over themselves; the organisms are trapped by some combination of tentacles and nematocysts. Probably the ambient movement of water is more important in getting food into their vicinity than is either the action of their tentacles or the active swimming of the prey. They are mostly "passive" suspension feeders in Jørgensen's (1975) sense, and passive feeding (like photosynthesis) favors forms unlike those which might minimize drag. Indeed, almost the opposite may be the case. Leversee (1976), for instance, found that in bidirectional tidal currents the gorgonian coral, *Leptogorgia,* assumed a relatively two-dimensional branching pattern, presenting maximal area to the current (Figure 6.11e). Even where a colony wasn't fully flat, branches were never downstream from other branches. Even bending over in the current, which these colonies do at speeds above 0.1 m s^{-1}, may be a device to regulate flow across the individual polyps instead of or as well as a drag reduction scheme. Still, drag is unlikely to be irrelevant to *Leptogorgia* colonies, especially in habitats where the footing is uncertain.

Leptogorgia, by the way, may have another trick. At current speeds too low to bend the branches, both lateral and downstream polyps appear to feed on organisms in the wake of their branch. Leversee noted that brine shrimp could remain in the eddies of the wake for 15 to 10 seconds. His data suggest that the Reynolds number under such circumstances is about 50: perhaps the dimensions of the system have been selected to utilize a pair of bound vortices as occur behind cylinders at low Reynolds numbers. It might be noted that *Leptogorgia* occur in tidal currents, so wave period is of little importance.

The sea fans such as *Gorgonia* are even more completely two-dimensional. Their branches anastomose, giving a planar and rather uniformly fenestrated structure (Figure 6.11f). In wave-induced reciprocating currents across reefs, as mentioned earlier, the sea fans are all oriented normal to the prevailing directions of flow (Wainwright and Dillon, 1969). In response to

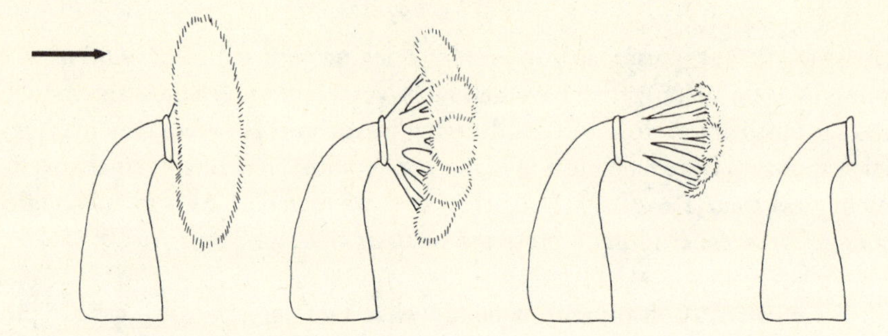

Successive changes in the appearance of a large sea anemone, *Metridium*, as current increases.

FIGURE 6.12

currents they bend, much as does *Leptogorgia;* it's likely, though, that the first response of *Gorgonia* is actually deflection of the individual polyps which protrude into the fenestrations, as suggested by Riedl and Forstner (1968). Again we'd predict that drag should increase at something less than the second power of velocity, and again, no data exist on either drag or drag coefficient.

The hydrostatically-supported sea anemones often bend in currents, but they have another device as well; they can deflate and, in extreme cases, retract down against the substratum to form a hemispherical lump. Deflation is only part of what may be a more complex behavioral response to drag (Koehl, 1977a). A tall anemone, *Metridium senile,* protrudes well into the mainstream, bends downstream at a constriction just beneath the oral disc, and filters with its outstretched tentacles (Figure 6.12). A rigid model equipped with flexible tentacles has a relatively constant drag coefficient of about 0.9. As the oral disc separates into lobes, the C_D drops to 0.4; collapse of the oral disc in the manner of an overstressed umbrella reduces the drag coefficient to 0.3, and retraction of the tentacles reduces C_D to 0.2 (all based on original projected area). Further bending and deflation reduces the drag still further.

Another large anemone, *Anthopleura xanthogrammica,* lives in less sheltered habitats, but it's short and wide instead of tall and thin. In fact, the more exposed the site, the shorter and wider are the *Anthopleura.* A tall model has a drag coefficient of 0.3; a short one has a C_D of 0.2; adding tentacles to the former increases its drag coefficient to 0.7 (Koehl, 1977a). The interspecific comparison is interesting. *Anthopleura* is typically exposed to velocities ten times greater than is *Metridium;* but, in spite of differences in height, shape, and location, the actual drag on each is about the same. Which nicely illustrates the pitfalls of using velocity alone as an indicator of drag (Koehl, 1976). A retracted *Anthopleura* is normally covered with bumps (verrucae), but they don't seem to have any appreciable effect on drag: the drag coefficients of bumpy anemone, smooth model, and bumpy model are essentially the same up to 1 m s^{-1} (*Re* = 100,000). The bumps are more likely to be significant at slightly higher Reynolds numbers, but the size and speed of the flow tank precluded testing. Since it was my flow tank that Koehl used, it would surely not be sporting of me to complain! I might mention that the absence of drag data for all but retracted anemones is no accident: live anemones seemed offended by the taste of the flow tank and just sat there like bumps on a log.

Similar results, with a few nicely quantitative fillips and flourishes, emerged from Koehl's (1977b) work on two species of zoanthids (*Palythoa*) living in overlapping habitats on a reef. It looks like a good thing to be shorter, stiffer, and more attached to your neighbors if you're in a microhabitat characterized by rapid flow.

All of these organisms, from trees to polyps, respond to rapid currents by bending, reorienting, or reversibly collapsing. The stony corals, by contrast, endure typhoons with hardly any deflection at all: they are perhaps the most rigid of large biological structures (Wainwright et al., 1976). Vosburgh (1977) towed coral heads of *Acropora reticulata* instrumented with strain gauges through a tank of water (Figure 6.13a). Depending on how some experimental artifacts

FIGURE 6.13

(*a*) The specimen of the rigid coral, *Acropora,* used by Vosburgh (1977).
(*b*) The relationship between form and exposure to currents in hard corals as described by Hubbard (1974).

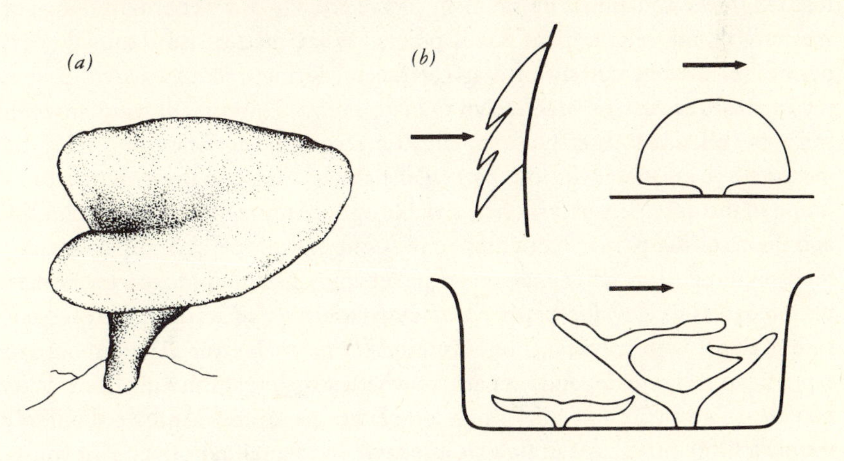

are viewed, drag varied either with the second or just under the second power of velocity; and the drag coefficients were constant or nearly so at around 0.8. From these figures, he extrapolated to the drag at the maximum water velocities predicted for 1 m above the substratum at varying depths. He then measured the forces necessary to break corals and was rewarded with the neat and tidy conclusion that the distribution of the species at Enewetak is limited by the drag forces of the largest waves. Corals do break in typhoons, but there may be surprisingly little large scale damage (Randall and Eldredge, 1977), especially when one considers how a coral stands unbending in the surge—a magnificent pièce de résistance.

Hubbard's (1974) observations provide further evidence that drag forces are important determinants of coral distribution. On walls facing major currents, the corals are usually in the form of downward-facing shingles, close against the substratum and each other (Figure 6.13b). Where currents are minimal, they proliferate into domes, discs, digitate forms, and vertical clusters. On exposed rock platforms, with currents parallel to the substratum, they take the form of round-topped mounds, not far different from limpets, chitons, deflated anemones, or Hoerner's (1965) best protrusions. Stony corals are high-current specialists. They grow fastest in rapidly moving water (Jokiel, 1978) and adjust their morphology to maintain appropriate internal flows (Chamberlain and Graus, 1975). Their stiffness may not help to keep down high-speed drag, but it does provide a protected habitat for the individual polyps in the face of continual currents which may be of substantial strength.

I've quite deliberately given much space to the business of the drag of sessile organisms because I consider the subject important whether one is interested in the functional morphology or the ecology of an enormous range of species. Let me reiterate the main points. What matters isn't usually "exposure" or average current or drag coefficient during everyday activities; it is the extremes of current which seem to matter most. Extremes of wind or waves can't be obtained directly from climatic recording agencies but must be measured or calculated for the particular habitat. Shape matters, but streamlining in the usual sense isn't often a viable option in irregular flows. Flexibility and the resulting passive reorientation in strong flows is perhaps the most common response. The latter may be an increasingly useful approach for human technology as we develop better flexible materials; we seem intuitively to confuse strength with stiffness, but nature makes no such error. The response to rapid flows seems to depend, in part, on whether the organism can afford, in its habitat, to go hypofunctional when local flows are strong. And we shouldn't simply assume that shape in flows is just some accidental morphogenetic consequence of the flow. Whether for alpine shrub or sea fan, it is a better working hypothesis to assume that the adjustment of shape in wind or currents is an adaptively important functional response. We need better information on maximal environmental flows; we need to know the drag of real organisms under conditions of naturally varying flows; and we need to know more about the drag of abstract but organism-like models under biologically relevant conditions.

Models have been useful and will continue to prove so: it is easiest to tell what is ordinary from what is special by comparing the behavior of organisms to models of varying degrees of idealization. But there is a special caution which must pervade any study using models. Organisms are selected to do well what they must. To the extent that we deviate in our tests from the natural conditions within which the design of an organism was selected, to that extent we are likely to encounter substandard rather than normal performance. Flow tanks and wind tunnels are such a deviation; the use of models for organisms is a more extreme one. Still, a model is a hypothesis about the relative importance of different aspects of life, and without such simplifying hypotheses it's hard to imagine making any progress at all.

CHAPTER 7

shape and drag: motile organisms

WHERE AN ORGANISM ALWAYS meets the flow head on, streamlining should be highly effective. Most organisms which consistently face into currents are those which propel themselves in continuous fluid media—the swimmers and fliers of the world. Humans have put a lot of effort into designing streamlined shapes; and, if Cayley's trout is no fluke, nature has done the same. For a sessile organism, resisting drag means resisting only force; deflections can be restored elastically and little work need be done once the system is constructed. For motile organisms, the price of excessive drag may be more immediate and serious: to move, animals must expend energy at a rate which is the product of drag and velocity. If the drag coefficient could be halved, then surely power output and perhaps also power input could be halved. Alternatively, an animal might be able to swim 40% faster.

CHAPTER 7

streamlining and drag coefficients

Some problems immediately cloud the skies and muddy the waters. First is the familiar one of lack of physical data. Between Reynolds numbers of 10^4 and 10^7, we have good information on the design of streamlined shapes. This ought to be a good base for work on birds, medium and large fish, and so forth. Between 10^2 and 10^4, we have almost no precise information about the shape of objects designed for minimal drag. One may guess that with lower Reynolds numbers and increasingly delayed separation (recall the data of Seeley et al., 1975, for spheres), the best shapes will be more rotund. Also, the increasing skin friction at low Reynolds number will favor shapes which expose less skin, again favoring the fatter. Of course "best" depends on the obejctive. Nature is probably more concerned with moving a volume of organism rather than frontal or surface area, again favoring less elongate shapes.

The use of drag coefficients (and to a lesser extent Reynolds numbers as well) encounters some unfortunate awkwardness when different streamlined shapes are compared. Part of the difficulty comes from the different conventions used in defining the reference area in the drag coefficient; the rest comes from the fact that just what shape gives least drag coefficient depends on the reference area used. As mentioned in Chapter 5, four different areas are in use: (1) frontal or projecting area, mainly with bluff bodies; (2) plan, profile, or maximum projecting area, for wings; (3) wetted or total area, for streamlined bodies; and (4) volume to the two-thirds power, for airships. Consider a streamlined body, symmetrical about its long axis. It's clear that a change in shape can affect the drag coefficient with no change in drag, and that how drag coefficient changes depends on how we define the area in it. If drag coefficient is based on frontal area or $V^{2/3}$, then stubbier shapes will look better; if it is based on plan or wetted area, then more elongate shapes will give lower drag coefficients.

Due to the diversity of conventions in the literature, it is useful to be able to convert one sort of drag coefficient into another. Table 7.1 gives some conversion factors; in particular, these may help in using the data of Figures 5.3 and 5.4 for comparisons with drag data for biological objects. For a sphere or cylinder moving normal to its long axis, frontal and plan areas are the same. Streamlined objects are a heterogeneous collection, but the factors for prolate spheroids may prove to be adequate approximations for crude comparisons. From the figures given by Pennycuick (1968) and Tucker and Parrott (1970) it looks as if the 4:1 spheroid (one of length four times its thickness) provides a

TABLE 7.1

Factors for converting drag coefficients from one reference area to another. Flow is perpendicular to the long axis of the infinitely long cylinder; flow is parallel to the long axes of the prolate spheroids, described by the relative lengths of their axes.*

			prolate spheroids		
	sphere	*cylinder*	*2:1*	*3:1*	*4:1*
Frontal area	1	1	1	1	1
Plan form area	1	1	0.500	0.333	0.250
Volume$^{2/3}$	1.208	0.932	0.762	0.581	0.480
Wetted area	0.250	0.318	0.146	0.102	0.078

* To convert a drag coefficient, multiply its value by the ratio of the factor for the area in which you want it to the factor for the area in which you already have it.

reasonable facsimile of a pigeon and perhaps of other birds. None of these figures should be used for bilaterally-flattened fish.

A good low-drag body of revolution (an elongate, radially symmetrical body) at a Reynolds number in the low hundred-thousands will have a drag coefficient, based on frontal area, of about 0.04 (Mises, 1945). Using his figures and those of Hoerner (1965) such a body may be approximated by a 3:1 spheroid, so the drag coefficient based on wetted area will be about 0.004, that based on plan area of 0.013, and on $V^{2/3}$ about .02.

These figures, though, are not a sufficient general guide to the quality of streamlining. How low is the lowest achievable drag coefficient? Recall just what streamlining accomplishes: it prevents separation aside and behind an object. The ideal nonseparating flow is that across a flat plate parallel to the flow, so the drag coefficients for such flat plates provide a convenient standard against which to compare other such coefficients. Fortunately, convenient and fairly trustworthy formulas are available for the drag of long, flat plates parallel to flow, based on wetted area or both surfaces of the plates. For laminar flow:

$$C_D = 1.33\, Re^{-0.5} \qquad (7.1)$$

and for turbulent flow:

$$C_D = 0.072\ Re^{-0.2} \qquad (7.2)$$

(Goldstein, 1938; Hoerner, 1965. Hertel, 1966, a widely cited but untrustworthy source, misquotes equation 7.2). As mentioned earlier, the transition to turbulent flow occurs somewhere between 5×10^5 and 1×10^7, depending on the circumstances. Formulas (7.1) and (7.2) generate Table 7.2, which nicely illustrates the advantage of postponing the turbulent transition to the highest possible Reynolds number. This is very nearly the same thing as postponing transition as far back as possible from the leading edge of a streamlined object, since the Reynolds number is proportional to that distance. The formulas and table also point out that drag coefficients typically drop with increasing Reynolds numbers for objects which experience mainly skin friction. Thus the common assumption that drag is proportional to the square of speed is even further from reality for streamlined objects than it is for bluff bodies.

A motile animal, of course, must do more than just resist drag. In particular, it has to produce thrust; and, at steady speed and horizontal motion in otherwise still fluid, thrust and drag must be equal. If you can measure one, you know the other, which is why figures for drag take on some significance in assessing power requirements for locomotion. If thrust-generating and drag-resisting structures are separate, as are the screw and the hull of a ship, the

TABLE 7.2
Calculated drag coefficients (on wetted area) for a long, flat plate parallel to flow.

	coefficient of drag	
Reynolds number	*laminar*	*turbulent*
10^2	0.133	-
10^3	0.0421	-
10^4	0.0133	0.0114
10^5	0.00421	0.00720
10^6	0.00133	0.00454
10^7	-	0.00287

situation is simple. The thrust producer must produce (whatever its own drag) a net thrust equal to the "parasite" or "extra-to-wing" drag—that of the rest of the craft.† The separation is satisfactory in birds and rowing water beetles; it is vexingly unsatisfactory for many fish, where the whole body may be involved in propulsion. At best, moreover, the nonthrust-producing part of the organism is subjected, not only to a smooth oncoming flow, but to a more complex and temporally-varying flow from wings or fins as well. But despite all these disclaimers, it's still worth looking at some drag coefficients for motile animals.

some swimmers and fliers

Fish. Webb (1975) has collected a large amount of data from his own and other measurements. For a 30-cm-long rainbow trout (*Salmo gairdneri*), freshly killed and with paired fins amputated, he measured drag coefficients of around 0.015 at Reynolds numbers between 50,000 and 200,000, where the coefficients were based on wetted area. From Table 7.1 and Figure 5.3, this is about 25 times less drag than that of a cylinder normal to the flow. That's impressive, and it begins to explain why trout seem to glide so effortlessly through a pond. But it's also about 3.5 times more than the drag of a flat plate in laminar flow (Table 6.2), which makes one wonder why nature does no better. Webb's fish were tested in a flow tank; some measurements based on towing fish in otherwise still water give lower values for drag. He quotes figures of 0.0043 for a small mackerel (*Scomber scombrus*) at Re = 100,000, which is essentially perfection. But there is no good evidence that at Reynolds numbers above 100,000 fish can maintain values of drag quite as low as those for laminar flow across a flat plate. The same mackerel, for example, has a C_D of 0.0052 at 175,000, not the ideal 0.0032. Yet these scombroid fish (including the tunas) look more like laminar flow airfoils than almost any others, and they live a high-speed pelagic existence.

Webb (1975) also provides a nicely critical review of the difficulties in getting reasonable measurements of drag, along with the various mechanisms

† The term "parasite drag" has been used in two quite different senses. As used here, it is the fuselage or body drag, excluding the drag of paired fins (swimmers) or wings (fliers), and it is based on area measurements excluding wings or fins. But the term has also been used (Tucker and Parrott, 1970, for instance) to include one part of wing drag, the "profile drag." The literature should be approached with due caution.

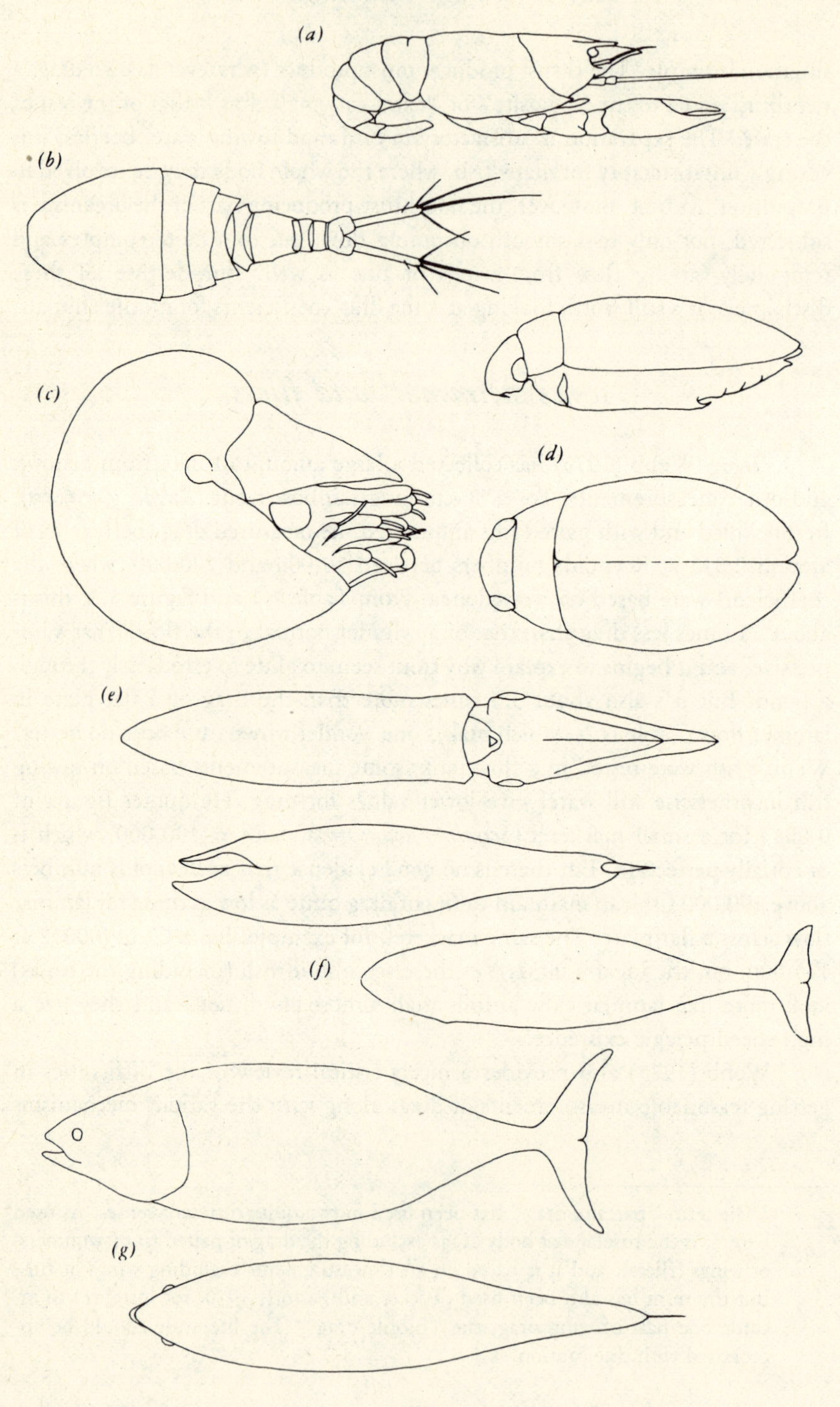
(a)
(b)
(c)
(d)
(e)
(f)
(g)

which have been suggested as possible ways by which fish might reduce drag. In none of these cases is the evidence for the use of a particular mechanism particularly persuasive, but the following have been suggested: (1) distributed viscous damping of disturbances in flow near the surface of the fish; (2) distributed dynamic damping via a pore canal system to prevent the initiation of turbulence; (3) periodic skin oscillation, for the same reason; (4) ejection of gill effluent into the flow to prevent transition; (5) heating the water at the surface of the skin to reduce viscosity; and (6) various schemes making use of surface roughness. All of these mechanisms are discussed with less skepticism by Aleyev (1977). He also gives a large number of figures (to one significant figure and of uncertain credibility) obtained on models modestly described as "exact;" the figures are at or above the values for flat plates in turbulent flow. Aleyev also supplies some amusing photographs of naked human swimmers.

Water beetles. Many beetles swim under water essentially by rowing with one or more pairs of legs. Nachtigall (1974, 1977a) reports extensive studies of the rowing locomotion of the medium-to-large dytiscid beetles. These have what look like fairly well-streamlined bodies, but at first glance their drag coefficients are not impressive. Values of 0.23 to 0.43, based on frontal area, in a Reynolds number range of about 50,000 to 20,000, are only about three to five times lower than the drag coefficients of cylinders. Still, at lower Reynolds numbers, the best and worst cases converge (recall Table 5.1), so their drag coefficients are only two or three times worse than that of a flat plate parallel to a laminar flow; the latter is about the same factor we noted for birds. Dytiscid beetles may be as short as 2 mm (from 35 mm in *Dytiscus*) with a Reynolds number based on body length of about 90. No drag coefficients have been determined for the smaller ones; they are increasingly round and less angular as their size decreases, which is what we'd expect as skin friction becomes more important and separation and pressure drag less so.

FIGURE 7.1

(*Facing page*) A variety of swimming organisms which presumably have been selected for low drag. (*a*) A crayfish making a rapid rearward evasion response; (*b*) a copepod; (*c*) the shelled cephalopod, *Nautilus*; (*d*) a large beetle, *Dytiscus;* (*e*) a squid; (*f*) a gray whale; (*g*) a tuna.

Cetaceans. The drag of dolphins and porpoises has been a matter of no minor interest since Gray (1936) suggested that a dolphin had to have drag corresponding to laminar rather than turbulent flow. Otherwise, its muscles would be unequal to the task of moving it at its estimated maximal swimming speed, the latter corresponding to a Reynolds number of about 1.6×10^7 (Alexander, 1977). The problem has achieved real status—it's called "Gray's paradox." The passage of time and fashion has greatly complicated the matter. On one hand, we now have figures for the power output of muscle much in excess of the values used by Gray (Weis-Fogh and Alexander, 1977); porpoises and dolphins may be much stronger than he believed. Also, Lang and his collaborators have estimated drag coefficients for coasting, decelerating, trained porpoises: from Lang's work, Newman and Wu (1975) cite a figure of $C_D = 0.0036$ (on wetted area) at $Re = 6.5 \times 10^6$ as typical. For comparison, a flat plate in turbulent flow (equation 7.2) gives $C_D = 0.0031$ and one in laminar flow (equation 7.1) gives $C_D = 0.00052$.

On the other hand, dolphins evidently do swim with remarkably little disturbance of the fluid around them. Moreover, the animals on which drag coefficients were estimated were not actively swimming. Also, these drag coefficients are really quite low, lacking that factor of two to five by which the parasite drag of birds, fish, and swimming beetles commonly exceeded the flat plate values. Finally, it is not clear that the animals can sustain high speeds for very long, and the maximum output of most animals is notoriously dependent on the length of time over which it must be sustained. So it appears that these smallest of whales need not maintain laminar flow as a *sine qua non* of going as fast as they do, but they probably do maintain laminarity over much of their surface to keep drag low. How they might do the trick is equally controversial, the compliant skin effect claimed by Kramer (1965) being in disfavor at present (Newman and Wu, 1975). Alexander (1968), Gray (1968), Leyton (1975), and Webb (1975) all have their say about the situation.

Shelled cephalopods. All of the animals we've mentioned are probably well arranged to minimize drag. Locomotion is important and expensive; and, even if we don't understand the details, we've fairly sure that drag minimization was accorded a high priority in their design. What drag coefficients characterize relatively poor swimmers? One group of cephalopod molluscs certainly looks poor—the shelled nautiloids and ammonoids, now represented only by the genus *Nautilus.* Chamberlain (1976) towed a *Nautilus* shell and 36 model shells in a tank, calculating drag coefficients for each. The coefficients, based on $V^{2/3}$, dropped only slightly between Reynolds numbers of 12,500 and

300,000. The real shell had a C_D of 0.48, and the models varied from about half to twice that value. A drag coefficient of 0.48 is a bit better than that of a sphere; at $Re = 100{,}000$, the latter is 0.56 when based on $V^{2/3}$. (Chamberlain errs in using C_D's based on frontal area for comparisons.) It's much better than a cylinder, for which C_D is 1.12. If this data had referred to wetted or frontal area, the *Nautilus* shell would probably have looked relatively better yet. Nevertheless, *Nautilus* is far from a good low drag design. One wonders whether these creatures lived (and live) an existence in which swimming is a minor matter (as in scallops, for instance); whether they are trapped by the inflexible geometry of a conical or spiral shell; and what effect the soft parts, extending downstream, might have on drag. Perhaps we should remind ourselves that moving fast isn't all that matters, now or in the paleo scene; after all, the shelled gastropods are slow, but surely successful.

Birds. They certainly look slick and smooth in flight posture, whether tiny hummingbirds or large, soaring eagles. And, according to Tucker (1973), overcoming the drag of the body absorbs a large part of a bird's power output at high speeds of flight. Pennycuick (1968, 1971) has measured the drag on wingless, frozen birds in a wind tunnel; he calculated drag coefficients, based on frontal area, of 0.43 for pigeons and vultures. These species (*Columba livia* and *Gyps ruppelli*), which are an order of magnitude apart in frontal area (36 and 300 cm^2), were tested over a wide range of speeds, corresponding to Reynolds numbers between 200,000 and 800,000. The figure of 0.43 is therefore likely to be reasonably typical for medium and large birds. The figure may be a little less than the drag coefficient of a sphere or cylinder before the transition to turbulence, but by any other comparison it is surprisingly high. It is very much worse than the drag of either laminar or turbulent flat plates. The very constancy of the drag coefficient with changes in Reynolds number suggests the pressure drag of bluff bodies rather than the skin friction of streamlined shapes. The question to be asked is whether parasite drag, measured in this way, actually reflects the situation in normal flight. I've faith enough in natural selection to suspect there's something going on here which we just don't understand.

Pennycuick also measured the drag of the dangling legs of both pigeons and vultures; the legs give drag coefficients of 1.08 and 1.25 (on frontal area). As he suggests, huge values are perfectly reasonable for structures ordinarily kept out of the way but called upon to act as airbrakes for creatures with neither the space nor running speed to land on runways. It's uncertain, however, whether legs and feet (except perhaps webbed feet) are large enough for their drag to make much difference.

Two large insects. Making estimates where necessary, I've calculated parasite drag coefficients for two relatively large flying insects, a migratory desert locust and a cockchafer beetle. The locust (data from Weis-Fogh, 1956) is more elongate and approximates a 4:1 spheroid; the beetle (data from Nachtigall, 1964) is shorter, roughly a 2:1 spheroid. Table 7.3 compares the two, along with data for a cylinder and flat plate. Neither locust nor beetle look like great prizes (Figure 7.2); the beetle appears more persuasively streamlined than the locust, but the beetle carries only half the weight of the latter. In status between "best" and "worst" (plate and cylinder), both are a little worse than the swimming beetles. A photograph of a locust in flight posture in smoke streamers (Weis-Fogh, 1956) shows a large and messy wake. The beetle, with its fairly blunt behind, is unlikely to do better.

Why do these excellent fliers suffer so much parasite drag? One can't easily invoke special properties of living skin and feathers for rationalizing poor results with dead insects. One possible answer was suggested by Weis-Fogh, who pointed out that, at cruising speed, less than two percent of a locust's metabolic expenditure is invested in overcoming parasite drag. The main costs are staying aloft and losses to drag in the wings theselves, not so much in the air

TABLE 7.3

Airflow across the fuselage of a locust and a beetle

	locust	beetle
Body length	3.8 cm	2.8 cm
Speed	3.4 m s^{-1}	2.25 m s^{-1}
Re (on length)	8000.	4000.
Frontal area	0.785 cm^2	1.54 cm^2
Drag	0.0008 N	0.00023 N
Weight	0.02 N	0.01 N
C_D (frontal)	1.47	0.48
C_D (wetted)	0.12	0.070
C_D (wetted) cylinder	0.318	
C_D (wetted) flat plate, parallel to flow	0.017	

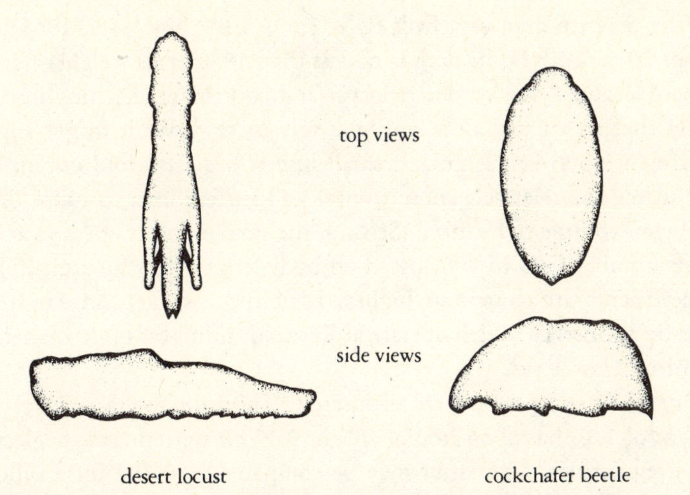

Silhouettes of a desert locust, *Schistocerca gregaria,* and a cockchafer, *Melolontha vulgaris,* in approximate flight posture.

FIGURE 7.2

resistance of the fuselage. Some index of the problem is given by the ratio of drag to weight. For the locust it is about 4%, for the beetle, about 2.5%. Although one can't simply convert force to work, it's evident that a lot more force must be generated to support the insects' weight than to offset their drag. Achieving a really splendid low-drag design just may not be a high priority matter in the design of these organisms.

shape and drag of a small insect

We turn now to a very different sort of flying insect, the fruit fly, *Drosophila virilis.* Its mass is only about 2 mg, its length about 3 mm. Whether it flies rapidly or not depends on how it's viewed. A cruising speed of 2 m s^{-1} can hardly be called fast and certainly puts a creature at the mercy of even slight breezes. Equally, though, it represents the fairly phenomenal speed of about 650 body lengths per second, a unit perhaps better termed "length-specific speed." The length-specific speeds of the locust and beetle mentioned previously are only about 80 or 90 s^{-1} and for a pigeon about 40. At cruising speed, the parasite drag of a fruit fly is 3.6×10^{-6} N (Vogel, 1966); with a weight of 20×10^{-6} N, its drag is no less than 18% of its weight—far higher than the 4% and 2.5% of desert locust and cockchafer. For tiny insects, it's relatively cheaper to stay aloft and relatively more difficult to get anywhere. This difference between large and small insects has a twofold origin. First, a smaller insect has relatively more frontal and surface area to cause drag but relatively less volume to be lifted. Second, the smaller insect operates at a lower Reynolds number (300 to 400, based on body length, for this fruit fly) where drag coefficients are somewhat higher. Fruit flies, in fact, are far from the smallest flying insects, which operate at Reynolds numbers more than an order of magnitude lower yet.

Our usual comparisons are of interest. The fruit fly has a parasite drag coefficient of 1.16 based on frontal area or 0.17 on wetted area, neglecting its surface irregularities. The latter may be compared to 0.382 for a cylinder or 0.068 for a parallel flat plate: the fly is right in between—2.3 times better than the former and 2.5 times worse than the latter. The full difference between "best" and "worst" has narrowed to 5.6 times instead of the 19-fold in the locust range or the 300-fold at a Reynolds number of 100,000. As the Reynolds number drops, skin friction relative to pressure drag increases and the value of streamlining therefore decreases.

I'd like to discuss this fruit fly for a few pages and to describe a set of experiments on the relationship between its shape and drag which I carried out around 1963. They were begun as an exploration of what I could do with a newly built microanemometer and the wake traverse system (see Chapter 4) for measuring drag, but they acquired a momentum of their own until their great tediousness proved an excessive drag. The anemometer (Vogel, 1969) is described in Appendix 2.

In certain critical respects these experiments were seriously flawed. I could measure drag with an imprecision no lower than about 3 to 5%; each fly after killing took a shape slightly different from its shape in life and from the shapes of its peers; and my dexterity was unequal to routinely mounting legs in flight posture. Furthermore, the fly normally encounters a complex airflow from the wings and from its forward progression, an airflow which I had no reasonable expectation of matching on an isolated fuselage. I meant to repeat this work with better techniques before publication rather than hang heavy conclusions from flimsy data; many readers will know personally the perils of postponing publication.

All of the measurements of drag were made with the wake traverse system, using only the half-wake on the side of the fly away from the mounting wire. A wind tunnel speed of 1.5 m s^{-1} represented a compromise between fast forward flight and wing-induced wind during hovering. (Actually, I just didn't know at that point how fast fruit flies could fly.) In any case, a sharp optimum in operating conditions is unlikely for an animal which normally operates over a wide range of conditions. Flies were killed with chloroform vapors, which gave minimal distortion of body shape, and only the least distorted animals were used. Wings and legs were cut off (with practice, youth, and profanity, it's much easier than it sounds). Animals lacking legs have about the same drag as animals with legs fixed roughly in flight posture.

The basic question is whether (and how) nature has arranged the fruit fly to minimize its parasite drag. One quick test is to compare the fly to an object of different shape, a shape which should not be the direct product of selection for low drag but which has the same volume and the same surface and frontal areas as the fly. The object meeting these criteria is, naturally, the same fly, now facing downwind instead of upwind. The reversed fly proved to have about 10% more drag between 1 and 2 m s^{-1}. This is nothing like the twofold difference noted in the previous chapter for a streamlined object at high Reynolds number, nor would the latter be expected; but it does suggest some specialization. The fly, of course, doesn't *look* very good in terms of drag; it veritably bristles with protuberances.

The main set of experiments involved both additive and subtractive alterations of body shape, some quite grotesque, with drag measurements both before and after the Procrustean procedures. A selection of the alterations are listed below and illustrated in Figure 7.3. Along with each alteration is the

FIGURE 7.3

Wingless, legless, fruit flies, *Drosophila virilis*, subjected to further paring.

normal fuselage

without thoracic setae

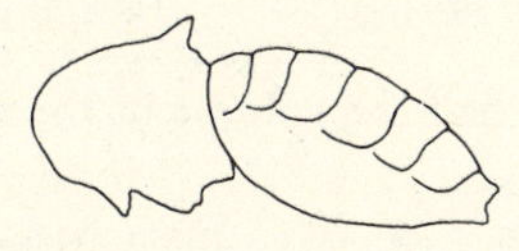

without setae or head

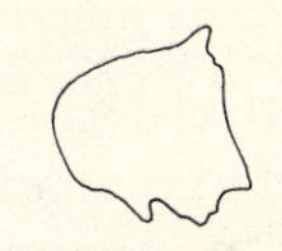

without setae, head, or abdomen

change in drag it caused when performed on a previously "normal" (dead, wingless, and legless) fly:

Removal of head	−0.1%
Removal of abdomen (including halteres)	−0.5%
Removal of 12 large thoracic setae (hairs)	−8.6%
Fairing groove between head and thorax with wax	+5.4%
Fairing groove between thorax and abdomen with wax	+3.2%

The first surprise was that rather drastic operations had only minor effects. Removal of head or abdomen raised the drag per unit volume or surface area, although to my eye, the headless fly looked more nicely shaped. Similarly, the waxed flies looked slicker, since hot wax wets cuticle and flows on very smoothly; but their drag was elevated. Only the least drastic operation, removing a dozen bristles, gave an appreciable reduction in drag. It was all quite counterintuitive.

The second surprise was what happened when precisely the same operations were performed on animals which had previously been operated on. One could, in fact, construct a sort of meat market matrix or (as below) pick a few illustrative data. Consider, first, removal of the head, which reduced drag only 0.1% on a normal animal. If the animals had been previously operated on, the following data were obtained:

On animal lacking abdomen	−11.3%
On animal lacking setae	−18.3%
On animal with thorax-abdomen fairing	− 7.1%

Once the animal is sufficiently abused, our intuitive view of what decapitation should do becomes reasonably predictive! About the same results occurred after removal of the abdomen, which reduced the drag only 0.5% in an intact animal:

On animal lacking head	−16.5%
On animal lacking setae	−18.3%
On animal with head-thorax fairing	−14.8%

Of course, removing the abdomen, about half the volume of the fly, *ought* to decrease the drag; and here, at least, it did.

There's quite a lot more of this kind of data, all consistent with the supposition that an altered animal behaves as an object not expressly designed for minimal drag: it experiences a decreased drag if its surface are is reduced or its surface is smoothed. By contrast, a normal fly shows an emergent property one is tempted to call "optimization." If its shape is changed, then drag rises relative to surface area. No single structural feature of the fly appears to be responsible for this optimization; instead, the animal seems to be a complexly integrated design, the result of which is a shape which incurs a minimum of parasite drag.

Two peculiarities deserve comment. First, there is that drop in drag following removal of the thoracic setae. Clearly, the setae cannot be special devices for reducing drag—more likely the opposite is true. Perhaps they serve a function which requires that they suffer a significant drag. One attractive possibility is that they are local airspeed detectors working as part of the flight control system. Whatever their role, their removal does render a fly non-intact with respect to further operations just as surely as the removal of head or abdomen, so they must still be regarded as elements in the integrated low-drag design. Shifting the location of the thoracic bristles might indicate whether their position is aerodynamically significant—an important question in a group of animals whose taxonomy is often based on the location of setae. While the critical operation cannot be conveniently performed, the bristles may be functionally relocated by changing another structure. If, for example, the head is removed, then subsequent removal of the setae reduces the drag of the fly considerably more (15.3%) than does their removal from an intact animal (8.6%). While an addition to total drag may be an unavoidable consequence of the function of the setae, their location appears important in minimizing this contribution.

Another peculiarity is the problem of the ventral surface. Legs were routinely removed before measurements of drag were made. Since the flies showed this "optimized" design nonetheless, it would appear that the ventral surface does not display the same crucial specialization as the rest of the body. In fact, application of wax to the ventral surface lowers the drag; elsewhere it increases drag. Perhaps the difference can be explained by the fact that these flies are convex dorsally and flat ventrally, with the ventral part of the thorax normally in the wake of the head. Or perhaps the ventral surface is less important since the wings meet at the top of the stroke and the airspeeds above the fly in normal flight are greater than those below. Or perhaps removal of legs does "deoptimize" the fly's bottom—hence the drag reduction attending waxing—but that optimization of the ventral surface simply doesn't interact strongly with optimization of parts along other streamlines.

This notion of an optimized design, if it holds up, is potent. A search for the structures responsible for low drag has led to the conclusion that nearly everything is important and nearly everything works in concert. A change in a downstream component can affect the character of a subsequent change upstream. It's obvious that there are serious implications for evolutionary studies and for both classical and numerical taxonomy. Certainly, this intriguing business ought to be pursued, and not just with fruit flies, which were a rather accidental choice in the first place; but with simpler, larger, and therefore more tractable organisms. The only absolute requirement for such an organism is that drag minimization be important to its biology.

CHAPTER 8

velocity gradients and boundary layers

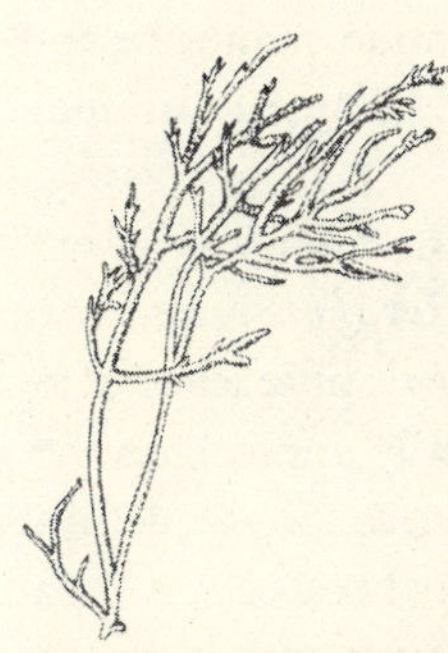

AT THE INTERFACE BETWEEN a stationary solid and a moving fluid, the velocity of the fluid is zero. This is the "no-slip" condition mentioned early in Chapter 2. The immediate corollary of the no-slip condition is that there is a gradient in the speed of flow near every surface. Speed changes from zero at the surface to what we'll call the "free stream" velocity some distance away. And it is in velocity gradients that viscosity works its mischief and gives rise to drag, power consumption, and so forth. The gradient region is associated with the term "boundary layer," which I've quite deliberately postponed mentioning until it could be defined carefully. Most biologists seem to have heard of the boundary layer, but they have the fuzzy notion that it is a discrete region rather than the discrete notion that it's a fuzzy region.

Velocity gradients are as universal as the no-slip condition and were undoubtedly recognized much earlier. The boundary layer wasn't so much discovered as it was invented, in the early part of this century, as the great stroke of genius of Ludwig Prandtl. In the basic differential equations for moving fluids, the Navier-Stokes equations, some terms result from the inertia of fluids and some from their viscosity. As we've seen, the Reynolds number gives an indication of the relative importance of inertia and viscosity, and it can be used to determine just which simplifying assumptions in the basic equations might be appropriate to a given situation. At Reynolds numbers well below unity, inertia can be ignored and highly accurate rules can be derived such as Stokes' law for the drag of a sphere. At high Reynolds numbers, one might expect to get away with neglecting viscosity, in which case the Navier–Stokes equations reduce to the Euler equations which are the three-dimensional analog of Bernoulli's equation. It may sound neat, but it doesn't work; results diverge from physical reality, drag vanishes, and d'Alembert has his paradox.

Prandtl reconciled practical and theoretical fluid mechanics at high Reynolds number by recognizing that viscosity could never be ignored in the gradient region near surfaces even if it could be disregarded everywhere else. That region might be small at high Reynolds numbers; but as long as the no-slip condition held, there had to be a place where shear rates were high and viscosity was significant. He called the place in question the "Grenzschicht" or "boundary layer." In general, a higher the Reynolds number implies a smaller boundary layer, but a higher shear rate in that boundary layer.

Shear, naturally, entails dissipation of energy, so another way of viewing the boundary layer is as the region near a surface where the action of viscosity produces an appreciable loss of total head. Bernoulli's equation cannot be applied to the differences in velocity within a boundary layer because it assumes constant total head.

The crucial question is, of course, just how thick is the boundary layer? Here's trouble. The inner limit is the solid–fluid interface. ***But there is no outer limit in nature to the gradient region.*** The speed of flow approaches zero almost linearly at the interface; by contrast, it asymptotically approaches the free-stream velocity with increasing distance from that interface. So the definition of the outer limit of the boundary layer must be arbitrary because it is not a naturally discrete region and is bounded by no physical discontinuity. It should be emphasized that the choice of an outer limit depends very much on the function for which one is invoking a boundary layer. The most commonly quoted outer limit is the curved plane at which the local velocity, U_x, has risen to 99% of the free-stream velocity, U:

$$U_x = 0.99U \tag{8.1}$$

Biologists should recognize that the constant, 0.99, is an accident of our bipedal and pentadactylic lineage.

the boundary layer on a flat surface

Consider a thin, flat plate oriented parallel to a flow as shown in Figure 8.1, a plate for which the Reynolds number based on the distance from leading to trailing edge is under about half a million. We'll call the distance downstream from the leading edge of the plate "x", the thickness of the boundary layer as defined above "δ" (delta), the velocity component along the plate U_x, and the velocity component normal to the plate U_y. Now, if fluid slows down upon entering the boundary layer, other fluid not entering the boundary layer must swing wide around it by the principle of continuity; slowing down means that streamlines diverge. If the boundary layer thickness, δ, is the order of magnitude of the distance downstream, x, then U_y will be significant. Conversely, if δ is much less than x, then U_y will be effectively zero. We'll consider only this latter situation. Note carefully that while U_y may be zero for all practical purposes, the rate of change of velocity along y, dU_x/dy, is always significant within the boundary layer.

Assuming that $U_y = 0$, an approximate solution to the Navier-Stokes equation gives a formula for the thickness of the boundary layer on one side of a flat plate. For a laminar boundary layer, defined as in equation (8.1):

$$\delta = 5\sqrt{\frac{x\mu}{\rho U}} \tag{8.2}$$

FIGURE 8.1

(*a*) The distribution of velocities above a point on a flat plate parallel to flow. The point is a long distance from the upstream edge of the plate. (*b*) The boundary layer thickness is defined as the distance normal to the surface of the plate which it requires for the local velocity to reach 99% of the free-stream value.

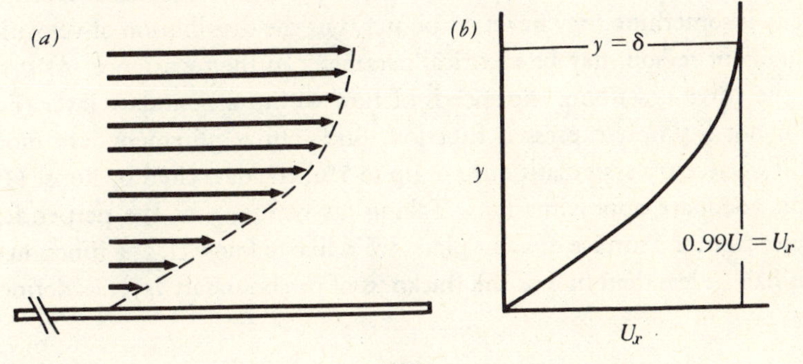

Notice that the boundary layer thickness increases in proportion to the square root of the distance from the leading edge; its outer limit thus forms a parabola (Figure 8.2). Increases in density or free-stream velocity thin the layer while increases in viscosity thicken it. Authors differ somewhat on the value of the constant in equation (8.2): I've found values ranging from 4.65 to 5.47, so 5 should be a reasonable approximation (Leyton, 1975).

Equation (8.2) can be expressed in terms of what is usually called the "local Reynolds number," Re_x, where the characteristic length is taken as the distance downstream from the leading edge. The result of this substitution, equation (8.3), shows that the relative thickness of the boundary layer depends solely on the Reynolds number:

$$\frac{\delta}{x} = 5\, Re_x^{-1/2} \qquad (8.3)$$

Thus at a local Reynolds number of 10,000, the thickness of the boundary layer is a twentieth of the distance from the leading edge while at a local Reynolds number of 100,000, it is about a sixtieth of that distance. Low Reynolds numbers mean relatively thick boundary layers.

Equations (8.2) and (8.3) are usable up to Reynolds numbers of 500,000. To how low a Reynolds number can they be applied? In my experience, they're safe down to where the boundary layer thickness is as much as 20% of the distance from the leading edge, in which case the Reynolds number is 625. For a cruder estimate of boundary layer thickness, they can be used down to 100 or a little lower (Vogel, 1962). They work for slightly curved surfaces almost as well as for flat surfaces, but there must be no separation of flow.

The most common use of these equations is in the calculation of forces; in fact, equation (7.1) was originally derived from (8.2). The biologist, though, should have a profound interest in the specific conditions within the boundary layer. Many organisms live partly or entirely within the boundary layers of either inanimate objects or other organisms. For many organisms, free-stream velocity is something they never encounter, but the distribution of velocities in the gradient region may be a critical parameter in their existence. As it turns out, the curve describing the speeds of flow within a boundary layer (Figure 8.1) is not a simply-expressed function. Still, our requirements are modest; and, if we accept a systematic error of up to 5%, the data cited by Rouse (1938) permit adequate approximations. Taking (as before) y as the perpendicular distance from the surface of a flat plate, we'd like to know U_x as a function of y. If y is half or less than half of the thickness of the boundary layer as defined in equations (8.1) and (8.2), then the following linear approximation is adequate:

$$U_x = 0.32Uy\sqrt{\frac{\rho U}{x\mu}} \tag{8.4}$$

If y can be anywhere within the boundary layer, then a slightly more complex parabolic distribution is necessary:

$$U_x = 0.39Uy\sqrt{\frac{\rho U}{x\mu}} - 0.038\frac{U^2y^2\rho}{x\mu} \tag{8.5}$$

Either is a simple matter for a programmable hand calculator.

What if our thin, flat plate sports warts? How rough can a surface be without materially affecting the flow over it? Goldstein (1938) cites formulas based on the finding that the flow is unaffected if the Reynolds number of a projection is 30 or less for a pointed projection or 50 or less for a rounded one. The permissible heights of roughness (ε) are then

$$\frac{\varepsilon}{x} \leqslant 9.5\ Re_x^{-3/4} \qquad \text{(pointed)} \tag{8.6}$$

$$\frac{\varepsilon}{x} \leqslant 12.2\ Re_x^{-3/4} \qquad \text{(rounded)} \tag{8.7}$$

where x is the distance downstream from the leading edge as well as the characteristic length in the local Reynolds number. Organisms are a bumpy, bristly bunch, and totally smooth surfaces are exceptional in nature. These formulas should provide a handy test for the applicability of the equations given earlier.

Evidently, roughness of a given size is more likely to have a disturbing effect if it is near the leading edge. It is also evident, by comparing the exponents in the equations above with that of equation (8.3), that the higher the Reynolds number, the deeper within the boundary layer a protuberance must be to prevent its disturbing the flow. As an example, consider a flat plate in water moving at 10 cm s^{-1} and a point 1 cm from its leading edge. The local Reynolds number is 1000, and a pointed projection of only about 0.5 mm is tolerable; this is one-third the thickness of the boundary layer as defined by equation (8.1). At a Reynolds number of 100,000, the protrusion must be only about a tenth of the thickness of the boundary layer to be without effect.

All of the above calculations work only for laminar boundary layers. Just as we needed a different description of the drag of a flat plate at high Reynolds numbers (equation (7.2)), so we need another equation for situations above Reynolds numbers of about half a million where the boundary layer is turbulent:

$$\delta = 0.376 \sqrt[5]{\frac{x\mu}{\rho U}} = 0.376x\,Re_x^{-1/5} \tag{8.8}$$

In practice, a boundary layer may be laminar near the leading edge and become turbulent somewhere downstream, the particular location being dependent on the local Reynolds number (Figure 8.2). Within the turbulent boundary layer, right near the surface, there remains a so-called "laminar sublayer." It is most important to recognize that a turbulent boundary layer is not the same kind of beast as a laminar layer, one differing only in the formula for how it happens to thicken. A laminar boundary layer is a semi-stagnant region in which wastes are likely to accumulate and in which nutrients may be depleted; it may represent a substantial barrier to diffusive exchange of heat or materials as well as a refuge from the forces of the free stream. A turbulent boundary layer may provide a refuge but, being well-stirred, it provides much less of a barrier to transport.

We should bear in mind the arbitrary definition of the thickness of the boundary layer stated in equation (8.1) and implicit in equations (8.2), (8.3), and (8.8). Considering all points where the speed is below 99% of the free-stream value is safe and conservative if you're interested in shear forces: one

FIGURE 8.2

The thickness of a laminar boundary layer on a flat plate. The *y*-direction is much exaggerated. Notice how the velocity gradient near the surface gets gentler with increasing distance downstream.

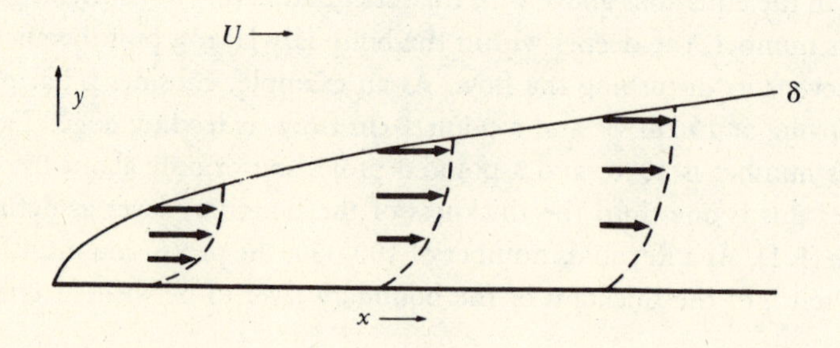

wouldn't want to leave out any major contributions to the shear forces and the resulting drag. For an organism living in a boundary layer and for a biologist interested in how much of the organism protrudes from the boundary layer, a slightly thinner definition may be more realistic. For such application, a flow of 90% of free-stream speed is probably functionally equivalent to full free-stream velocity. Thus if one is interested in what part of a sessile creature sticks up into the free stream, a definition based on 0.90 U may be appropriate. If so, the coefficient of 5 in equations (8.2) and (8.3) should be replaced by 3.5.

The trouble with all of these formulas is that they apply strictly to situations which are rarely found in nature. A stone in a stream certainly has a boundary layer, but it's rarely flat and rarely has a discrete leading edge. An organism sticks up into a boundary layer; by doing so it distorts the velocity distribution in a boundary layer and, of course, has a complex boundary layer of its own. A layer of organisms effectively raises the level of the surface, so pervasive is its influence. The obvious response is to measure velocities rather than trying to calculate them. Flowmeters and anemometers can be quite small (Appendix 2), and boundary layers are thick at low Reynolds numbers.

Even where the geometry seems appropriate, one must exercise caution and skepticism. Perhaps a few examples will stress the point sufficiently. Most leaves are fairly flat, and the wind might appear to blow smoothly over them. But..., (1) in a comparison of soybean leaves and metal models it turned out that the structure of the boundary layer was comparable only at very low wind speeds (Perrier et. al, 1973). (2) Turbulence occurred at much lower wind speeds on a poplar leaf than on a flat plate, and evaporation rates were more than twice those predicted (Grace and Wilson, 1976). (3) The turbulence of the natural wind decreased the heat transfer resistance of the boundary layer of a leaf to just 40% of the resistance of a leaf in a smooth and steady air flow (Parlange and Waggoner, 1972). So don't perpetuate the practice of equation-grabbing predecessors. Don't blindly adopt the 99% definition; don't use the formulas unless they demonstrably apply; don't be intimidated by the prospect of measuring low flows in small places.

A reasonable compromise between full faith in some revealed truth of physics and the complete agnosticism of empirical measurements is the generation of semi-empirical formulas. These can be derived from measurements on models and then tested for their applicability to organisms in nature. Good (and useful) examples are the equations of Nobel (1974, 1975) for effective, average boundary layer thicknesses on various parts of plants for the prediction of water loss and gas and heat exchange.

(1) For a flat, leaflike object, Reynolds numbers 300 to 16,000:

$$\delta = 0.40\sqrt{\frac{x}{U}} \tag{8.9}$$

(2) For a cylinder, Reynolds numbers from 1300 to 200,000:

$$\delta = 0.56\sqrt{\frac{d}{U}} \tag{8.10}$$

(3) For a sphere, Reynolds numbers from 400 to 40,000:

$$\delta = 0.33\sqrt{\frac{d}{U}} + \frac{2.9}{U} \tag{8.11}$$

These formulas, by the way, are not dimensionally homogeneous as given—a poor practice. They yield boundary layer thicknesses in centimeters if one expresses lengths in cm and speeds in cm s^{-1}.

forces at and near surfaces

Avoiding the full wind on a beach by lying against the sand is an experience familiar to most of us. As we've seen, the closer one gets to the surface, the lower is the local wind, with no wind at all right at the surface. Does this mean that the force of the wind, the drag, also approaches zero as an object flattens itself against the substratum? The drag quite clearly diminishes, but it doesn't actually drop to zero. Rather, it drops to a value which reflects the local skin friction on the object. There may be no velocity at the surface, but most certainly there is still a velocity gradient there.

This minimal drag of a flat object on a surface can be easily calculated if an appropriate formula for the velocity gradient is available. Consider, as an example, a small spot (not a point!) on a flat plate parallel to a flow, a spot well back from the leading edge lying beneath a laminar boundary layer. The velocity gradient near the surface is just the derivative of equation (8.4) with respect to y:

$$\frac{dU_x}{dy} = 0.32U^{3/2}\rho^{1/2}x^{-1/2}\mu^{-1/2}$$

The drag per unit area is, of course, the shear stress, so:

$$\frac{D}{S} = \mu \frac{dU_x}{dy}\bigg|_{y = 0} \qquad (2.4)$$

And the local drag coefficient is defined as:

$$C_{DL} = 2\left(\frac{D}{S}\right)\rho^{-1}U^{-2} \qquad (5.4)$$

Substituting the first of these expressions into the second and the second into the third, we get

$$C_{DL} = 0.64\ Re_x^{-1/2} \qquad (8.12)$$

a formula for the local drag coefficient on a part of a surface. (Hoerner gives the same formula, but with a coefficient of 0.664. If formula (8.12) is integrated from the leading edge rearward, equation (7.1) results.) For turbulent flow Leyton (1975) cites an analogous formula:

$$C_{DL} = 0.058\ Re_x^{-1/5} \qquad (8.13)$$

The skin friction of a part of a surface provides a useful base line with which the measured drag or the force needed to dislodge an organism can be compared. If no explicit formula is available for the velocity gradient, it may be measured directly by determining the speed of flow at a few points near the surface. Alternatively, one might use the shear on a force platform to indicate the velocity gradient of the inner part of the boundary layer or even the boundary layer thickness.

Naturally, it will be a rare organism which can get sufficiently intimate with substratum to have a drag as low as that given by equations (8.12) and (8.13). Most creatures will protrude to some degree above any surface of attachment, and most creatures are themselves equipped with protrusions—eyes, pinnae, nares, and other such excrescences. Aircraft and ship designers have given some attention to the drag contribution of minor surface protuberances. Hoerner (1965) gives values of the drag coefficient for bolt heads and pegs as between 0.74 and 1.20 for Reynolds numbers of 20,000 to 50,000 based on the frontal area exposed to the flow (in C_D) and the diameter of the protuberance (in *Re*). In general, the higher the head, the worse are both the drag and the

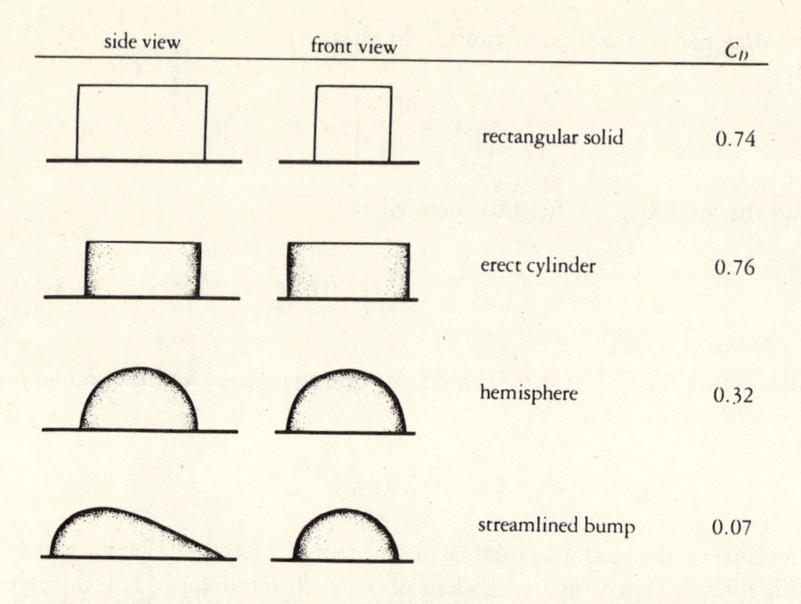

Drag coefficients of protuberances on flat plates, based on frontal area. The values have only relative significance since, in practice, they depend on the heights of the protuberances relative to the thickness of the adjacent boundary layer.

FIGURE 8.3

drag coefficient. Protuberances with rounded tops are better than flat-topped cylinders. If the "plan area" or top view is taken as S, a hemispherical protuberance has a drag coefficient of 0.32. But a half-streamlined body is better, with a typical drag coefficient of 0.07. The latter, of course, requires that the direction of flow be either predictable or invariant (Figure 8.3). In fact, the half-streamlined body is still far from ideal because it experiences quite a lot of interference around its base, especially in front. It helps to lengthen and flatten the leading edge of the protuberance; by such maneuvers the drag coefficient can be reduced to a minimum of about 0.03 at these Reynolds numbers.

These drag coefficients give some idea of the effects of shape on drag for objects within a boundary layer, but they shouldn't be strictly applicable to very many biological situations. The Reynolds number may be relatively low (less than 50,000) but the local Reynolds number based on the distance downstream from the leading edge of the surface of attachment is about two orders of magnitude larger. It is likely, however, that the same general rules apply at more modest Reynolds numbers: (1) Round tops are better than flat, sharp-edged tops. (2) Half-streamlined bodies are better yet, and it's better to have maximum thickness near the upstream rather than the downstream extremity to avoid separation. (3) Smooth fairing of the edge of the protuberance into the surface of attachment improves matters. (4) With such fairing, the distinction in shape between upstream and downstream ends of a protuberance should be slight—certainly less than that distinction for well-streamlined bodies in a free stream. Thus good low drag shapes for unidirectional, bidirectional, and even omnidirectional flows may converge. In any case, altogether too little hard practical information is available.

the earth's boundary layer

What happens if our flat plate lacks a leading edge? Columbus (1492) and others have asserted that the surface of the earth is quite without such an edge. Some sort of gradient region is obviously present; the no-slip condition isn't easily violated, and one can escape most of the wind by lying down even on an unobstructed beach. In fact, only at altitudes of 1000 m or more above the ground can the wind be treated as unaffected by friction with the surface (Geiger, 1965).

For the most part, air movement near the ground is highly turbulent and temporally irregular, so, at best, one can speak of average velocities and variation about averages. For applications such as wind throw of trees, maxima are of most consequence (see Chapter 6); for investigations of peak temperatures in convective heat dissipation, minima are most important (see Chapter 14); for estimation of evaporation rates, average speeds are of greatest interest. Superimposed on an overall horizontal wind, eddies of various sizes transport momentum vertically, and the roughness of the surface has a major influence on the distribution of velocities well above it.

Both theory and measurements indicate that the variation of average horizontal wind speed with height above the ground is a logarithmic function, at least if the effects of temperature variations are negligible. And the logarithmic relationship persists up to the winds of destructive gales (Oliver and Mayhead, 1974). The following general formula is usually cited:

$$U_x = \frac{U_*}{k} \ln\left(\frac{z-d}{z_0}\right) \qquad (8.14)$$

It will stand a bit of explanation. We're interested, most often, in how the horizontal wind, U_x, varies with height, z,[†] above the ground. Von Karman's constant, represented by k, is dimensionless and has an empirically determined value of 0.40. The so-called "roughness parameter," z_0, adjusts the steepness of the velocity gradient because it is related to the size of the eddies generated at the surface. The variable d is called the "zero plane displacement"; it accounts for the fact that the logarithmic profile extrapolates to zero velocity somewhere above the ground, especially on vegetated surfaces. U_* is called the "velocity of shear" or the "friction velocity"; it indicates the amount of turbulence, and its value is independent of height for a given surface and free-stream wind. More precisely, U_* is the square root of the shear stress divided by air density. Clearly, the higher the wind, the greater the shear stress it exerts on the ground; but the shear stress and friction velocity depend on the character of the surface as well as the wind speed.

With three not-quite independent unknowns, equation (8.14) isn't terribly convenient. It does indicate that a plot of $\ln(z - d)$ versus wind speed will be linear, which is helpful for interpolation and extrapolation. And it provides a base line against which thermal effects can be contrasted. Monteith (1973) gives some rough-and-ready formulas for getting d and z_0 from the vegetation height, h: d is about 63% of h and z_0 is 13% of h. Lowry (1967) gives values of z_0 ranging from 0.01 to 0.1 cm for fine sand up to between 4 and 10 for tall grass. Figure 8.4 gives a pair of graphs of equation (8.14), using the values of U_*, z_0, and d obtained by Oliver and Mayhead (1974) from measurements within and above a pine forest during an especially intense storm. These intrepid investigators, incidentally, found that z_0 and d in the storm weren't very different from the values at lower wind speeds, so apparently it is safe to use low speed data and to extrapolate to profiles and forces which occur during storms.

Within stands of plants, the situation is still more complex. Wind speeds are lower than those above the stand and, as a rough rule, are inversely related to the local density of foliage (Grace, 1977), although they naturally drop off near the ground. In open forests with little understory, winds beneath the canopy may exceed those within the canopy. A completely analogous situation was found by Mark Patterson (personal communication) within a "forest" of

[†] z here is what we have previously called y or h. In switching fields, we're caught between the Scylla of internal consistency and the Charybdis of common practice.

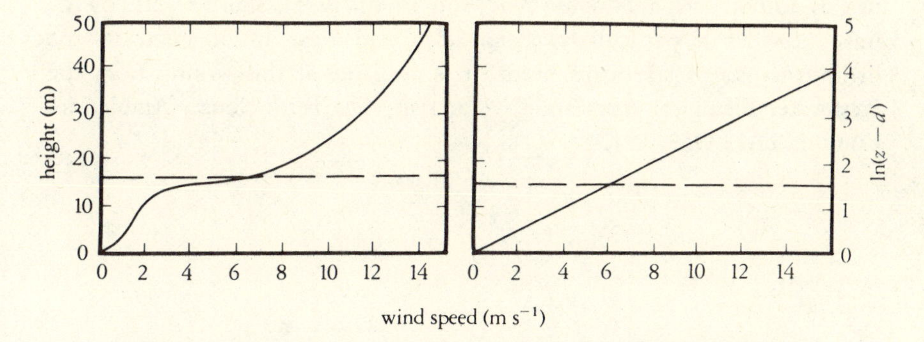

Two views of the distribution of velocities within and above a pine forest in an intense storm. The dashed lines correspond to treetop height, and the lines below 2 m s^{-1} are extrapolations. (From data of Oliver and Mayhead, 1974).

FIGURE 8.4

tall anemones (*Metridium senile*) on a subtidal rock wall. For further information on winds just above the earth's surface, the reader should consult Geiger (1965), Lowry (1967), Monteith (1973), Leyton (1975), or Grace (1977), and should peruse the large number of related articles in the journal, *Boundary Layer Meteorology*.

To bring the matter full circle, there are edge effects to be considered. A wind which passes from one sort of surface to another will thereby alter its velocity profile and turbulence. The distance required to achieve a steady-state velocity profile is called the "fetch" and is about 200 times the thickness of the velocity profile (Monteith, 1973). Thus is takes a fetch of about 200 meters to get a logarithmic profile a meter thick. Related to fetch is another kind of edge effect—the action of windbreaks and shelterbelts. These latter provide a means of manipulating the velocity gradient near the ground and are of great agricultural importance. They are the subjects of books by Jensen (1954) and, less formally, by Caborn (1965) and of various papers, such as that of Plate (1970), in journals such as *Agricultural Meteorology*.

The phenomenon of turbulent flow over an unbounded surface is just as applicable to flows of water as of air; events on the bottoms of rivers and oceans have merely had less practical application. With only a mild permutation of notation, Defant (1961) uses equation (8.14) to describe the effects of ocean currents near the bottom. Even von Karman's constant, 0.40, is unchanged. Similarly, the equation is invoked for flows near the bottom of rivers and channels by authors such as Simons (1969) and Smith (1975). Smith's work, by the way, provides a particularly cheap (£1) and clear introduction to the hydrodynamics of lakes and rivers. It is available at this writing from the Freshwater Biological Association's Librarian, The Ferry House, Ambleside, Cumbria LA22 OLP, U.K.

CHAPTER 9

life in velocity gradients

THE DRAG OF ATTACHED ORGANISMS comprises the entire content of Chapter 6. In that discussion, it was assumed either that such organisms were exposed to free-stream velocity or that the spatial variations of velocity impinging on them could be adequately represented by some simple average. We now turn to the world of organisms whose heights above the substratum are less than or of the same order of magnitude as the thickness of the boundary layer. (Of course, no organism in the logarithmic boundary layer of an unbounded surface will really experience mainstream current.) If we want to afflict biology with yet another name, we can easily coin one for this assemblage. We might call them *craspedophilic* creatures, from the Greek *craspedo-*, an edge or border, and *phile,* fond of (Jaeger, 1930): they are the interfacial plants and animals.

A boundary layer is both good news and bad news. It is a hiding place from drag, but it is a barrier to the exchange of materials and energy. A barrier may be useful as insulation; it's bad when exchange is necessary. A boundary layer may afford some mechanical protection for a moving organism; it may also constitute an extra mass to be accelerated. Current-dispersed propagules are likely to travel less far if released deep within a boundary layer, especially in air; the same propagules are more likely to remain where they settle if the boundary layer is thick and the velocity gradients are gentle. On a filter-feeding device, a boundary layer may cause deflection of edible particles around the filter, particles which would, in an ideal fluid, impinge upon it. It is clearly important that one consider not just free-stream velocity, but the currents which actually affect an organism.

drag of the flat and not-so-flat

The torrential insect fauna. Even the swiftest streams have stones covered with vegetation and carry suspended comestibles. Despite currents as high as 3 m s^{-1} (Nielsen, 1950), a specialized fauna of both grazers and filterers manages to make a living in such streams without being swept downstream. Most of the macroscopic members of this "torrential fauna" are immature insects: mayfly nymphs (Ephemeroptera); caddisfly larvae (Trichoptera); beetle larvae and adults (Coleoptera), especially of the family Elmidae; black fly larvae (Diptera: Simuliidae); net-winged midge larvae and pupae (Diptera: Blepharoceridae); larvae and occasional pupae of some moths (Lepidoptera: Pyralidae); and a few others. Their diversity is far less than that of the animals which live in the same streams but avoid the current by hiding in crevises, by burrowing, or by other means. The true torrential fauna live in a physically harsh habitat which demands substantial structural, functional, and behavioral adaptations. Merritt and Cummins (1978) provide a recent general account of who lives where and how.

Early in this century, Steinmann (1907) suggested that the insects of rapid waters are dorso-ventrally flattened to reduce drag and to help them stay attached. The idea sounds reasonable, and it penetrated several generations of textbooks, but it is a gross oversimplification and has since been questioned by almost every serious paper on the subject. Some torrential forms are indeed flatter than their calm-water relatives, but others are less flat than closely related non-torrential forms. Thus among the mayflies, *Iron* and *Rhithrogena* are notably flattened, whereas *Baetis* and others have adopted the fusiform shape of a fully streamlined body (Figure 9.1a and b; Dodds and Hisaw, 1924; Needham et al., 1935; Resh and Solem, 1978). Still, *Baetis* can flatten itself against the substratum to some extent (Ambuhl, 1959), and at least one species of the genus tolerates currents (presumably free-stream) of 3 m s^{-1} (Dodds and

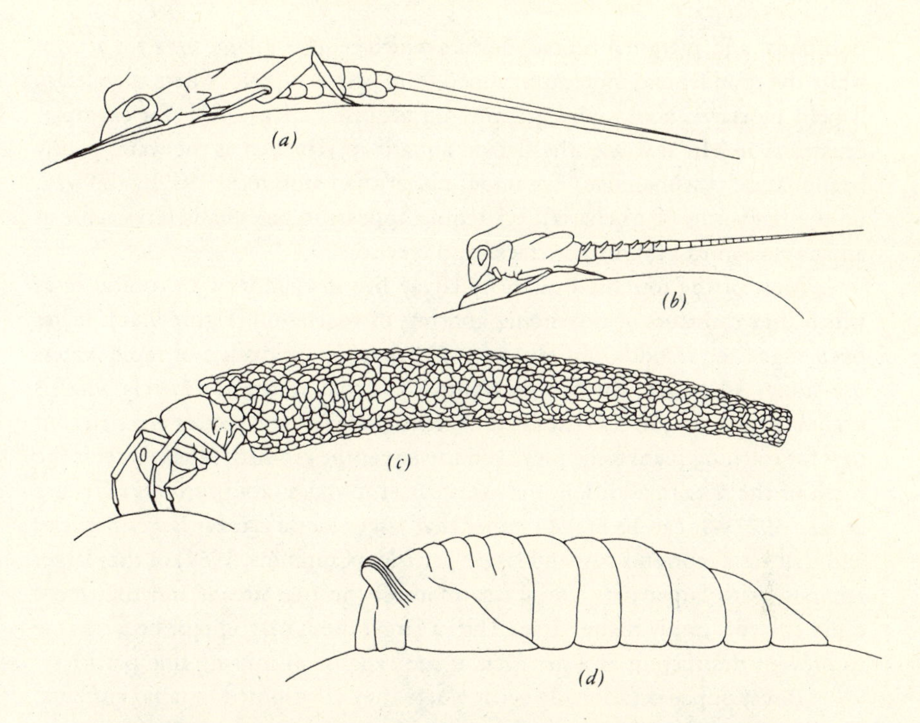

Some insects which live atop rocks in swift streams. (*a*) The mayfly nymph, *Rhithrogena;* (*b*) The mayfly nymph, *Baetis;* (*c*) the caddisfly larva, *Neothremma;* (*d*) the blepharocerid pupa, *Bibiocephala.*

FIGURE 9.1

Hisaw, 1924). *Baetis* carries streamlining even to the cross sections of its stout legs; they are flattened overall, but are thicker and rounder at the anterior margins and thin and sharp at the trailing edges. All of these torrential mayfly nymphs orient to face upstream.

Steinmann viewed the flow as pushing the flattened insect down against the substratum by striking its inclined, upstream-facing surface. Not only are torrential insects not consistently flattened, but (as Ambuhl, 1959, has pointed out) Steinmann's physical rationale was almost certainly incorrect. More likely, a convex upper surface would generate lift (recall the plaice in Chapter 4), and the use of a more fusiform and less flattened shape might be a way to avoid excessive lift. Flattening is better viewed as a way of getting down within the boundary layer and increasing the area of contact with the surface. Perhaps a flattened form is appropriate when attachment is maintained by suckers, friction pads, and marginal contact devices which require a large area for action, while the rounder and more streamlined form proves superior where the surface is held by claws, hooks, and grapples on an insect's legs. (See, for example, Stuart, 1958.) In any case, the flattest aquatic insects such as the water-penny beetle larvae (Psephenidae) live under rather than atop rocks. So, by the way, do the flatworms (Turbellaria). Flattening appears to be equally serviceable as an adaptation to life in tight cracks and crevises.

Most of the torrential caddisfly larvae live in cylindrical or conical cases which they construct of tiny stones and bits of vegetation (Figure 9.1c). It has been suggested (Dodds and Hisaw, 1925) that the caddisflies of rapid waters use stones as ballast; the counterargument is that stones are merely what is available in faster currents (Resh and Solem, 1978). Some species erect elegant nets for catching planktonic prey, and the larger the caddisfly, the coarser is the mesh of the net. In addition, nets are coarser in more rapid currents (Wallace et. al., 1977). It can be argued either that larger larvae can eat larger particles and that faster currents suspend larger particles (Cummins, 1964) or that larger animals, with larger nets, must face more of the free stream and that more rapid currents imply higher drag. Thus a larger mesh may in part be a scheme to prevent destruction of a net even at the expense of missing fine particles.

Insect pupae occasionally occur where they are exposed to rapid currents. They don't take on particularly flattened shapes (although some Pyralid pupae live in very flat cocoons; Nielsen, 1950); probably this is due to the morphological constraints of metamorphosis. But pupae are usually compact and fusiform and thus, assuming appropriate orientation, reasonably streamlined. Some morphological adjustments to life in a velocity gradient are evident. Pennak (1978), for example, provides a drawing of a Blepharocerid pupa which looks almost identical to one of Hoerner's streamlined rivet heads (Figures 9.1d and 8.3).

A rapid current enhances gas exchange, and it is not uncommon for torrential insects to have reduced gills. Smaller gills, of course, incur less drag, so, for these structures at least, the presence of a current permits a lower-drag design. Perhaps species exist in which gills are progressively less exposed as current increases and in which the drag coefficient thereby decreases with velocity. Reduced gills, of course, lead to an absolute current dependence. Another sort of current dependence has been described for one Simuliid larva. It cannot keep the cephalic fans with which is feeds open at currents below about 0.2 m s^{-1} (Harrod, 1965).

Current dependence is certainly a multidimensional parameter. Living in swift water may require a specific attribute, but possession of that attribute need not restrict an organism to rapid currents. Conversely, living in swift water may permit omission of some other attribute where the omission does restrict the range of the organism. Field studies that merely document the distribution of organisms are unlikely to explain distributions in any truly causal sense. Equally, though, a laboratory study is prone to focus too exclusively on a single facet of multidimensional adaptations such as drag. Ambuhl (1959), for example, found that some stream insects preferred rapid flows in an experimental channel, others avoided currents by hiding among stones along the bottom, while still others seemed almost indifferent to current velocity. However, in nature these latter insects might have sharp preferences with respect to currents as an indirect consequence of factors such as predation and trophic preference.

All of this information on stream insects is essentially anecdotal. The literature is very large, but I've not discovered a single actual measurement of drag under reasonable or even unreasonable conditions. Only when velocity profiles are measured at the location where creatures are collected and these profiles are then duplicated in experimental flow tanks for tests of both organisms and models will we be able to speak with any confidence about the shape adaptations of this torrential fauna.

Drag and tenacity of limpets. Currents on rocky coasts are less regular than those of streams, although the common bidirectional surge represents a fairly predictable flow. The adaptations of erect, attached marine organisms to wave forces were discussed in Chapter 6; let's now look at some relatively flat animals which live attached to a substratum. In particular, several auspicious studies have been done on limpets. These ordinary gastropod mollusks have single shells of modified conical form; growth is by addition to the exposed edge of the cone as well as by thickening from within. Limpet shells are low and unspiralled cones, several times wider across the aperture than their height. They vary from almost radially symmetrical to rather bilaterally compressed, often with the apex forward of center, and they run from almost smooth to elaborately sculptured (Figure 9.2).

The proper limpets, (*Docoglossa*), even more than the other limpet-shaped gastropods, are high-current specialists. Warburton (1976), for example, mentions that *Patina pellucida* reaches greatest density at currents around 1.0 to 1.4 m s^{-1}. This particular species lives on the fronds of a large alga, *Laminaria,* and the individual limpets orient with their long axes parallel to the long "fingers" of the algae (Vahl, 1972). Since the algal fronds move somewhat like weathervanes, the limpets should thus avoid the higher drag of a broadside current.

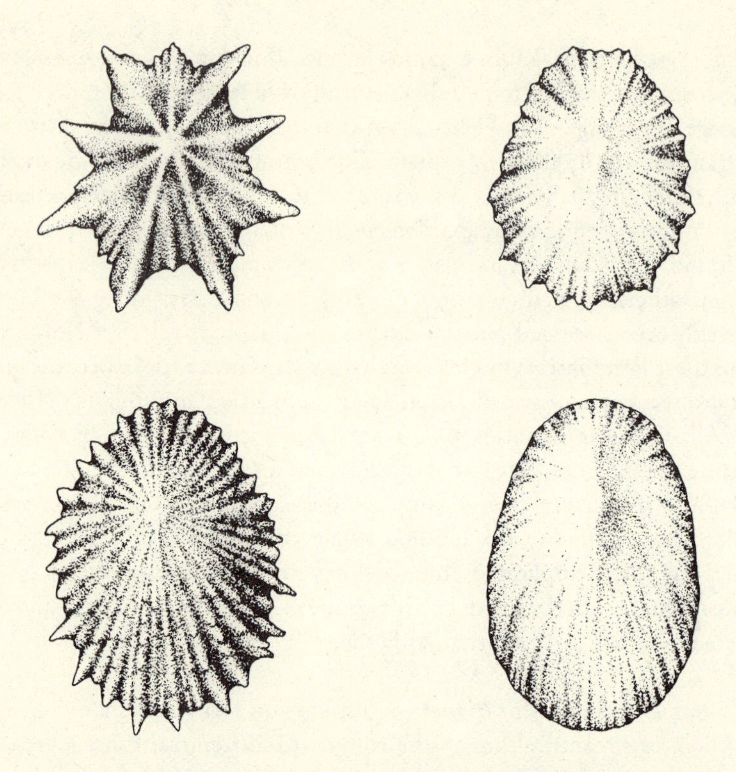

A potpourri of limpet shells (Genus *Patella*), showing the variations in surface texture. What adaptive significance might be attached to this diversity? The limpet shape occurs in many other mollusks which live on rocks in moving water, such as keyhole limpets (Fissurellidae), false limpets (Siphonariidae), slipper limpets (Crepidulidae) and chitons (Order Polyplacophora) in sea water, and the Ancylidae in fresh water.

FIGURE 9.2

Patina is a fairly asymmetrical, erect-sided, and roundspired limpet which looks, in side view, very much like a half-streamlined body. Living on algae, it faces currents more typical of stream organisms and, indeed, rather resembles the torrential, freshwater limpets (Nielsen, 1950) with whom its taxonomic affinities are remote. At or above 1 m s^{-1} dislodgment becomes a serious problem, at least in steady flows in the laboratory. A sudden current can dislodge a limpet at a lower speed, but with gradually increasing currents it clamps down in anticipation and tolerates higher flows. In any case, there seems to be little safety margin, so drag minimization might be crucial. Just prior to dislodgment, the front of the shell elevates (Warburton, 1976), which is just what we'd expect, based on the normal distribution of pressure around a sphere or streamlined object. It also is precisely the opposite of the anterior depression which Steinmann's mechanism would predict. Interestingly, Vahl (1971) finds some evidence that conditions of exposure are reflected in growth pattern and the consequent shape of the shell; the matter would certainly bear further investigation.

I'd like to take a gentle poke at both of these studies on *Patina.* Vahl asserts that larger limpets are relatively flatter; this is certainly of interest, if so. But my analysis of his data suggests that this is not the case; his height-to-aperture area ratio is not dimensionless and is therefore misleading. Warburton presents a number of graphs involving the drag of shells; basing figures for the drag of a naturally selected organism of irregular shape living in a velocity gradient on textbook drag coefficients seems to be quite unrealistic.

Patella is a genus of limpets which have more pointed and central apices and more sloping and less rounded side walls than *Patina.* (Figure 9.2). The individual species vary considerably in size, shape, and surface ornamentation. Branch and Marsh (1978) studied the shape, drag, and tenacity of six species and found that the tallest and fanciest were those living in the most protected places. *Patella* live on rocks rather than algal fronds, so currents are presumably less regular than those striking *Patina.* Nevertheless, the drag on a limpet broadside to the flow is as much as three times that on one facing the current. The tenacity with which such a limpet attaches is nearly proportional to the area of its foot. Thus if the animals are geometrically similar and if drag is proportional to area, a big limpet should be neither more or less likely to dislodge than a small one. However, if a large limpet protrudes further through the boundary layer, then it would be more likely to dislodge unless, at the same time, its drag coefficient were lower: this may in fact happen.

Branch and Marsh (1978) determined the drag coefficients for shells of their six species. The one from the most exposed habitats, *P. cochlear,* had the lowest drag coefficient, 0.28 (based on frontal area) in a current of 1.0 m s^{-1} over a plate with a minimal boundary layer. For a 5.5-cm-long shell, the Reynolds number is 55,000, so the drag is far above that for a spot on a flat plate (drag corrected to plan-form area and compared to equation (8.12)) but a bit lower than that of a locust fuselage (Table 7.3). Surprisingly, they found that the drag coefficient varied inversely with speed: at 0.81 m s^{-1} it was 0.34; at 0.58 m s^{-1} it was 0.47. This corresponds to a situation in which the drag coefficient is almost exactly proportional to the reciprocal of the Reynolds number. Drag should thus be proportional to the first, not to the second, power of velocity: useful, but most peculiar. Since limpet shells are rigid, the relationship should apply as size, instead of speed, is varied: larger limpets should have less drag per unit area than small ones and might therefore be able to tolerate more of the free-stream current. Again, I emphasize the peculiarity. For a bluff body, the drag coefficient is nearly constant in this range, or proportional to Re^{0}; for a streamlined object, the drag coefficient is proportional to $Re^{-0.5}$. For these limpets it is proportional to $Re^{-0.95}$. We saw such a relationship for a tree in a wind, but a tree can reshape itself in ways a shell cannot. Someone should look into the matter. Measuring the drag of limpet shells should be simple.

diffusion through boundary layers

The rate at which a type of molecule diffuses from one place to another is proportional to the concentration gradient between those places. Flow might appear to be of little direct relevance, but in practice it is often of great importance in determining the magnitude of that concentration gradient. Consider an air-water interface through which water is evaporating into the air. At the interface, the air will be water-saturated, while at some distance away the concentration of water will be lower. If the air has some free-stream concentration of water vapor, the rate at which water can diffuse away from the interface will depend on how far it has to go to reach that lower free-stream concentration. "How far" must, of course, refer to the thickness of the boundary layer, which will have a humidity gradient as well as a velocity gradient. Anything which makes the boundary layer thinner should promote movement of vapor from the surface, such as higher wind speed, greater proximity to a leading edge, and so forth.

Water loss in air with wind. These situations should be familiar. Greater air movement lessens the likelihood that liquid water from perspiration will accumulate on the skin. Wind often has a deleterious drying effect on plants. In fact, the phenomena are slightly more complex than they appear. At least three processes are involved: transport of water to the surface by whatever means; vaporization; and, finally, diffusion. The overall rate of water movement depends on a set of resistances, mostly in series; each of these resistances depends on different geometrical and physical parameters. If the resistance of the boundary layer is a major element in the total resistance, then alterations in its value will have a noticeable impact on water loss. By contrast, if the boundary layer resistance is low compared to other series resistances, than increases in wind speed will have little effect on water loss. For example, Bange (1953) found that a moderate wind could cause a fourfold increase in the transpiration of a *Zebrina* leaf, provided that the leaf had fully open stomata. Otherwise, stomatal resistance was a large component in the series. On the other hand, Tracy and Sotherland (1979) showed that the boundary layer of bird eggs is of little consequence in water loss; its resistance is less than 10% of the total resistance, including shell and membranes, under any circumstances. Still, situations are quite common in which the boundary layer resistance is a major element, even where the surface of an organism isn't obviously wet or grossly perforate.

Does the boundary layer viewed as a resistance to water vapor diffusion correspond to the boundary layer we've previously described in terms of flow? Using garden snails as material, Machin (1964a, 1964b) compared aerodynamic and diffusional boundary layers at low speeds. On a 5-cm-long snail in a wind of 26 cm s^{-1}, the aerodynamic boundary layer was only 1.5 mm. Depending on the circumstances, diffusion boundary layers were three to six times thinner than aerodynamic boundary layers, but they varied in much the same manner. One probably shouldn't make too much of the difference: recall just how arbitrary was the definition of the thickness of the aerodynamic boundary layer. A water molecule needn't reach the free-stream wind to reach free-stream vapor concentration. Even more important, the diffusional boundary layer invoked by Machin represents an imaginary stagnant layer with the same net effect as the real gradient region. In any case, evaporation from either a free-water surface or a snail proves to vary with the square root of wind speed or, in effect, with the reciprocal of the boundary layer thickness—diffusion and concentration gradients are what matter. Indeed, Ramsay et al. (1938) measured humidity gradients on single transpiring leaves in a wind tunnel and found that the gradients very nicely fitted the expected parabolic form of a boundary layer's outer "limit." In practice, thicknesses of diffusional boundary layers are not often measured or calculated (but recall the references to Nobel, 1974, 1975). The measurements are awkward and the calculations are based on figures for resistance; resistance itself is usually the datum desired.

Diffusion of water vapor through boundary layers has been of especial concern to plant physiologists, who produce periodic reviews of the state of the art. For further information, I suggest consulting Slatyer (1967), Meidner and Mansfield (1968), Monteith (1973), or Grace (1977). For animals, a good source is Schmidt-Nielsen (1969).

Other physiologically important gases such as CO_2 and O_2 move around in response to the same rules as does water vapor; their diffusivities are merely a little lower as a consequence of their higher molecular weights.

Aquatic systems. Boundary layers are even more important in diffusion processes in water than in air. If a transport system is viewed as a series of resistances, the largest resistance is the most important rate-limiting component. The diffusion constants of small molecules through water are about 10,000 times lower than their values in air. Thus a diffusional link through water in a transport sequence is much more likely to be seriously rate-limiting. Concomitantly, currents are more important in water, and it takes a lower current to have a noticeable effect on exchange between an organism and its environment. I should emphasize that pure diffusion is a poor way to move materials over macroscopic distances in any medium; the common example of the spread of perfume in a room really demonstrates convection and flow, not molecular diffusion.

A simple consequence of diffusion through a boundary layer is commonly seen when a photographic print is developed. As the print is agitated in a tray of developing solution, the image first appears near the edges (Figure 9.3) because the necessary chemical reactions proceed more rapidly where boundary layers are thin. The phenomenon could probably be employed as a semiquantitative way to judge the steepness of velocity gradients, the onset of tur-

FIGURE 9.3

A photographic image appearing, edges first, in a tray of developer.

bulence, and the location of separation. The test object could be made of photographic paper or covered with emulsion and immersed briefly in dilute developer.

Since very low water currents may have a substantial effect on transport and exchange, still water may be no simple thing to achieve as a control. Schumacher and Whitford (1965) measured both respiration rates (in the dark) and photosynthetic rates (with light) on four classes of attached algae; in all cases currents greatly increased these rates. In *Spirogyra,* a filamentous green alga, a current of 1 cm s^{-1} increased the photosynthetic rate by 18% compared to the rate in "still water." Even lower speeds were investigated by Westlake (1967), who was able to detect an increase in photosynthetic rate in *Ranunculus pseudofluitans* (an aquatic dicot) when the current was raised from 0.02 to 0.03 cm s^{-1}. The latter is about a meter per hour, well below what we'd normally notice as moving. I find no evidence which places a lower limit on currents that augment diffusive exchange; probably it's more a matter of accuracy and the practical definition of still water than any particular threshold: eventually one would reach, at zero speed, a baseline rate reflecting pure diffusion.

Just how limiting diffusion through a boundary layer can be is clear from the records of Schumacher and Whitford for higher speeds. An 18 cm s^{-1} current could produce more than a fivefold increase in photosynthetic rate compared to still water. Here, however, one ought to expect an upper, high-speed asymptote. Eventually, diffusive resistance becomes trivial compared to other rate-limiting processes. It's not easy to distinguish an approach to a true asymptote from the gradual diminution of the current effect at higher speeds due to the dependence of boundary layer thickness on the inverse square root of speed. Also, other complications enter. High-intensity water movement has usually been associated with an increase in productivity in large marine algae due to enhancement of nutrient uptake. But Gerard and Mann (1979) found the opposite effect in a pair of *Laminaria* populations. In exposed locations the algae developed thicker and narrower blades (perhaps for strength and drag reduction), which more than offset the diffusional advantage of the current.

In talking about drag, the issue of morphological adaptations was obviously central. It is only a little less certain that another suite of adaptations reflects the problems of diffusion through boundary layers. Small leaves, which would have thinner boundary layers on the average, are variously reported to have higher photosynthetic rates in crop plants than do large leaves. Aquatic plants often have smaller, thinner, or more highly dissected leaves than their terrestrial relatives (see Sculthorpe, 1967, or Haslam, 1978). The possibility of reduced damage from flow forces has already been mentioned, but many of these plants never experience rapid currents. So their shape may be primarily an adaptation to the much greater difficulty of diffusive exchange through a boundary layer in an aqueous medium. But current, as we've seen, helps exchange. As mentioned earlier, at least some of the insects of torrential currents have gills of reduced area (Dodds and Hisaw, 1924); probably the augmentation of gas exchange in such currents makes gills of "normal" size an extravagance and a drag.

Diffusion through boundary layers is important in the world of cell biologists as well. I can't resist citing one iconoclastic experiment (Stoker, 1973). On a culture of fibroblasts a strip was removed, leaving a denuded patch. Cells at the border of the denuded zone increased their growth and invaded it. Conventional wisdom would call this a reduction of contact inhibition between cells. But Stoker showed that one needn't invoke anything so biological as an explanation; the phenomenon is caused merely by higher concentration gradients in the stirred medium at the edges of the patch due to locally more rapid fluid movement.

So far our discussion has tacitly assumed that boundary layers were laminar or that events took place only within the laminar sublayer. Turbulence provides an additional agency of transport normal to the surface of the local direction of overall flow, and it leads us into quite a different physical world. Similarly, I've not mentioned heat transfer through boundary layers; the subject of convective heat transfer will get brief attention in Chapter 14. For now it might just be noted that conductive heat transfer and diffusion follow very much the same formal rules. The only major difference is that the density of the medium is rarely much affected by the concentration gradients which "drive" diffusion. By contrast, thermal gradients quite commonly affect density enough to generate local flow—so-called "free convection"—even in the absence of a free-stream current.

suspension feeding and local currents

Animals which remain attached to the substratum and which extract food from the passing water have appeared from time to time in these pages. They are representatives of the trophic group called "suspension feeders" or "filter feeders," (Jorgensen, 1966). Each of these forms must represent some compromise between, on one hand, extension into the free stream and the consequent drag and, on the other, hiding in the local boundary layer but intercepting less water. Their situation is not drastically different from that of an organism which exchanges molecules by diffusion through a boundary layer.

Some suspension feeders are completely dependent on ambient currents; most of the net-erecting trichopteran larvae are good examples. Others actively pump water and make their own currents—bivalve mollusks, ascidians, encrusting bryozoans, and others. A third group—represented (as far as we know) mainly by sponges and some brachiopods—can pump but are capable of using environmental currents as well (Vogel, 1978a). The difference in feeding habits is reflected in choice of habitat. Hughes (1975) measured the currents around erect colonial hydroids (*Nemertesia antennina*) in about 1 m of water. Velocities near their tops were generally above those near the holdfasts. He then made a very thorough catalog of the numerous species which live on the hydroid, noting where each was found. The more passive feeders (some smaller hydroids and others) were most distal or apical; the more active feeders were more proximal or basal. Really active pumpers—some bryozoans, bivalves, ascidians, and sponges—were usually found only on the holdfast. Furthermore, the suspension feeders on the holdfast treated food differently from the higher and more passive sorts. The former were much less selective in initial capture but had well-developed mechanisms for subsequent rejection of unwanted stuff. Clearly there is a secondary disadvantage of being on the bottom. Suspended plankton is more likely to be nutritive than particles which have dropped out of the water column and are tumbling along in the boundary layer.

Even where organisms do not seem to take advantage of local currents to induce internal flow, filtration rates can depend quite strongly on the rate of ambient flow. The mechanism is unclear—somehow the current must stimulate more active pumping—but the general phenomenon seems established from the work of Walne (1972) on five species of clams, mussels, and oysters. Perhaps more rapid flow signals the likelihood of a richer food resource and the desirability of a greater investment of energy in pumping. Kirby-Smith (1972), by the way, found almost the opposite effect in bay scallops, in which currents (up to 12 cm s^{-1}) inhibited growth. Again, the mechanism in unknown.

Barnacles are high-current, non-pumping suspension feeders. Local current seems to be one of the major factors determining the course of adult growth. They commonly orient with the rostrum facing the predominant source of current (Crisp, 1960). Often they grow in low "hummocks" about 5 cm across; about 100 individuals make up a hummock, with the largest in-

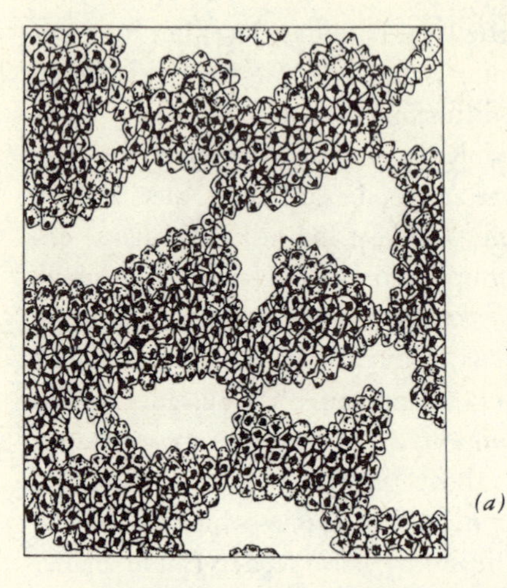

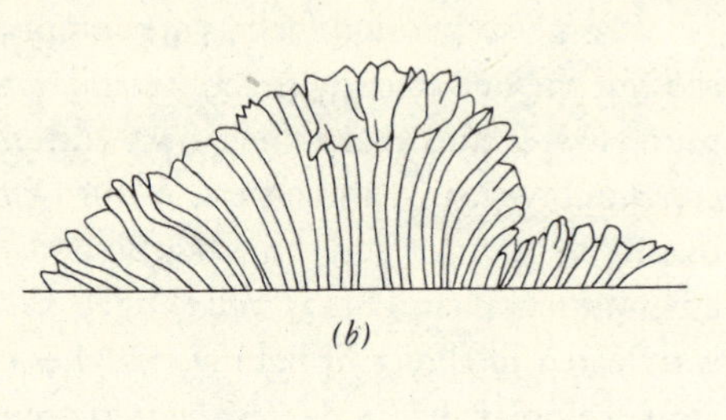

(*a*) Hummocks of barnacles on a rock; (*b*) a cutaway of an individual hummock showing individuals of varying heights.

FIGURE 9.4

dividuals at the summit (Figure 9.4). The pattern of hummocks and furrows is quite independent of the contours of the underlying rock (Barnes and Powell, 1950), nor does it depend on variations in the age of the individuals. Knight-Jones and Moyse (1961) suggested that faster growth gets an individual out into faster-moving water; it can then capture still more food and increase its relative advantage. Such individuals then form the centers of these domeshaped clusters. They further suggested that the hummocky surface benefits the whole population by increasing turbulence and preventing the build-up of a food-depleted laminar boundary layer. I like the idea but would rather reserve judgment on whether it truly explains the phenomenon of hummocks until someone investigates the matter under more controlled conditions.

Many species of bryozoans (''moss animals'') form thin colonial encrustations on the flat surfaces of algae. These animals are highly active suspension feeders but live out their adult lives quite thoroughly enveloped by the boundary layer of the host's surface (Figure 9.5a). How then do they keep from wasting their time continually reingesting their neighbor's effluent? One neat device has been described by Cook (1977) and investigated in detail by Scott Lidgard (personal communication). While the individual zooids have no gaps between them, a colony of *Membranipora villosa* has regions of non-feeding, degenerate individuals every few millimeters. These regions act as chimneys for the colony-wide currents. Water from near the surface of the colony is pumped downward by cilia on the tentacles of the individuals. Water then moves laterally toward the chimneys and emerges from them as high-speed jets (2.5 cm s^{-1}) which broaden and coalesce well above the zooids (Figure 9.5b). Excurrent water is therefore located above incurrent water and is as well separated from it as if physical chimneys were present. So no zooids can reprocess water previously used by other zooids except, perhaps, in the stillest of still waters. Encrusting Bryozoa certainly care about local currents. Norton (1973) found that colonies of *Membranipora membranacea* expand proximally on algal fronds. They seem to prefer the portions where boundary layers are thinner and, as the alga grows in one direction, they expand in the other. And Buss (1979) found evidence of microcurrent interference between two species: as one species overgrew the other, the former managed to subject the latter to a pure diet of previously-filtered effluent.

This business of arranging a high-speed jet on one's output is fairly common. Bidder (1923) seems to have first recognized it (in sponges) as a scheme to minimize the chance of reingesting previously filtered water in the

FIGURE 9.5

Colony-wide currents in the encrusting bryozoan, *Membranipora villosa.* The rectangles in (*a*) are "chimneys"—zones of degenerate, nonfeeding individuals. A cross-sectional view of a slice (*b*) is indicated by the dashed line in (*a*). Dye goes through the tentacle ring (lophophore) of an individual, moves laterally, and emerges at a chimney. Taller individuals surround the chimneys, which are about 4 mm apart.

(*a*)

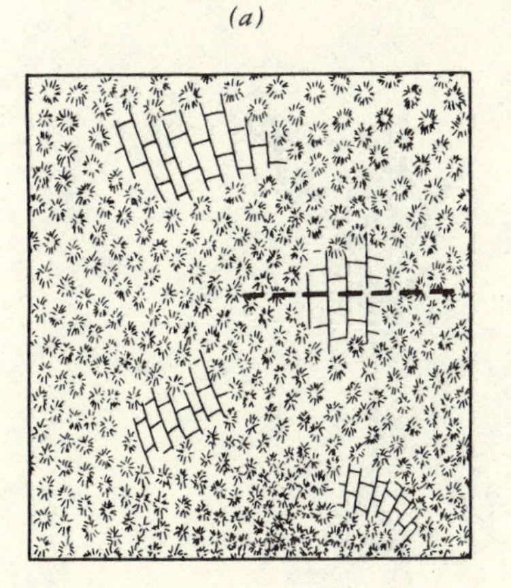

(*b*)

low currents near a surface. The siphons of ascidians and bivalves accomplish the same, and some of the colonial ascidians (Botryllidae) look as if they have a cooperative discharge system analogous to that of the bryozoans already mentioned (Figure 9.6). One might argue in favor of connecting the input rather than the output directly to the free stream; but, of course, it is easy to eject a jet by using a nozzle whereas ingesting a jet would require a long, tapered pipe.

An alternative to a jet is elevation of the filtration apparatus itself. Wherever external nets or tentacles are employed, this arrangement almost inevitably occurs. One can point to the various erect colonial coelenterates and bryozoa; sea anemones such as *Metridium;* several families of polychaetes living in tubes and extending feeding appendages upward (see Fauchald and Jumars, 1979); sea lilies among the echinoderms; goose barnacles among the crustacea; the stalked solitary or colonial protozoa such as *Vorticella* and *Carchesium;* the simuliid dipteran larvae which keep their cephalic fans uppermost and pick thin boundary layers (Wallace and Merritt, 1980); sessile rotifers; and even some ascidians which live at the tops of stalks (Figure 9.7). In some of this diverse assemblage the stalk or other elevation device is retractile, in others it is not. Some of the retractile systems operate fairly slowly (anemones for example); another involves the most rapid contractile mechanism known in any living system—the "spasmoneme" of *Zoothamnium,* a colonial stalked ciliate similar to the solitary *Vorticella* (Weis-Fogh and Amos, 1972).

For more information on filter and/or suspension feeding, the reader should consult Rubenstein and Koehl (1977) for basic mechanisms, Jørgensen (1975) for physiology, and Wallace and Merritt (1980) for ecology. The subject will also be discussed in Chapter 13.

FIGURE 9.6

A single output jet discharging the common effluent of a colony of ascidians (*Botryllus*).

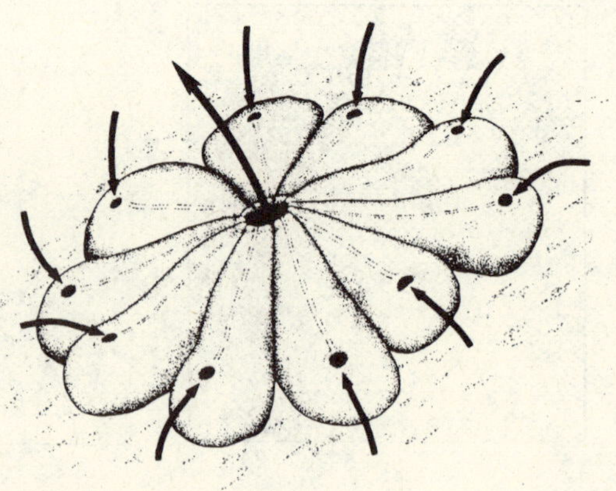

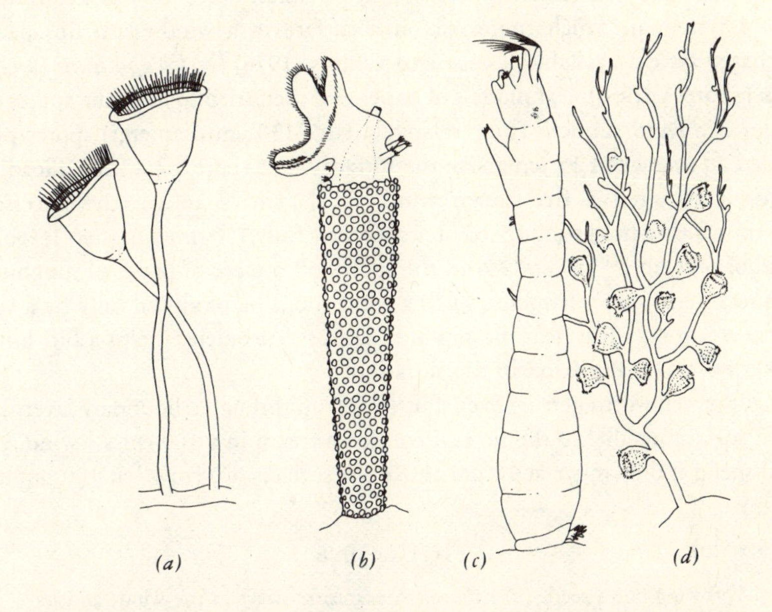

Various creatures which elevate their filtration apparatus. (*a*) The ciliate, *Vorticella;* (*b*) the sessile, tube-making rotifer, *Floscularia;* (*c*) the dipteran larva, *Simulium,* (*d*) the colonial ascidian, *Perophora.*

FIGURE 9.7

dispersal and the boundary layer as barrier

Passive dispersal of propagules through the use of winds and water currents is practiced by at least one member of every phylum of animals and division of plants. Aerosol particles, polluting and otherwise, likewise enjoy the benefits of the free transit system in the sky. It looks as if all a propagule has to do is to arrange a sufficiently low sinking rate through high drag, lift production, upcurrent detection, or density adjustment and to avoid becoming someone else's lunch en route (*sic transit gloria*). In fact the organism must do more. Somehow it has to get liberated into the current and it needs to deposit itself in some way on a surface out of the current. In both cases, boundary layers must be crossed, and nature has produced an extraordinary collection of devices for accomplishing these mundane but essential steps.

Liberation. The basic problem may be easiest to see if we first consider a model system in which spores sit on a surface in a wind, with no special discharge system available. Grace and Collins (1976) let *Lycopodium* (a club moss) spores settle on leaf models of paper and then tried to blow the spores off again in a wind tunnel. These relatively large (30-μm diameter) spores protruded an average of 15 μm above the surface. For a spore 4 cm from a leading edge, it required a free-stream wind of 270 cm s^{-1} to get the necessary 1.8 cm s^{-1}-threshold wind speed on its center—fully 150 times higher. It took a wind of 680 cm s^{-1} to get 50% of the spores off a piece of paper of roughness comparable to spore diameter. Such a wind could be produced only by a very severe storm on a leaf near the top of a canopy. Turbulence helps a bit, but it makes no major alteration in the situation.

It is well worth getting even a little way up through a boundary layer. For a flat surface parallel to the flow, 4 cm downstream in a 100-cm s^{-1} wind, the local speed is 0.4 cm s^{-1} at 10 μm above the surface, 4.0 cm s^{-1} at 100 μm up,

FIGURE 9.8

Throwing one's seed. (*a*) *Papaver somniferum* sways in the wind, and its seeds are thrown out, guided by grooves, as it accelerates. Wind blowing in one side and out the other probably assists. (*b*) The nutlets of *Salvia lyrata* are thrown after a raindrop has depressed the pedicel.

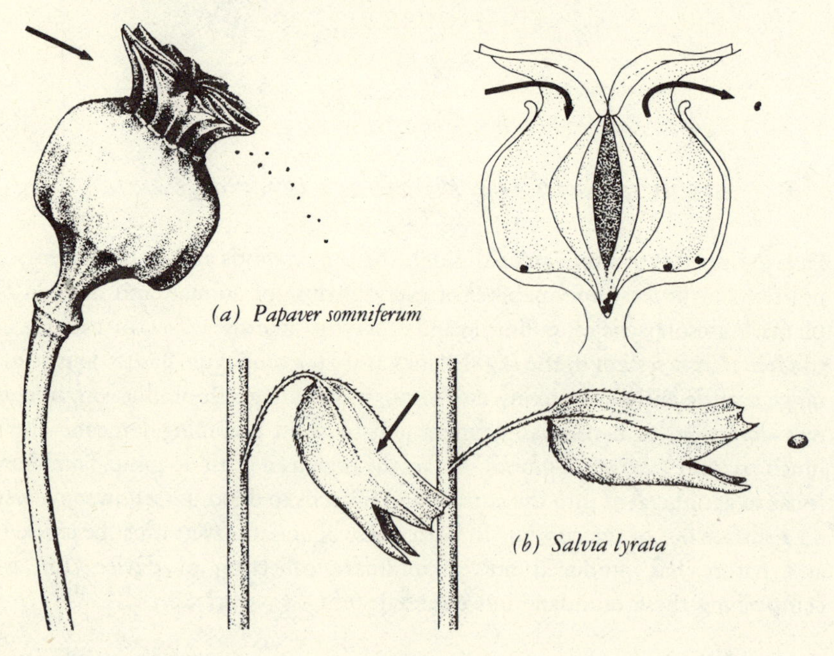

and 40 cm s^{-1} at 1 mm off the surface (equation 8.4). And elevation is the usual tactic. Spores, pollen, eggs, and sperm are commonly presented to the wind on some sort of elevation, whether we consider the spores of a *Penicillium* fungus less than 100 μm high or the elaborate arrangements to expose pollen to wind in many higher plants. When I mow my scruffy lawn, the highest items are the structures bearing wind-dispersed seeds of grass, dandelions, and plantains. And both lycosid spider young and gypsy moth caterpillers are reported to climb as high as they can before spinning their silk threads and letting themselves go in the breeze (Tolbert, 1977). The examples are endless.

More fun to recount are the more bizarre schemes (Figure 9.8). A relatively mild one is employed by a fungal smut of barley (*Ustilago nuda*). Smutty plants grow a little taller than healthy ones and diseased heads don't bend over in the wind. Thus the smut spores are held above the rest of the crop in the best possible position to be picked up by the wind (Sreeramulu, 1962). There are, furthermore, the so-called "wind-ballists" (Van der Pijl, 1972). One of these, *Papaver somniferum,* the opium poppy, has long elastic pedicels which swing in the wind; apical capsules throw relatively heavy seeds up to 15 m like fine old Roman artillery. And there are the "rain-ballists." In *Salvia lyrata* (a sage) the force of a falling raindrop depresses a springy pedicel which causes nutlets to be thrown up to 2 m, much like a swimmer projected from a diving board (Brodie, 1955). "Splash cups" occur in several groups of plants, sometimes, as with the pedicel of *Salvia,* just transmitting a blow, and other times, as in the moss *Polytrichum,* producing the propagating units as well.

One general problem in employing ballistics is the conflict between the design of a projectile which can be thrown far, one dense and low in drag, and a projectile which once aloft can be carried by the wind, one buoyant and high in drag. The problem is evident from a look at the performance of the famous ballistic fungus, *Pilobolus* (Figure 9.9a). It elevates its sporangiophore on a stalk, takes aim (a positive phototropism) and launches its sporangium atop a jet of sap. Buller (1934), in a clever experiment, determined that the "muzzle velocity" is as high as 14 m s^{-1}. Nevertheless, the sporangium is less than half a millimeter in diameter and the initial Reynolds number less than 400. The projectile, so well launched, goes only about 3 m horizontally in still air, so great is the effect of drag on small items at low Reynolds numbers. In a breeze it might be carried as far again. Shooting individual spores, which are much smaller, would be less effective; but, if they were aloft, they would fall out again at lower speeds. The optimization is not simple.

Guns of one sort of another have developed many times among plants. *Sphaerobolus,* another fungus, puts a glebal mass about 1 mm across in a

Fungal artillery. (*a*) *Pilobolus;* (*b*) *Sphaerobolus.* These are small organisms, only a few millimeters tall.

FIGURE 9.9

hollow cup; the inner wall of the cup suddenly everts, and this minimortar heaves the sphere a distance of several meters (Buller, 1933). (See Figure 9.9b.) *Sordaria,* yet another fungus, seems to hedge its bets with between one and eight spores per projectile. The larger projectiles go further (nearly 10 cm) from the initial shot, but they are probably less suitable for dispersal by wind (Ingold, 1965). Higher plants, as well, have weaponry. *Arceuthobium,* a dwarf mistletoe, has the same muzzle speed as *Pilobolus;* but, by virtue of a larger projectile (ellipsoidal, 2.9 × 1.1 mm), its seed goes about five times as far (Hinds et al., 1963).

A partial evasion of the problem of optimal size of projectile is obtained by ejecting fluid as well as projectiles. A puffball (*Lycoperdon*) works this way. Indenting its thin and flexible wall forces out a jet of air and spores; in nature, raindrops are probably the immediate stimuli (Gregory, 1973) in this bellowslike arrangement. And most of the filter-feeding animals which produce a jet of water manage to include their eggs, sperm, gemmules, or other propagules in the jet. In water, of course, drag forces are much larger, and jets have an overwhelming advantage over guns.

Some lichens and slime molds bear their propagative elements on the insides of upright cups. Brodie and Gregory (1953) showed that such cups not only kept soredia and spores from falling onto the substratum but also have a distinct aerodynamic role in dispersal. A wind of 2 m s^{-1} removed soredia of the cupulate lichen, *Cladonia;* the same wind was inadequate to blow them off a flat surface. A wind of 0.5 m s^{-1} blew spores from the cupulate sporangia of a myxomycete.

At least one system involves wind dispersal entirely within the velocity gradient at the earth's immediate surface. Several families of higher plants have produced "tumbleweeds," seeds with outgrowths sufficiently long for them literally to roll along the landscape. The lowest part of a roller is in contact with the ground and is at ground speed; the highest moves at twice the speed of the package as a whole but is, in a velocity gradient, exposed to the highest wind. The main disadvantage of the scheme (among some lesser ones) is probably its restriction to rather barren steppes with a minimum of obstacles. Such vehicles aren't much good at steering.

Deposition. Getting down can be almost as much of a problem as getting up. In air, the issue would seem trivial: what goes up must come down, as the saying goes. And the terminal velocities of a large number of pollens and spores are matters of public record (see, for example, Gregory, 1973). In still air, spores settle randomly on a horizontal, flat surface. But with increasing wind, the efficiency of capture is progressively reduced, with the least deposition on surfaces parallel to the flow. From the photographs of Hirst and Stedman (1971) it looks as if deposition is inversely related to the thickness of the local boundary layer. Several phenomena are probably involved. First, with a thick boundary layer, fewer particles will enter the inner reaches from which they may be captured in a given time. In addition, in a boundary layer of appreciable thickness, the downstream thickening involves a downstream reduction in speed of flow near the surface, and that, in turn, implies a divergence of streamlines from the surface. Such fluid particle trajectories will tend to deflect dropping spores or pollen from the surface. Also, there's some evidence that dropping spores can bounce off again, at least from leaves, although, as we saw earlier, a pollen grain or spore once settled doesn't easily blow off. Low turbulence seems to promote capture just as it inhibited blow-off (Forster, 1977).

In water, getting up is perhaps less difficult than in air, but getting down again is certainly more so, given the increased effects of buoyancy and drag. Rapid currents may enhance algal productivity in streams, but they retard the initial attachment of colonizing cells. On rocks diatoms are a relatively more important component of the flora at high currents, and they seem to settle better than filamentous green algae on experimental glass slides (McIntire, 1966).

For actively swimming larvae in marine situations there is evidence of active selection of attachment sites. Barnacles, for example, perfer rocks in high current areas. Crisp (1955) passed cyprids of the barnacle, *Balanus balanoides*, through glass pipes in which the velocity profiles were known. From these and other experiments he showed that free-stream velocity was largely irrelevant; it was the current 0.5 mm from the surface which determined whether attachment would occur. In practice, that meant that what mattered was the velocity gradient at the surface. A shear rate of 50 s^{-1} was the minimum which stimulated attachment, 100 s^{-1} was optimal, and the cyprids were unable to attach above 400 s^{-1}. The phenomenon is probably general among motile marine larvae and is of obvious relevance ot the ecology of fouling communities. Perhaps studies of attachment to artificial substrates should pay more attention to velocity gradients and less to free-stream current.

the mass of the boundary layer

What must be the least pervasive role of a boundary layer is one which I turned up as my first attempt at science (Vogel, 1962). The frequency of wingbeat of many insects seems to be determined largely by the natural frequency of thorax and wings as a mechanical oscillator. For most insects, this frequency does not change as air density is altered. However, for a certain number of smallish insects, frequency drops in a slight but regular manner as air density is increased. This inverse relationship can be explained if it is assumed that changing air density changes wing mass (or, more precisely, moment of inertia). And this odd assumption becomes plausible when one realizes that a wing, beating over one hundred times each second, "carries" with it a boundary layer or some portion of one which must have some mass. Insect wings are extraordinarily light. A kilogram of the wings of a large fruit fly, *Drosophila virilis*, laid end to end, would extend from Boston to Washington; wing mass is two-thousandths of body mass. So the mass of a fraction of the boundary layer can be significant. The scheme works quantitatively. The assumption of an inertial boundary layer correctly predicts the shape of the frequency vs. air density curve for *D. virilis*, where sufficient data exist for a good test. It also correctly predicts which insects will and which won't show this variation in frequency. In fact, I'd stumbled across something called "virtual mass," about which more will be said in Chapter 14. But virtual mass in the usual treatment requires that an object displace fluid. The displacement of a fruit fly wing is trivial on its own; only its boundary layer endows it with a decent volume.

CHAPTER 10

flow through pipes

PIPES WERE PUT ASIDE AFTER THE discussion of the principle of continuity and Bernoulli's principle; with the aid of streamlines, imaginary pipes could be created at will and real pipes could be left on the back burner. Now armed with notions of drag, skin friction, and boundary layers, let us return to pipes. We'll consider for a start a portion of a long, unbranching pipe of circular cross section and rigid walls. Fluid has entered this pipe a long way upstream and flows steadily through it in a fully laminar fashion; at each point the instantaneous and time-averaged velocities are the same. Continuity works with a vengeance; the amount flowing through any cross section is the same as that passing any other cross section.

CHAPTER 10

basic rules for laminar flow

What formulas would we like to have concerning flow through pipes? For one thing, it would be useful to know the way in which velocity varies across the pipe. An engineer usually uses a velocity distribution as a step toward some other calculation; for a biologist, the distribution itself is of no slight interest, a point we rather dwelled upon in Chapter 9. Obviously, the velocity at the walls of the pipe will be zero, and therefore the velocity along the axis will be maximal. Far enough down the pipe, the no-slip condition will have exerted its influence throughout the cross section; thus, in effect, there will be within the pipe nothing but boundary layer. Still, the skin friction of the walls can't continually slow the flow or continuity would be violated. So we can reasonably expect that some velocity distribution across the pipe will, once established, persist further downstream. This contrasts with the behavior of the boundary layer on a flat plate: it continues to thicken, if at a slower and slower pace.

In addition, we'd like to have a formula to describe how flow varies with the dimensions of the pipe and the pressure applied. With the continual skin friction of the walls, one might expect that flow would gradually slow down as it moved down the pipe, but it's already been pointed out that continuity forbids such sloth. At all cross sections, the total flow must be the same. But to keep things going, force must be applied and work done, and the longer the pipe, the more force will be needed to maintain a given rate of flow. We'll talk about pressure drop, not force, but the distinction is just a minor convenience. By pressure drop we mean the difference in pressure between two ''stations'' or cross sections; it is analogous to the momentum drop in the fluid from front to back of an object experiencing drag. A formula for pressure drop would give a measure of the cost of pushing fluid through, or the rate at which the presence of material walls causes energy to be dissipated. If the velocity distribution and the total flow remains constant from station to station, then neither the pressure drop nor energy dissipation per unit length of pipe should vary.

Velocity distribution. Consider a section of a pipe of length l and radius a (Figure 10.1). What keeps the fluid moving through it? It is a drop in pressure, Δp, between one end of the section and the other, exerted as a uniform push across the cross section. We needn't worry about variations in pressure across the pipe; Bernoulli, remember, is strictly applicable only along streamlines, so pressure, not total head, will be constant, just as in a boundary layer. What will resist the motion? Only the skin friction of the walls; this is dependent on the shear stress at the walls, again just as in a boundary layer. In a

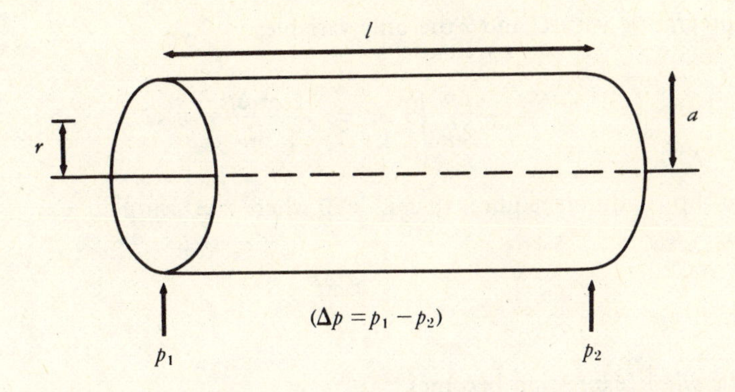

Conventions and symbols for describing flow through a length of cylindrical pipe.

FIGURE 10.1

steady flow the push and the resistance must be the same. For a circular pipe, flow will be axisymmetric, so we can consider the pipe as filled with concentric cylinders, each with a characteristic radius, r, measured from the center. These cylinders will slide past one another, with the center one going fastest and the outer one not moving. The force pushing a cylinder and each cylinder within it will thus be the pressure drop times the cross-sectional area:

$$F_p = \Delta p(\pi r^2)$$

The force resisting the push will be the shear stress (τ) times the surface area of the side walls of the cylinder:

$$F_r = \tau(2\pi rl) = \mu\frac{dU}{dr}(2\pi rl)$$

These two forces will be equal and opposite, so:

$$\Delta p\pi r^2 = -\mu\frac{dU}{dr}(2\pi rl)$$

Cancelling and rearranging, we get:

$$dU = -\frac{\Delta pr\,dr}{2l\mu}$$

And integrating with U and r the only variables:

$$U_r = -\frac{\Delta p}{2l\mu}\int r\,dr = -\frac{r^2\,\Delta p}{4l\mu} + C$$

The no-slip condition requires that $U = 0$ where $r = a$, so:

$$C = \frac{a^2\Delta p}{4l\mu}$$

And the whole expression becomes

$$U_r = \frac{\Delta p}{4l\mu}(a^2 - r^2) \qquad (10.1)$$

Equation (10.1) is *exactly* what we sought—a near formula for how the velocity varies across the pipe. The distribution is parabolic (Figure 10.2); at the center ($r = 0$) the velocity is maximal, while at the walls ($r = a$) it is zero. The velocity approaches the zero value at the walls almost linearly, not asymptotically. There is no nonsense about arbitrary constants, empirical coefficients, Reynolds numbers, or other bits of *deus ex machina.* Indeed, the main reason for presenting a derivation (others are available) was to expose this uncharacteristic lack of chicanery. Assuming only uniform pressure and viscosity, we got a unique and simple solution. In fact, equation (10.1) says other agreeable things. It asserts that if the pressure drop per unit length ($\Delta p/l$) doesn't change, then the velocity distribution won't either and, of course, vice

FIGURE 10.2

Velocities in laminar flow at a series of points across the diameter of a long, straight, circular pipe, a long distance from the entrance to the pipe. Lengths of the arrows are proportional to flow speeds at the points.

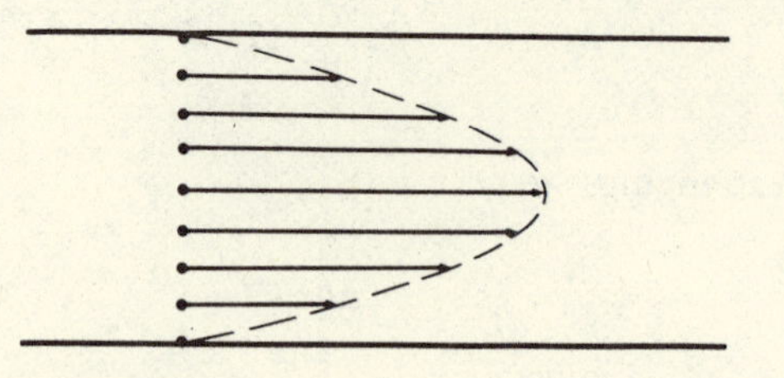

versa; as we anticipated, the parabola retains the same shape as fluid moves downstream.

Total flow. Let us think of the same pipe filled with concentric cylinders, but now with a face area, dS, on each cylinder in the cross-sectional plane of the pipe. Each face will then have an area

$$dS = 2\pi r \, dr$$

and the total flow through each face will be, by continuity:

$$dQ = U_r \, dS = U_r(2\pi r) \, dr$$

The total flow is then

$$Q = \int_0^a U_r(2\pi r) \, dr$$

This integral would be awkward, but we have equation (10.1) and can simply insert it for U_r, plug 'n chug, and get:

$$Q = \frac{\pi \, \Delta p a^4}{8\mu l} \tag{10.2}$$

This expression is known as "Poiseuille's equation" or the "Hagen–Poiseuille equation" after its independent discoverers. As Prandtl and Tietjens (1934) pointed out, Hagen has precedence by a year (1839 vs. 1840), but he expressed his results in obscure units, and for many years missed recognition. There's a lesson in that. As for Poiseuille, opinion in the English-speaking world varies on an appropriate mispronunciation.

Equation (10.2) makes the famous statement that total flow is proportional to the *fourth* power of radius or diameter. A larger pipe of the same length carries *much* more fluid for a given pressure drop than does a smaller one. The larger one has both a greater cross-sectional area and a lower area of wall per unit area of cross section. Another way of viewing the situation is to note that two equal, small pipes in parallel will carry only one-fourth the flow of a single large one of the same total cross-sectional area for a given pressure drop.

Resistance. In laminar flow through a pipe, the total flow is directly proportional to the pressure drop (equation 10.2). The constant of proportionality can be called the resistance (R), as in its analog, Ohm's law, for electrical conduction. Thus

$$R = \frac{\Delta p}{Q} = \frac{8\mu l}{\pi a^4} \tag{10.3}$$

This parameter, resistance, can be used to characterize some pipe or array of pipes just as the electrical resistance characterizes an ordinary (ohmic) conductor. In general, it will be independent of the particular pressure drop, total flow, or velocity at which it was determined. For a circular pipe and fully developed flow (see page 170), the resistance can be calculated from the last expression in equation (10.3). Notice the fourth power of radius in the denominator: resistance, like total flow, is exceedingly sensitive to small changes in pipe bore. For arrays of pipes, the usual (electrical) rules for series and parallel hook-ups apply: resistances in series add directly; the sum of the reciprocals of resistances in parallel gives the reciprocal of the system resistance. Finally, it's worth noting the peculiar dimensions of pipe resistance: force times time divided by length to the fifth power.

Power. As mentioned, work has to be done to push a fluid through a pipe that has material walls. It's a simple matter to calculate the power required either from the dimensions of the system and velocity, total flow, or pressure drop, or it can be obtained from some prior resistance measurement and a datum for either velocity, total flow, or pressure drop. As with resistance, the case is formally the same as conduction of electricity:

$$P = Q\,\Delta p = Q^2 R = \frac{(\Delta p)^2}{R} \tag{10.4}$$

Average and maximum velocities. Still other useful relationships emerge in a straightforward manner from equations (10.1) and (10.2). The average velocity ($\bar{U}$) of flow in a pipe is the total flow, Q, divided by the cross-sectional area, S. From (10.2), then:

$$\bar{U} = \frac{\Delta p\, a^2}{8\mu l} \tag{10.5}$$

The maximum velocity, occurring at the axis of the pipe, can be obtained from (10.1) by setting $r = 0$:

$$U_{max} = \frac{\Delta p \, a^2}{4\mu l} \quad (10.6)$$

These are truly splendid results! They tell us that the maximum velocity is precisely twice the average velocity, something not only neat and tidy, but exceedingly useful. If you mount a calibrated flow probe at the axis of a pipe you can obtain average velocity, total flow, resistance, and power without the bother of a lateral traverse or integration. Conversely, if you can measure total flow, perhaps by catching the output of the pipe for a measured time, you can calibrate a flow probe located along its axis.

There's an even more potent application of equation (10.6) for maximal velocity. In a transparent tube of liquid, the progress of an advancing front of dye can be followed with ease to obtain axial velocity. From the dimensions of the tube and the viscosity of the liquid, one can determine the pressure drop by equation (10.6). This, then, is a kind of pseudo-manometer, useful for measuring very low pressures in liquid systems. We need only connect two points in a flow by an appropriate tube and inject a dye. As an example, assume a tube of 0.2-cm internal diameter and 20-cm length with a measured axial flow of 0.2 cm s^{-1}. The corresponding pressure drop is 1.6 Pa (or 16 microatmospheres). (In a water current of 5.7 cm s^{-1} this would be the difference between dynamic and static pressures.) As a general rule, the Reynolds number, based on the diameter of the tube, should be kept under about 50 for reasons related to entrance phenomena (see page 170). Also, the flow in the tube should be less than a tenth of the external current being measured or else flow in the tube might appreciably relieve the pressure difference being measured. Of course, the flow in the measuring tube can be slowed at will by using a longer tube. Presumably, some degree of automation might be achieved with photodetectors and timing circuits.

Roughness. The relative irregularity of the inner wall of a pipe has only a small effect on the total flow or pressure drop in laminar flows. The same kind of criteria for tolerable roughness apply as were cited for flow over a flat plate (Chapter 8). A critical Reynolds numbers is based on the heights of projections; 30 is used for pointed projections and 50 for rounded projections. Above these values, roughness appreciably affects the shearing stress and character of the flow. The permissible heights of roughness (ε) are then, according to Goldstein (1938):

$$\frac{\varepsilon}{a} \leqslant 4\, Re^{-1/2} \qquad \text{(pointed)} \qquad (10.7)$$

$$\frac{\varepsilon}{a} \leqslant 5\, Re^{-1/2} \qquad \text{(rounded)} \qquad (10.8)$$

The Reynolds numbers in these formulas are, of course, based on the diameters of the pipes and not on the heights of projections. These are very generous limits; at a Reynolds number of 1000 (as with a water flow of 10 cm s^{-1} through a 1-cm pipe) a pointed protrusion can be up to an eighth of the pipe radius without making much difference. For lower Reynolds numbers, even large protrusions are tolerable.

The "entrance region." All of the previous formulas assume that a steady-state parabolic velocity distribution has been established somewhere upstream. But the matter can't be so blithely evaded indefinitely, and an "entrance region" has to be considered. As fluid enters a pipe from say, a reservoir, its velocity will be nearly uniform across the pipe; the situation has been called *plug flow* or *slug flow.* Gradually, a gradient region develops, beginning at the edges of the pipe and thickening just like a boundary layer until the final parabola is achieved; then the flow is said to be "fully developed" (Figure 10.3). In this entrance region, mass and momentum are transferred from periphery toward the axis of the pipe; the kinetic energy of the fluid increases with distance from the entrance, and the pressure drop per unit length is therefore greater than that predicted by the Hagen-Poiseuille equation.

The most immediate question about the entrance region is its length. Here we have to resort to an approximation, and a relatively stringent criterion lies behind the usual one. The flow may approach full development asymptotically, but it is said to be fully developed for practical purposes when it has achieved 99% of the final axial velocity. Development, as mentioned, is much like the growth of a boundary layer. Just as the thickness relative to downstream distance of a boundary layer depended only on the local Reynolds number, so does the entrance length (L') expressed in units of pipe diameter (d):

$$\frac{L'}{d} = 0.058\, Re = 0.058\, \frac{\rho U d}{\mu} \qquad (10.9)$$

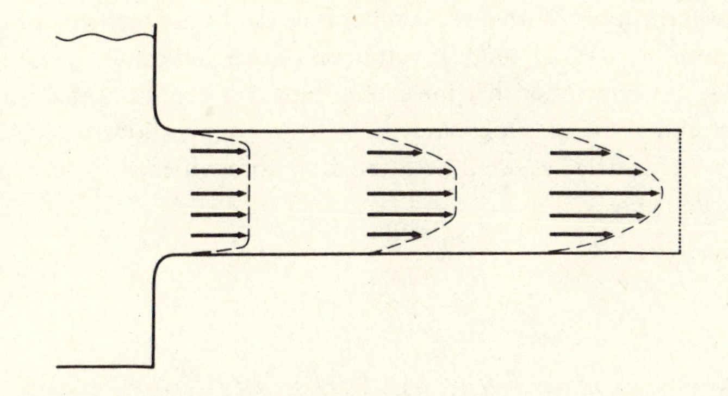

Gradual development of the parabolic profile of velocity in the entrance region of a cylindrical pipe.

FIGURE 10.3

(Some sources give slightly different numerical constants.) For many purposes it is not necessary to get so close to full development, and a constant about half that of equation (10.9) may be ample. In fact, the Hagen–Poiseuille equation gives a good general estimate of the pressure drop almost to the very entrance. Conversely, whatever the outcome of applying equation (10.9), it is best to allow at least one diameter-length downstream from an entrance for the parabolic profile to develop.

Another way of looking at development in the entrance region is in terms of the time needed for the process. Since length is average velocity times time:

$$t = 0.058 \frac{\rho d}{\mu} \qquad (10.10)$$

The time it takes for a flow to become fully developed is independent of its velocity and depends inversely on the kinematic viscosity of the fluid and directly on the cross-sectional area of the pipe.

The limit of laminarity. The classic experiments of Osborne Reynolds, although mentioned in Chapters 3 and 5, are more strictly relevant to the present topic. Reynolds established a limit of around $Re = 2000$ for the sudden transition from laminar to turbulent flow in a circular pipe. In fact, the transition may be sudden, but it doesn't necessarily happen at that value. Below 2000, a disturbance will not persist but will damp out through viscous action. Above 2000, a disturbance, once started, propagates throughout the fluid as it travels down the pipe. With care, laminar flow can be maintained at Reynolds numbers well in excess of 2000. If you need a quick indication of the limit of laminarity, just remember that for a 1-cm pipe it is about 0.2 m s^{-1} in water and about 3 m s^{-1} in air. Only rarely in biological applications are these limits exceeded; fortunately, as we'll see, the flow through most of our pipes is decently laminar.

basic rules for turbulent flow

Turbulent flow is less straightforward than laminar flow, both physically and mathematically. In turbulent flow, momentum is continually being transported across a pipe, so the notion of a set of concentric sliding cylinders applies only to a time average of the instantaneously jumbled motion. The precise parabola disappears, and the velocity distribution even as an average can no longer be described simply. The speed of flow is still zero at the walls, but flow along the axis is less than twice the average speed. And the roughness of the inside wall of a pipe, of little effect in laminar flow, takes on a major role in determining the resistance offered by a pipe.

Friction factor. To deal with turbulent pipe flow it is most convenient to lump the eccentricities into a dimensionless parameter analogous to the drag coefficient. C_D, you'll recall, can be defined as the ratio of the drag per unit of surface to the dynamic pressure. Similarly, the *friction factor* (f) can be defined as the ratio of the shear stress to the dynamic pressure. The former is what impedes flow in pipes; the latter is what promotes it. Alternatively, the friction factor can be viewed as the ratio of the pressure drop per unit length times pipe diameter to the dynamic pressure:†

†Taken literally, the first definition of friction—shear stress over dynamic pressure—generates a coefficient which has a numerical constant 1 rather than 4. Some sources use the constant of 4, as here, but others don't. If it isn't obvious which is intended, find the coefficient which applies to laminar flow: if the latter is $16/Re$, then the constant is 1; if it's $64/Re$, the constant is 4. The names vary also—pressure drop coefficient, pipe resistance coefficient, and so on. Most recent American sources use the 4 and call the result the friction factor.

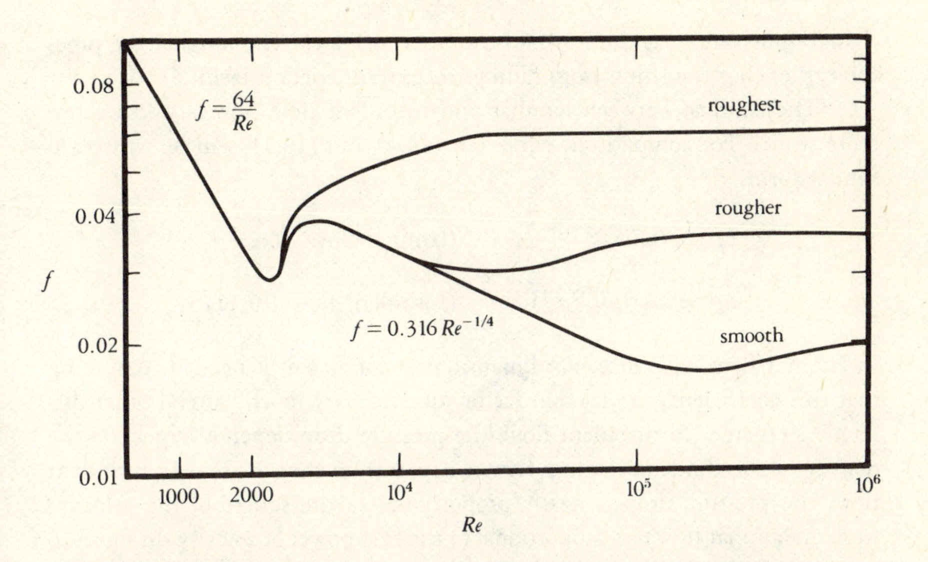

Graph of friction factor, f, versus Reynolds number for flow through a cylindrical pipe. Note the transition at $Re = 2000$ and the influence of wall roughness at Reynolds numbers above this value.

FIGURE 10.4

$$f = \frac{\Delta p(2a)}{l} \div \frac{\rho \bar{U}^2}{2} = 4\,\frac{\Delta p a}{\rho l \bar{U}^2} \qquad (10.11)$$

Just as the drag coefficient is a function of the Reynolds number and the shape of an object, the friction factor is a function of the Reynolds number and the roughness of the lining of the pipe. And in just the same way, we can plot the friction factor as a function of the Reynolds number (Figure 10.4).

This plot of friction factor against Reynolds number has several noteworthy features. The left-hand line refers, of course, to laminar flow, and the friction factor is simply $64/Re$, a result which follows directly from equations (10.2) and (10.11). The right-hand lines are a little more complicated. They begin well above the line for laminar flow, indicating that the transition to turbulence involves an abrupt rise in the resistance of a pipe to flow. Again, flow over a flat plate is analogous. For smooth pipes, the friction factor decreases slowly and steadily with increases in the Reynolds number. For rough pipes that decrease ceases at some point and the friction factor subsequently does not change much more as the Reynolds number increases. The rougher the pipe, the earlier that transition from falling to constant friction factor.

The contrast between laminar and turbulent pipe flow is worth a few more words. For comparison, equations (10.2) and (10.11) can be written in similar forms:

$$\Delta p = 8\mu l\bar{U}a^{-2} \qquad \text{(laminar flow; 10.2)}$$

$$\Delta p = \tfrac{1}{4}f(\rho l\bar{U}^2a^{-1}) \qquad \text{(turbulent flow; 10.11)}$$

In laminar flow, of course, no dimensionless coefficient is needed. Assuming that this coefficient, the friction factor, doesn't vary much, several other differences emerge. In turbulent flow, the pressure drop depends largely on the density of the fluid; in laminar flow it depends on the viscosity. In turbulent flow, the pressure drop is nearly proportional to the square of the velocity, while in laminar flow it is proportional to the first power of velocity. In short, in turbulent flow the resistance is not linear or "ohmic," and the resistance which might be calculated as in equation (10.3) is no longer particularly useful. These same differences will reappear when we consider open flow fields at very low Reynolds numbers (see Chapter 13).

Velocity distribution. The onset of turbulence in a pipe is accompanied by a drastic alteration of the distribution of velocities across the pipe. The regular and consistent parabolic distribution characteristic of laminar flow disappears; even worse, the actual distribution for turbulent flow depends on both wall roughness and the Reynolds number. For smooth pipes, $U_{max}/\bar{U}$ diminishes slowly with increases in the Reynolds number, eventually reaching an asymptote of about 1.25 (Prandtl and Tietjens, 1934). For rough pipes, $U_{max}/\bar{U}$ is nearly constant above the transition range. The subject of velocity distributions and friction factors is both complex and of great practical (if not biological) consequence; it's considered in detail by almost every fluid mechanics textbook.

Entrance length. The entrance length for turbulent flow is less than that for laminar flow. This results from the lateral transport of momentum in eddies and the fact that the steady-state velocity profile is closer to slug flow. Caro et al. (1978) give the following equation for estimating entrance length:

$$\frac{L'}{d} = 0.693\, Re^{-1/4} \qquad (10.12)$$

manipulating the velocity profile

As we've seen, the velocity distribution across a circular pipe in fully-developed flow is parabolic, with an axial velocity precisely twice the average velocity. In many biological situations, either material or energy is exchanged across the walls of pipes; a moment's consideration should persuade you that a parabolic profile isn't exactly ideal for exchange. Too much of the material or energy passes down the middle of the pipe and too little moves along the edges, near the site of exchange. This places a heavy burden on molecular diffusion or conduction, weak reeds in a macroscopic world. And nature repeatedly seems to have found it prudent to arrange something better than a parabolic profile. Her schemes are more subtle than mere reductions in the size of her pipes.

How far is a flow from a wall? Before revelling in adaptive tricks, I think it's worthwhile to develop an expression for the distance of flow from the walls of a pipe or channel. The most convenient would be a dimensionless and size-independent index which compares the average distance of the flow from a wall with the radius of a pipe or the depth of a channel. We'll call our expression the *distance index*, Di, and we'll consider, first, an axisymmetric circular pipe. To get the mean distance, the pipe again needs to be viewed as a series of concentric cylinders, each with an annular face. The area of each annulus ($2\pi r\, dr$) must be multiplied by the local velocity (U_r) and the distance from the wall ($a - r$) and divided by the total flow ($\pi a^2 \bar{U}$). The result is then integrated across the radius to get the mean distance, which is divided by the radius of the pipe. Written out as an integral, the procedure can be summarized as:

$$Di = \frac{2}{a^3 \bar{U}} \int_0^a U_r\,(ar - r^2)\, dr = \frac{2\pi}{Qa} \int_0^a U_r\,(ar - r^2)\, dr \qquad (10.13)$$

Both forms of this equation can be solved by either of two approaches. If an explicit expression is available for the velocity distribution, that is, for U_r as a function of r, the expression can be substituted for U_r and the integral evaluated by the usual mathematics. Alternatively, an experimentally determined set of measurements of velocity at different radii can be used. One by one, each is multiplied by its appropriate dr (now a Δr) and by ($ar - r^2$); the results are then summed and further multiplied by the factors outside the integral sign. (Note that this procedure is not universally applicable. The Romberg method, conceptually a little more complex, would require fewer measurements for a given level of precision; for details, see Pennington (1965).)

Plug, or slug flow implies a uniform velocity across the pipe, so $\bar{U} = U_r$. The remaining integral is easily evaluated, and everything cancels out after integration except a numerical coefficient. The latter is the desired distance index, 1/3, or 0.333.

The parabolic profile of fully-developed laminar flow is almost as simple. Equation (10.1) is substituted for U_r and equation (10.2) for Q in (10.13). After a little bookkeeping, a numerical coefficient again emerges; this time the distance index is 7/15, or 0.467.

With a uniform velocity across a pipe, the flow is thus, on the average, one-third of the way from the walls to the axis. By contrast, with a parabolic profile, the flow averages 40% further from the walls. This is one of the fringe problems of real, as opposed to ideal, fluids (plug flow will occur with an ideal fluid). This is the same problem we considered with regard to diffusive exchange across boundary layers in Chapter 9. Indeed, with the efficacy of diffusion as an agency of net transport dropping with the square of the distance, a 40% greater distance index may have profound practical effects. Fully-developed laminar flow is, in a sense, a worst case as well as the one which occurs in the absence of special devices.

As a further set of examples, I have determined the distribution of velocities across a real pipe with a microanemometer (see Appendix 2) and have calculated the distance indices from them for several arrangements. For an unobstructed pipe, the figure is 0.447—the flow is not quite developed fully. When the same pipe is loosely filled with wool fibers, the distance index drops to 0.387—much closer to plug flow. The effect on the filling is to distribute the resistance to flow across the entire volume of the pipe and to reduce the steepness of the parabola. With only a little more effort, one can do still better. If a set of wires are threaded diametrically across the pipe, each wire must cross the axis. The scheme therefore puts more of a barrier in the middle than at the periphery—in effect, an axial plug (Figure 10.5). With diametrical wires in the pipe, the distance index drops to 0.311—better even than plug flow. Although not specifically measured, the pressure drop is clearly less than that for the wool-stuffed pipe. It is uncertain whether any organism grows hairs which, by extending to or beyond the axis of a pipe, push the average flow towards the walls and increase the exchange between fluid and walls. It is also uncertain, to me at least, whether anyone has looked for such a phenomenon. Don't sniff at the notion: who knows what nasal hairs might do!

An analogous distance index can be created for an open channel of depth a and width far greater than the depth. Using h as the distance from the

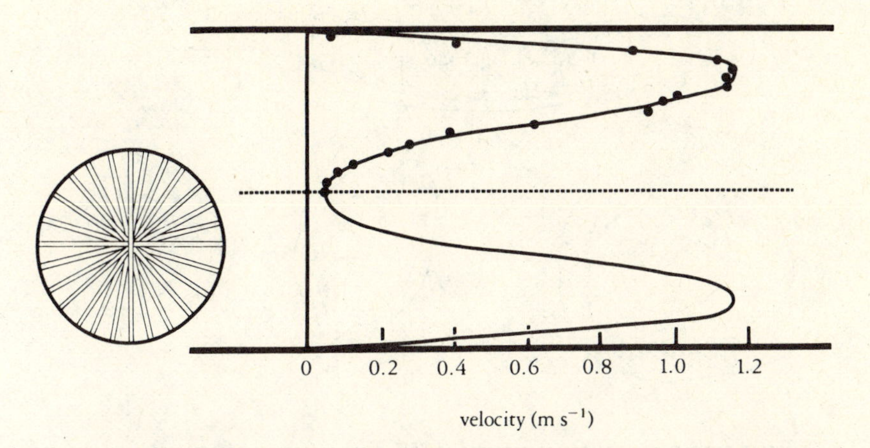

Flow profile across a cylindrical pipe with laminar flow where the pipe is crossed by diametric wires. A length of the pipe, looking axially, is shown on the left.

FIGURE 10.5

bottom and *dh* in place of *dr*, the index is:

$$Di = \frac{1}{a^2 \overline{U}} \int_0^a U_h h \, dh \tag{10.14}$$

If the velocity does not vary with the depth, then $\overline{U} = U_h$ and the distance index is 0.5. If the velocity is zero at the bottom and increases linearly to the surface, then $Di = 0.67$. If the velocity is parabolically distributed, that is, if it increases with the square root of the distance from the bottom, then $Di = 0.60$.

Many schemes are available to minimize the distance index. Some involve increasing the internal surface area of the pipes; and one usually thinks of the increased surface area as the sole benefit, ignoring the concomitant improvement in the intimacy of flow with the walls.

Use of non-circular cross sections. From the point of view of exchange, a circular section is the worst geometry; its only advantages are ease of construction, mechanical robustness in the face of pressure differences across the walls, and minimum pressure drop per unit of total flow. Large pipes through which

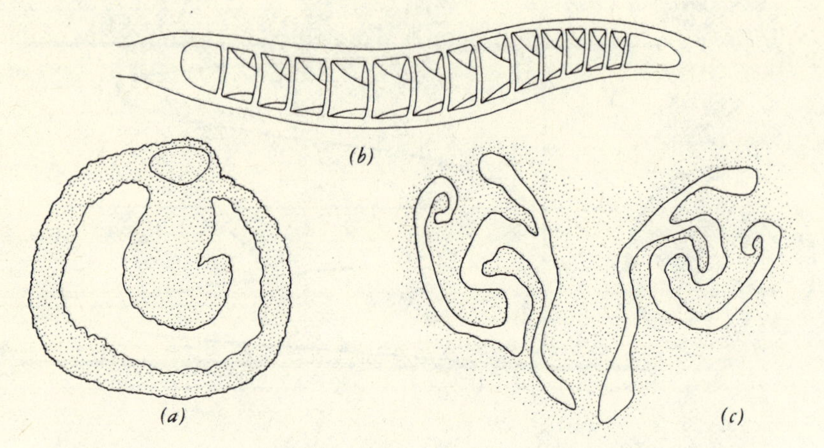

Pipes with noncircular cross sections. (*a*) The intestine of an earthworm with the large dorsal typhlosole protruding downward: (*b*) the "spiral valve", in cut-away view, which runs the length of a shark's intestine; (*c*) a cross section of the nasal passages of a kangaroo rat.

FIGURE 10.6

exchange takes place are quite often non-circular. A few examples (Figure 10.6) are nasal passages—especially those in which a countercurrent heat exchange system functions (Schmidt-Nielsen, 1972); the intestine of an earthworm with its large dorsal invagination, the typhlosole; the "spiral valve," a spirally-coiled fold filling most of the small intestine of sharks; and the passages of the open parts of the circulatory systems of the bivalve and gastropod mollusks and of the arthropods.

For some situations, such as flow between two closely-spaced parallel plates, formulas for velocity distribution, total flow, resistance, and so forth, are easy to derive: the procedure is analogous to that presented at the start of this chapter.

Making the pipes very small. Strictly speaking, this doesn't reduce the distance index since the index is size-independent, but it certainly does improve exchange. The use of very small pipes is the general solution evident in capillary beds, the parabronchi of bird lungs, renal tubules, and numerous other situations. At the same time, the total cross-sectional area of arrays of small pipes is generally huge, and therefore the velocity in them is low, which provides more time for exchange to take place. Of course, it also prevents the cost of pumping the fluid from becoming inordinately high.

Eddies and turbulence. Turbulent flow has a flatter velocity profile than does laminar flow and thus has a lower distance index. In practice, the distance index either becomes irrelevant or takes on quite a different meaning in the presence of lateral bulk transport of fluid. Exchange, though, should be greatly augmented by turbulence. A minor difficulty is the greater cost of turbulent flow associated with the greater pressure drop per unit length of pipe. A major difficulty is the impracticality, and sometimes impossibility, of getting turbulent flow at the low Reynolds numbers set by the size of the most biological pipes. Eddies can be generated by protrusions in a pipe, and these will improve exchange, but the Reynolds number still has to be reasonably high—at least above about 50. The nasal passages of mammals would appear to be the best candidates for such deliberate generation of turbulence; the latter would then improve olfaction and heat exchange.

Periodic boluses. If a solid plug is passed through a pipe, then flow immediately fore and aft of the plug should be more nearly linear than parabolic in profile. We send red blood cells through capillaries at a great rate, and the cells barely squeeze through. As Caro et al. (1978) point out, there must be some circulation generated in the space between the boluses, so perfect plug flow isn't achieved. Also, development is relatively rapid at low Reynolds numbers. Still, it's most likely that exchange between plasma and capillary walls is improved as a purely physical consequence of the presence of red blood cells.

Pumping at the wall. We've assumed all along that the walls provided the resistance to flow. It is entirely possible to use the walls as the location of the pump, whereupon the parabolic profile might be nearly reversed. The obvious agency is a lining of cilia propelling a liquid. (The cilia on gas-filled pipes are usually beneath and within a mucus layer and don't pump the main fluid.) Cheung and Winet (1975) analyzed the velocity profile at a ciliated surface and found the expected no-slip situation but with a very steep velocity gradient immediately above it. Free-stream velocity extends almost down to the tips of the cilia (in their case) of the protozoan, *Spirostomum,* about 9 μm. Under some circumstances they could detect a bit of back flow just above the surface, presumably as a result of the return strokes of the cilia. Besides their presence on the outside of swimming ciliates, cilia or flagella form a propulsive coating in the flagellated chambers of sponges, molluscan gills, the excretory tubules of some flatworms, oviducts of various animals, and other places. In general, they seem to be more common where exchange processes are taking place and less common where simple propulsion of fluid is the primary mission. Perhaps the very steep velocity gradients which must occur in ciliary propulsion in tubes improve the efficacy of exchange by moving fluid very near walls; but as a consequence of the high shear rates, they increase the cost of using cilia as pumps relative to the costs of other biological pumps.

the cost of pumping fluid through pipes

The price organisms pay for pushing fluid through their plumbing is of no slight interest. Not only may this price form an appreciable fraction of metabolic rate, but it may underlie much of the choice of structural arrangements made by the evolutionary process. From equation (10.4) it is clear that, for a given total flow, the power requirement is inversely proportional to the fourth power of the radius. So smaller pipes appear to have disproportionately high resistances. In fact, the fourth-power relationship can be a bit misleading. For a given total flow, by the principle of continuity, the velocity must be inversely proportional to the square of the radius. At a given velocity rather than a given total flow, therefore, the power requirement is inversely proportional to the second and not the fourth power of the radius. Nonetheless, there is still a high price paid for passage through an array of small pipes instead of a single big one of the same overall cross-sectional area. One common compensation is to make the total cross section very large where the individual parallel pipes are small and thus, by continuity, to minimize the velocities. We briefly touched on this matter in Chapter 3.

Our own systemic circulation is a good example. The total cross-sectional area of the capillaries is about 300 times that of the aorta, and the blood velocities in capillaries are correspondingly 300 times less. Nevertheless, most of the resistance to flow resides in the smallest vessels. The capillaries account for 27% of the total resistance of pressure drop. Of the arterial resistance, the arterioles account for 62%, and of the venous resistance, the venules account for 57%. In all, in these three kinds of smallest vessels resides almost three-fourths of the entire resistance of the systemic circulation. Minimization of resistance is probably not a trivial consideration in the design of a mammalian circulation. The power output of the human heart, which must balance the power required for circulation, is about 1.3 watts at rest. Assuming the heart to be about 10% efficient as a pump, 13 watts of input are required. As a comparison, the resting metabolic rate of a person is about 100 watts.

Our respiratory system has a rather different distribution of resistances than does our circulatory system. It is also a progressively branching array; after the first few generations of branching, the total cross-sectional area begins a dramatic increase, about 1.2 times for each dichotomous branch. By the twenty-third generation, the total cross-sectional area has increased from 2.5 to 11,800 cm^2 and the Reynolds number has dropped from around 2000 (in quiet breathing) to around 0.05 (Pedley, 1977). But here, in contrast to the circulation, the bulk of the resistance occurs in the larger pipes: 50% resides in the nasal passages and 20% in the rest of the upper airways such as the trachea and larynx (Hildebrandt and Young, 1965), which, by the way, is why it is noticeably easier to breathe via mouth than nose. From the figures in these sources, I calculate that substantially less than 1% of resting metabolism is devoted to the action of the respiratory musculature. Thus the losses in the upper airways are probably not important, especially in view of the functions of olfaction and speech in which these airways participate. Pedley's monograph (1977) is the best source I've seen on the fluid mechanics of lungs.

The cost of pumping fluid should be nowhere more significant than in some filter-feeding and/or tubicolous animals. Such animals often process vast volumes of water for a modest yield of food, and their pumping activities may be a major consumer of metabolic energy. Jørgensen (1966) gives a few cost-benefit calculations which suggest that filter-feeding is a marginal game under ordinary circumstances. But there is only a little information with which to proceed beyond Jørgensen's approximations. Pressures and volumes have been measured for a variety of worms, a clam, an amphipod crustacean, and a burrowing sea urchin, and the power output calculated for each (Chapman, 1968; Brown, 1977; Foster-Smith, 1978). The relationship between structure and mechanical performance is still largely a matter of guesswork due, as much as anything, to the morphological and ecological diversity these animals represent. For example, the lugworm *Arenicola* pumps water up through loose sand (Figure 10.7); it uses its body as a piston pump to produce high pressures (1200 Pa) and low flow rates (0.014 $cm^3\ s^{-1}$) (Foster-Smith, 1978). The parchment worm *Chaetopterus* pumps through a two-ended tube into which it fits more loosely. It processes more water (0.54 $cm^3\ s^{-1}$) but at a lower pressure (30 Pa) for a similar power output (Brown, 1977). In none of these cases is the geometry appropriate for Poiseuille's equation to give us much assistance.

Rather a different sort of pipes are the xylem elements which carry water upward in trees. Even in very large trees, these are small pipes: a 250-μm (diameter) vessel in an oak is reckoned large. Individual vessels may be as long as 0.6 m (grapevine, maple) or 3 m (ash). Neither the total cross-sectional area

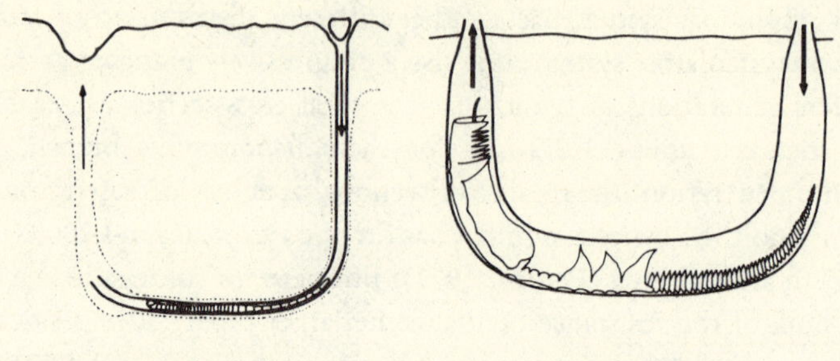

Two burrowing marine worms. (*a*) *Arenicola* pumps water up through sand, perhaps making quicksand, but in any case obtaining food by processing sand and sediment. (*b*) *Chaetopterus* lives in a parchment tube and feeds on suspended material in the water which it pumps through its tube.

FIGURE 10.7

nor vessel diameters change much with height above ground or location in trunk, branches, or twigs. Water moves from vessel to vessel through "pits," and a comparison of the conductivity as calculated from Poiseuille's equation with the experimental values gives an indication of the relative obstruction of such structures. For a grapevine, conductivity is fully 100% of that predicted; for an oak (a "ring porous" species) it is 53 to 84% of the predicted value, so the resistance of the pits can be appreciable. For trees with shorter xylem elements, this resistance is still higher (Zimmermann, 1971).

The functioning of these vessels makes a sharp contrast with all other biological fluid transport systems, and it's worth our attention on that account alone. First, it is now quite clear that the water isn't pushed up but is mainly pulled up; and normally it quite literally hangs from the top of the tree. Most of the water column is under negative pressure of a few atmospheres in a tall tree, but the columns don't "cavitate" or break into a series of vacuum voids. As a result there is a gravitational pressure gradient of about 0.1 atmosphere (10^4 Pa) per meter of height up the tree. If the pressure at the ground is $+1$ atmosphere, then 0 pressure occurs at 10 m, -1 atmosphere at 20 m, and so forth. That is, if there is no flow through the xylem. In fact, there is flow; speeds of over 1 cm s^{-1} have been measured in oaks. Flow is associated with an additional pressure drop; and, between the small vessels and the long distances, the pressure drop is substantial enough so that a tree gets measurably skinnier when rapidly transpiring. Using Poiseuille's equation and these data

(70% of ideal capillary conductivity, $\bar{U}$ = 1 cm s^{-1}, a = 0.0125 cm) the pressure drop comes to 0.072 atmospheres (7300 Pa) per meter of height. Although this is still less than the gravitational pressure drop, the two figures are similar. Their sum has been verified experimentally (Zimmermann, 1971).

Apparently, once the gravitational pressure drop has been endured, the pressure drop due to flow in long, small vessels isn't overwhelming. It is of some interest that just as the gravitational pressure drop is independent of species and both vessel and pore diameters, so, nearly, is the pressure drop due to flow. Trees which raise water more slowly do so with narrower xylem elements. It looks as if the xylem elements are made about as small as nature can get away with without being stuck with a pressure drop much in excess of that due to gravity. The advantage of using many small pipes is probably safety in numbers as well as further resistance to cavitation. There's a lot of water being stretched, and embolisms appear to be a significant hazard for trees. The cost, by the way, may be purely structural and not directly metabolic. The bits and pieces of tree must be built to tolerate these large negative pressures. But the pump is a solar engine—the evaporation of water—which incurs no immediate expenditure of the plant's energy. Broad-leaved trees would probably be out of the question if they had to use a muscular pump to lift all that water.

These xylem vessels aren't smooth-walled pipes; most have spiral, annular, or reticulate thickenings of their inner walls, so the lumen is nearly a series of apertures. Jeje and Zimmermann (1979) give some data on the effect of these irregularities in setting vessel resistance, but further work using physical models seems necessary. Similar reinforcements occur in insect tracheae, again with essentially unknown fluid mechanical function.

a few other phenomena

The applicability of odd physical phenomena to living systems is more likely to be recognized if investigators are aware of their existence. So here, for more than amusement, are a few phenomena, several of which play no known biological role.

Turbulent plugs and spouting. Consider a fluid being pushed through a long pipe by a pressure just sufficient to achieve a velocity in laminar flow which would raise the Reynolds number above the point of initiation of turbulence. As turbulence begins, the resistance of the pipe will abruptly rise, causing the speed of flow to decrease below the turbulent range. A turbulent "plug" will be pushed through the pipe, laminar flow will be reestablished, the resistance will drop, and the velocity will again increase toward the critical Reynolds number. The result at the end of the pipe will be an alternation of laminar spouts and turbulent dribbles.

Converging and diverging pipes. A gradual convergence of the walls of a pipe (6° or less) not only speeds up the flow, but also stabilizes laminar flow and elevates the critical Reynolds number for the onset of turbulence. Furthermore, it distorts the parabolic velocity distribution, giving a lower peak velocity and more flow closer to the walls. The convergence can be put to good use in wind tunnels and flow tanks to accelerate fluid near the walls and thus achieve a more nearly uniform distribution of speeds across some working section (see Appendix 1).

A gradual divergence has the opposite effect. It decreases velocity, destabilizes laminar flow, and promotes the transition to turbulence. The axial velocity is raised relative to the velocities near the walls. If the angle of divergence is more than about 6° (at fairly high Reynolds numbers) the flow will separate near the walls. Rouse (1938) gives a good account of these phenomena.

Bends and branches. Any change in the direction of flow in a pipe will involve a pressure drop beyond that for a straight and uniform pipe. Masses of data exist for bends and branches in commercial pipes, almost all applicable only at high Reynolds numbers. In laminar flow, the losses still occur: parabolic flow is always distorted at a bend or branch and energy is expended in reestablishing it. Pedley (1977) gives an elaborate account of the effects of branching in lungs. But the losses relative to the normal resistance per unit length are generally less in laminar flow and become relatively lesser still at lower Reynolds numbers.

Nature seems to have recognized the phenomenon. Large pipes have gentle bends, and their branches rarely diverge or converge fully normal to the parent pipe. In small pipes, abrupt branching and bending is a constant and shameless morphological indulgence. Horsfield and Cumming (1967) found that the bifurcation angle of the branches in our bronchial tree was only 64° when the pipe diameters are greater than 0.4 cm; the angle increases to 100° for bronchial diameters of less than 0.1 cm. For normal breathing, these figures correspond to Reynolds numbers of above 500 and below 25, respectively. The same situation is evident in the illustrations in any text on the structure of the mammalian circulatory system. And I've noticed that in sponges the larger excurrent pipes are often connected at acute angles while the smaller vessels connect abruptly and perpendicularly with sharp corners wherever convenient. The general rule seems to be to keep pipes as short as possible at low Reynolds numbers, but to keep bends and junctions gentle at higher Reynolds numbers even if the pipes have to be a little longer.

One other phenomenon worth noting happens at bends. The fastest fluid, near the center of the pipe, has more momentum that that at the periphery. Therefore, the former moves laterally toward the outer part of the curve (Figure 10.8). Its place is taken by fluid moving circumferentially toward the inside of the curve and medially toward the center of the pipe. The overall effect is to generate two equal and opposite helical streams which persist for a distance beyond the bend. Even without turbulence, a bend augments mixing.

FIGURE 10.8

Secondary flow just downstream from a bend in a cylindrical pipe. A swirling or doubly-helical motion is superimposed on the normally parabolic profile, and the most rapid flow is shifted toward the outside of the bend from the axis of the pipe.

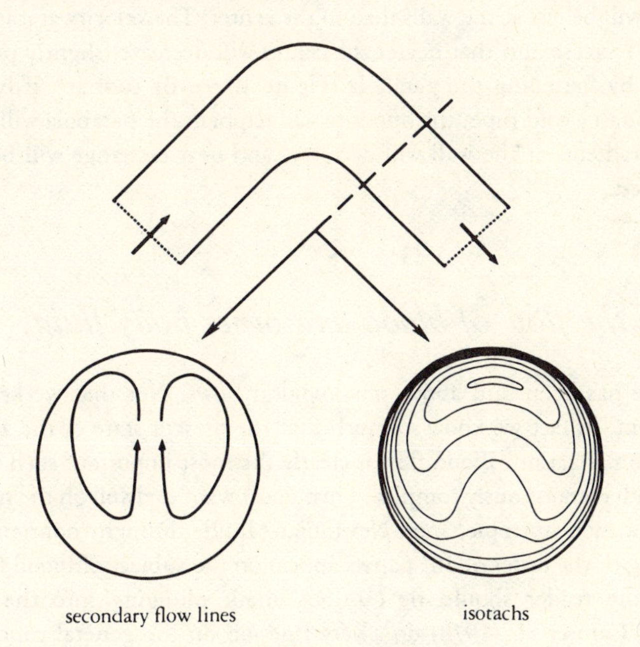

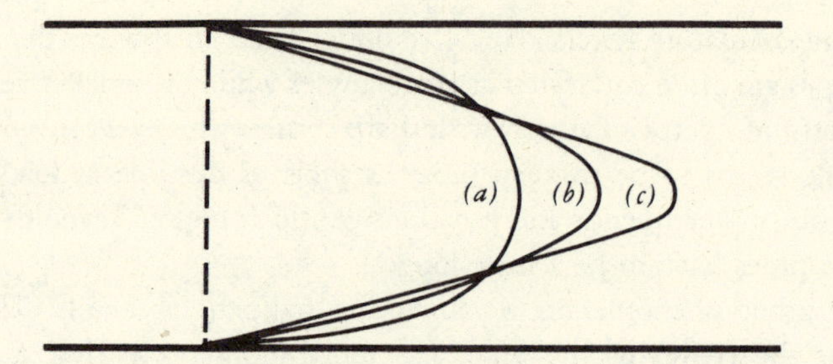

The effects of non-uniformity of temperature on the flow profile across a circular pipe: (*a*) cold fluid flowing through a hot pipe; (*b*) isothermal parabolic profile; (*c*) hot fluid flowing through a cold pipe. The local velocity is proportional to the horizontal distance from the dotted line to a point on a curve.

FIGURE 10.9

Non-isothermal water flows. The extreme variation of viscosity of liquids with changes in temperature was mentioned in Chapter 2. It gives rise to a curious sort of hysteresis in parabolic flow through pipes. If cold liquid is flowing through and exchanging heat with hot pipe walls, then the viscosity of the fluid will be less at the walls than in the center. The velocity gradient at the walls will increase and that nearer the center will decrease, slightly promoting exchange by flattening the parabola (Figure 10.9). By contrast, if hot liquid flows through a cold pipe, the opposite will happen: the parabola will steepen, velocity gradients at the wall will decrease, and heat exchange will be slightly discouraged.

the flow of blood and other body fluids

Very little has been said about physiological flows. Not that we know little about them; in fact we know so much that the present state of the art almost defies summarization. Blood flow is clearly the most important such transport system and is enormously complex. I just don't want to dwell on the nonsteady flow of a non-isotropic, non-Newtonian fluid through nonrigid pipes. Something of the order of 10^4 papers appear on the subject of blood flow each year, so the reader should be cautious about plunging into the primary literature. Caro et al. (1978) do a very fine job on the general concepts and phenomena; Lighthill (1972) examines the logic of the design; and Burton (1972) gives a graceful overview.

Flows of body fluids are frequent topics of articles in the *Annual Review of Physiology* and the *Annual Review of Fluid Mechanics.* Jaffrin and Shapiro (1971) provide a good treatment of persistaltic flow. There is even a symposium volume entitled *Hydrodynamics of Micturition;* I didn't find much of interest in it, though.

flow through a circular aperture

A circular aperture is essentially a pipe of negligible length. Flow through such orifices is not uncommon in biological systems: consider pits and stomata in plants, perforate vessel walls in various circulatory systems, and various filtration devices such as the dermis of sponges. For a single aperture, reasonably far from other apertures, a formula which is analogous to the Hagen–Poiseuille equation (10.2) will work:

$$Q = \frac{a^3 \Delta p}{3\mu} \tag{10.15}$$

The practical difficulty with equation (10.15) is that it is dependable only up to a Reynolds number of about 3. Above that, edge effects complicate matters; the shape and sharpness of the lip become important, and vortices are generated as fluid leaves the aperture. For higher Reynolds number, another formula is used:

$$Q = C_o \pi a^2 \sqrt{\frac{2\Delta p}{\rho}} \tag{10.16}$$

In (10.16) a dimensionless *orifice coefficient,* C_o, is needed. Its value varies with the Reynolds number but may be taken as 0.6 above $Re = 30{,}000$ as a reasonable approximation.

Notice that, just as in pipes, resistance in apertures depends exclusively on viscosity at low Reynolds numbers but more on density at high values. Also, no coefficient is needed at low Reynolds numbers, but one must be used when the Reynolds number is high.

CHAPTER 11

lift and airfoils

ALMOST WITHOUT EXCEPTION, THE FORCES considered thus far either have been omnidirectional (hydrostatic) or have acted parallel to the direction of flow. We now turn (90° to be precise) to forces normal to the direction of flow. Whatever their orientation with respect to the surface of the earth or to the direction of gravity, we'll call these "lifting forces." Just as drag did not originate as one's raw intuition might have suggested, neither does lift. Here, though, is a different if equally odd story. It is sufficiently peculiar that one wonders whether humans would ever have seriously entertained the notion of flight had nature not unmistakeably demonstrated its practicality.

The obvious way to produce an upward force on a body in an airstream is to arrange matters so that the speed of flow above the body is greater than that below under circumstances for which Bernoulli's principle (Chapter 4) holds. The pressure below will thereby exceed that above, and the body will experience lift. What is required, in effect, is creation of a new flow *around* the body, with the flow above in the same direction as the free stream and the flow below in the opposite direction. As we'll see, this net flow around the body isn't at all hard to accomplish. The only tricky part is explaining why it happens.

CHAPTER 11

the origin of lift

To begin, lets talk about vortices. One sort of vortex turns out to be of particular interest because, to a first approximation, it is the type which occurs naturally in fluids; this is the so-called *irrotational vortex.* The curious notion of an irrotational vortex is perhaps best presented in the form of a comparison between rotational motion and circular but irrotational motion. As an aside, it might be noted that the former is the motion of a bicycle wheel while the latter is the motion of its pedals; the former is what bacterial flagella do and the latter is what protozoan flagella do. For the comparison, we must bear in mind the difference between translation and rotation. In translation, an object changes its location—it moves from one place to another. That motion, in fact, may take it in a circular path in which it periodically revisits an earlier position. In rotation, an object changes its orientation over time, whether or not it ever comes back to a place it previously occupied. The "object" can, of course, be an element of a fluid just as well as a solid.

If we consider a small mass of water moving around and around, it is easy to see that there are two fundamentally different ways it can do so. If we spin a bowl of water by rotating the outer container (Figure 11.1a), the water clearly

FIGURE 11.1

Two ways of moving water in a circular path within a cylindrical container. In (*a*) the container as a whole is rotated; in (*b*) the container is fixed and a smaller cylinder is rotated coaxially within it.

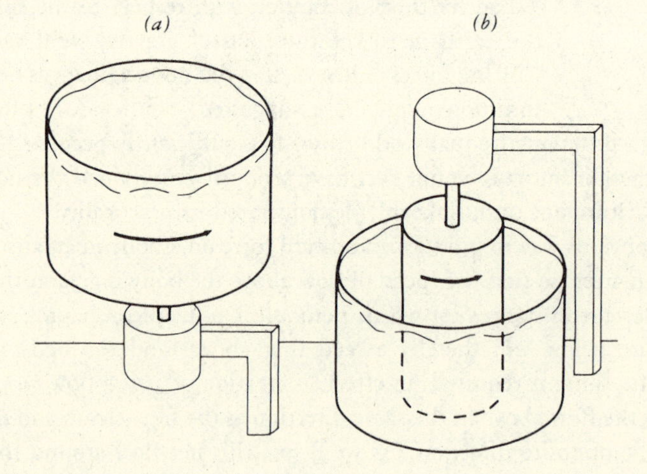

rotates both as a whole and as a set of separate bits. Each element of the water changes its orientation, rotating 360° with each turn of the bowl, as well as translating around the center of the bowl. The further from the center of the bowl, the faster must be the tangential velocity of an element of water; tangential velocity is directly proportional to distance from the axis. If, on the other hand, we take a large container of otherwise still water, insert a cylinder with a vertical axis into the water, and rotate the cylinder at a steady rate (Figure 11.1b), water again goes around the cylinder but in a much different manner. The water translates about the central cylinder, it circulates around the cylinder, but it does so with very little rotation. An object suspended in the water will not face east, then south, then west, then north; it will maintain its original orientation. We have made an irrotational vortex.

How are velocities distributed in an irrotational vortex? Consider an object some distance from the axis of such a vortex. For it to maintain its orientation, the water nearer the axis must be moving faster than the water further from the axis. More specifically, it can be shown that for the irrotational condition to exist, the product of tangential velocity and radius must be constant. An irrotational vortex, then, forms a set of concentric streamlines, with the inner ones representing higher velocities.

Of course, a constant product of velocity and radius implies that velocity approaches infinity as the radius approaches zero. In practice, any real vortex has a core which does indeed rotate as a result of the viscosity of real fluids. So our original driving cylinder in the bowl has some physical reality. At the axis one is quite literally in the eye of the storm (Figure 11.2).

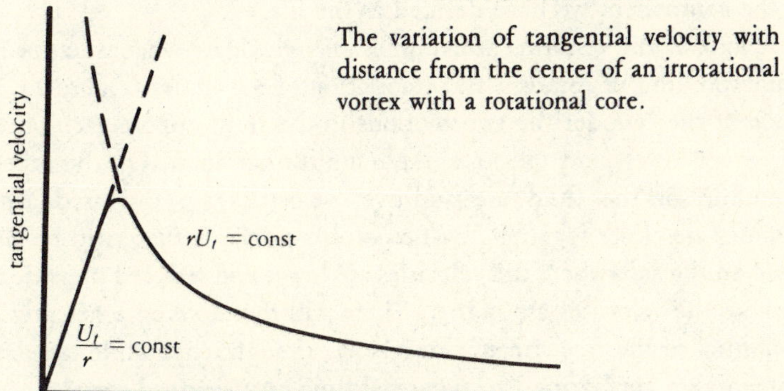

FIGURE 11.2

The variation of tangential velocity with distance from the center of an irrotational vortex with a rotational core.

If something is constant, it gets named, and this speed-distance product is called the *circulation*, (capital gamma, Γ). It is defined as the product of circumference and tangential velocity. Thus for all streamlines surrounding the core of an irrotational vortex:

$$\Gamma = 2\pi r U_t \tag{11.1}$$

Circulation has dimensions of distance squared per unit time. It turns out that the constancy of the circulation (at least in an ideal fluid) does not depend on staying on a streamline. So one can give a more formal and more general definition of circulation as the line integral of the component of the velocity on and tangential to a closed curve lying entirely within the fluid:

$$\Gamma = \int_l U_l \, dl \tag{11.2}$$

The value of the circulation defined in this way is the same for *any* closed loop enclosing the core of an irrotational vortex. And in inviscid fluids it turns out to be zero for any closed loop which doesn't encircle a core.

Let's return to our rotating cylinder. In an otherwise stationary fluid it will be surrounded by an irrotational vortex. Now let us superimpose upon this circulation an additional motion, a translation in a line, either by moving the cylinder through the fluid or by moving the fluid past the cylinder. To no one's surprise, the cylinder now incurs a certain amount of drag, but the force on it has an important difference from the drag forces we've previously discussed. The direction of the force is no longer exactly in the direction of motion, and it can be resolved into two components. One is the familiar drag, the force component in the direction of motion; the other is a component normal to the flow, the component we have defined as the lift.

A look at the resulting streamlines may elucidate what is happening in this superposition of rotation and translation of a cylinder (Figure 11.3). On one side of the cylinder the two motions in the fluid oppose each other, the velocities are lower, and the streamlines are further apart. On the other side, the motions of the fluid are additive, velocities are increased, and the streamlines are closer together. By Bernoulli's principle there will be elevated pressure on the side where the velocities are lower and reduced pressure on the side where the velocities are higher. There will therefore be a net pressure or force normal to the free-stream flow. Note, though, that while circulation is fundamental in generating lift, no actual fluid particle need travel all the way

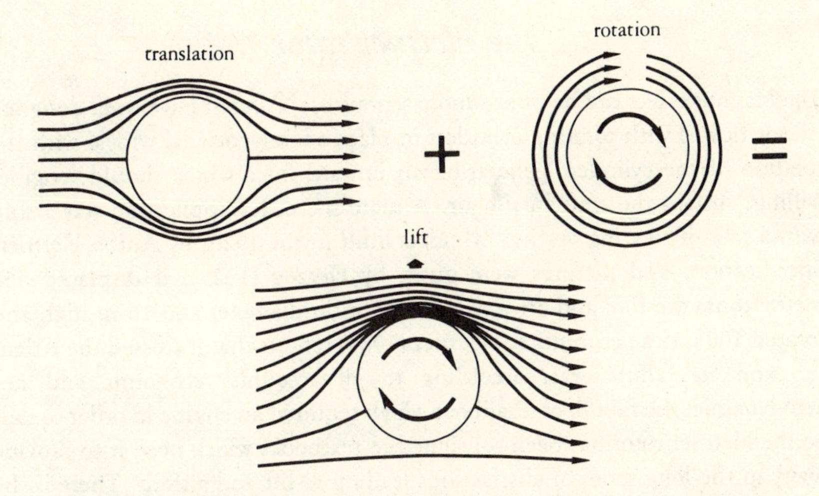

If a rotational motion of a solid is superimposed on a translational motion of that solid, the resulting asymmetry of flow generates a force normal to the free-stream flow. We call the force *lift*.

FIGURE 11.3

around the cylinder. (A limited amount will go around as a result of viscosity but not enough to merit much attention.)

Upon what does this lift depend for its magnitude? Within certain practical limits, it is directly proportional to the rate of rotation of the cylinder —in short, to the circulation. It is also directly proportional, again within limits, to the speed of the translational motion. A formal statement, the Kutta-Joukowski theorem, puts it in tidier form: "If an irrotational air stream surrounds a closed curve with circulation, a force is set up perpendicular to that air stream, and the force (per unit length) is the product of the fluid's density, the free-stream velocity, and that circulation":

$$\frac{F}{l} = \rho U \Gamma \qquad (11.3)$$

This phenomenon, the lift of a rotating cylinder moving through a fluid, is termed the "Magnus effect," after H.G. Magnus (1802–1870). It works for spheres as well as for cylinders and makes physically lawful the curved path of the sliced golf ball, the eccentricities of cut tennis balls, and other bits of nastiness.

CHAPTER 11

the Flettner rotor

The Magnus effect can be put to more serious use. Aircraft have been designed (if not flown) with rotating cylinders in place of conventional wings, with the rotation of the cylinders generating an upward force which should, engines willing, sustain the craft in the air. A more successful application was a ship with a pair of rotating vertical cylinders built in the 1920s by Anton Flettner. Specifications and pictures were given by Herzog (1925); it displaced 550 metric tons, the fore and aft rotors were 3 m in diameter and 16 m high and rotated 100 times per minute. Hoerner (1965) reports that it crossed the Atlantic; and its failure was, according to all accounts, economic and not aerodynamic: the failure of a sailboat which required an engine in order to sail. So the Flettner rotor has been consigned to textbooks which need it to provide levity in the long series of abstractions leading to lift and airfoils. There is, by the way, nothing special about the use of a cylinder; Hoerner says a rod of X-shaped cross section works as well.

McCutchen (1977) has recently pointed out that nature long ago and upon several occasions invented the Flettner rotor. Flat plates, not cylinders, are used; and the necessity of a rotating joint is avoided by having the entire craft rotate about its long axis. These rotorcraft are certain of the winged seeds (samaras), gliding down and away from the parent tree by autorotation. In the tree-of-heaven (*Ailanthus altissima*) the samara is a blade about 3 to 4 cm long, twisted along its length, with the seed in a thickening in the center (Figure 11.4a). When released, it begins to rotate about its long axis and either travels away from the parent or descends in a wide helix, slowly enough to achieve some wind dispersal. We'll talk about gliding later, giving a fuller account of how lift is usefully invested. You might at this point drop a small index or playing card, holding it with its long axis horizontal and flicking the top edge toward you as you release it.

Two other common autorotating samaras were described by McCutchen (1977): those of the ash (*Fraxinus*) and the tuliptree (*Liriodendron*). Each has its seed not in the middle but at one end. Each does a more complex, dual autorotation, combining the lengthwise Flettner rotation with a tightly-helical gyration about the seed as an axis. The former axis of rotation is nearly horizontal, the latter vertical (Figure 11.4b). These are, like maple samaras, autogyros, but they are Flettner-rotating autogyros, obtaining their lift in this unconventional manner. More will be said about autogyration in the next chapter.

Many long, thin leaves act as Flettner rotors when they're shed. In central North Carolina, the willow oaks (*Quercus phellos*) are particularly conspicuous

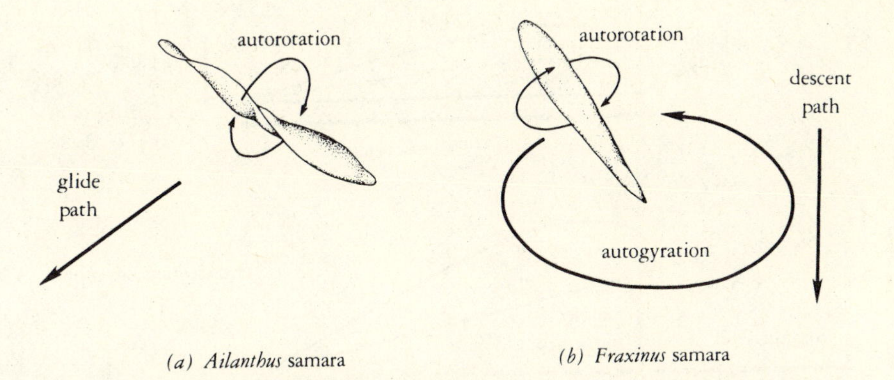

(*a*) *Ailanthus* samara (*b*) *Fraxinus* samara

(*a*) The samara of *Ailanthus* rotates about its long axis and thus glides instead of descending vertically. (*b*) The samara of *Fraxinus* autorotates similarly, but its glide path is a tight helix which certainly slows its descent and permits a long exposure to any ambient wind.

FIGURE 11.4

autorotators, and their leaves are carried well beyond the tree even with no apparent breeze. It's not clear whether the behavior has any adaptive significance or is merely an incidental consequence of a leaf shaped by selection for other functional characteristics.

airfoils

We've defined and produced lift, but we've done so in a way that seems far removed from the wing of a bird. Several sticky problems remain to be dealt with:

1. It is a common observation that means other than rotating devices can produce lift. I refer, of course, to airfoils or wings, to whose lift-producing action we commonly trust our nearest and dearest. Do these also work by the superposition of circulation and translation, or must we seek a different physical phenomenon?

2. It may seem intuitively reasonable for a flat, inclined surface to deflect an air stream and produce lift. But curved surfaces with their convex side upward can produce an upward force even if the leading and trailing edges are at the same horizontal level, with no inclination.

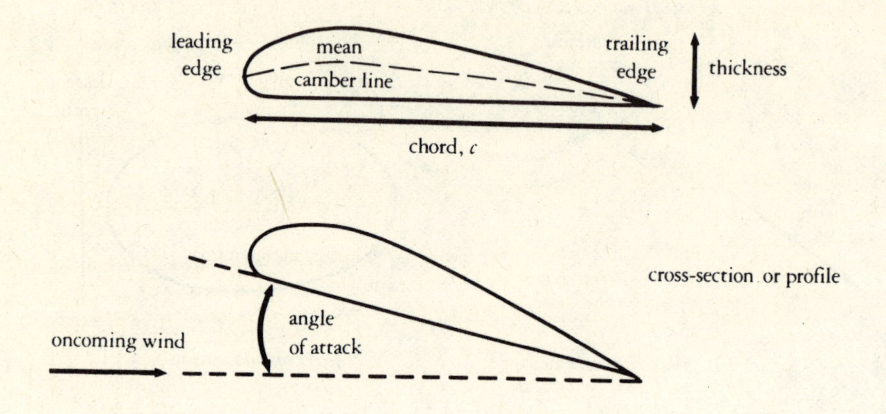

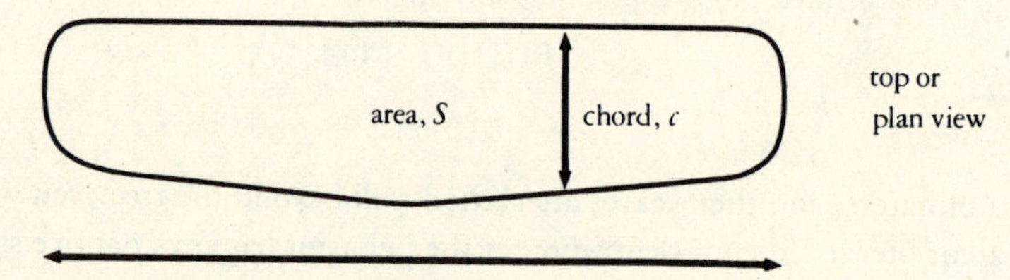

Aspect Ratio (AR) equals b/c or b^2/S or S/c^2

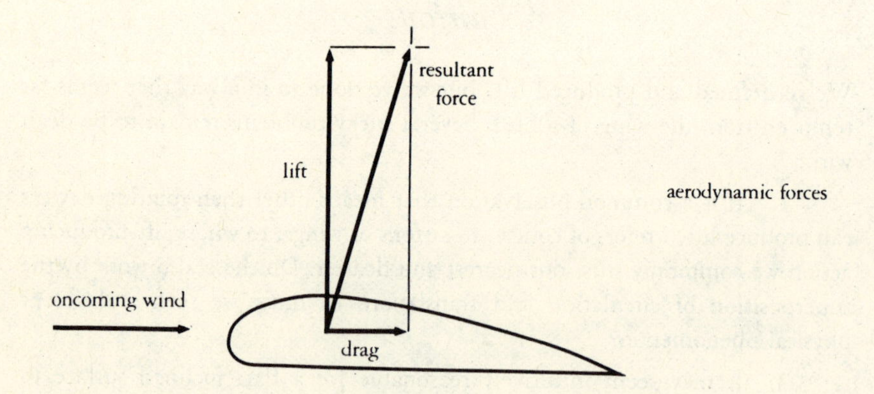

The terminology for lift-producing airfoils.

FIGURE 11.5

3. For rotating cylinders, lift is proportional to the first power of the translational velocity (equation 11.3), while for airfoils, lift is more nearly proportional to the second power of the translational velocity. Is the Kutta-Joukowski theorem applicable to non-rotating airfoils?

4. A theorem concerning ideal fluids, established by Kelvin, states that circulation can be neither created nor destroyed. Can we evade that theorem, or is circulation and lift, like drag, totally at odds with ideal fluid theory?

Circulation and airfoils. The route around these awkward points is the following. We'll explore in more detail the behavior of this something we're calling an airfoil, the relevant terminology for which is given in Figure 11.5. According to F.W. Lanchester (1868–1946), an airfoil is a device which can produce circulation in its vicinity without itself actually rotating. What is the origin of this circulation? The fact that the lift for an airfoil is proportional to the square rather than to the first power of the free-stream velocity suggests that the circulation itself must be proportional to the speed of the oncoming air stream. The interaction of the airfoil shape and orientation with the oncoming air must create circulation.

What determines the value of the circulation and hence the lift was first realized by Joukowski. He pointed out that for an airfoil with a sharp trailing edge there is only one pattern of flow in which the air slips off the rear of the airfoil without any discontinuity, or without having to turn the sharp corner at the trailing edge. This pattern determines the amount of the circulation and lift. In effect, it isn't practical for fluid to sneak around from a high pressure airfoil bottom to a lower pressure top to relieve the pressure difference; instead, everything else shifts, creating a net circulation (Figure 11.6). It's only a net circulation—no particles of fluid actually travel around the airfoil—but it's still a flow pattern equivalent to that which would be generated by the superposition of translation and circulation. In short, a wing can be viewed as a fictitious bound vortex (fictitious because it's only a *net* vortex), and the strength of that vortex (its circulation) is proportional to the airfoil's translational velocity. Hence the dependence of lift on U^2. One U comes directly from translation, the U in equation (11.3); the other U comes from the circulation which is itself proportional to the translation.

We still have to deal with Lord Kelvin's theorem. Prandtl and Lanchester solved that paradox by recognizing the existence of a "starting vortex," equal in strength and opposite in direction to the "bound vortex" of a wing (Figure 11.7). Thus there is no net circulation for the system as a whole. Prandtl even made movies showing starting vortices; these also showed that when an airfoil

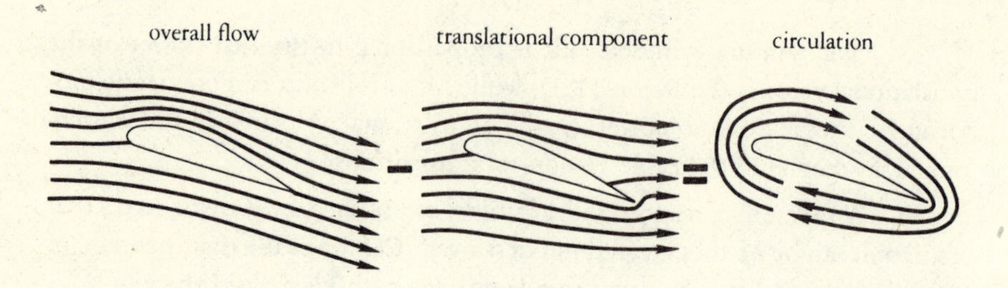

The overall flow across a lift-producing airfoil consists of a translational component and a net circulation around the airfoil.

FIGURE 11.6

was suddenly stopped, the bound vortex slipped out from around the airfoil and became a free and obvious "stopping vortex." These starting and stopping vortices have the unfortunate habit of hanging around airports until eaten up by viscosity, buffeting small craft. For reasons which will become clear shortly, wings of finite length shed "tip vortices" at their ends which are essentially continuations of the bound vortex or circulation of the wing. These tip vortices extend back to the starting vortex, completing a "vortex ring" much like a smoke ring. Naturally, the vortex ring is almost always much longer than wide, about 100,000 times so for a large plane on a transcontinental flight. As a result of viscosity, in the real world the starting vortex has been dissipated long before the stopping vortex is shed. But we needn't postulate any discontinuities in the flow, and Lord Kelvin's theorem isn't seriously abused.

With the role of circulation established, there's no great mystery as to why curved plates can produce lift even if leading and trailing edges are at the same level with respect to the free-stream motion. If a plate is convex on the side facing upward, then the path from leading to trailing edge is greater on the top than on the bottom surface. To avoid discontinuity and to satisfy the principle of continuity, the speed of flow has to be greater above the plate than below; thus it generates a net circulation without any inclination. This also explains why the contour of the upper surface of a lift-producing airfoil is generally of more consequence than that of the lower surface, and why flat lower surfaces work about as well as concave ones. With a curved upper surface and a flat lower one, a wing may have an enclosed volume which can be put to use housing fuel tanks, muscles, control equipment, and so forth. The sharp trailing edge is crucial for the functioning of an airfoil as a lift-generator as well

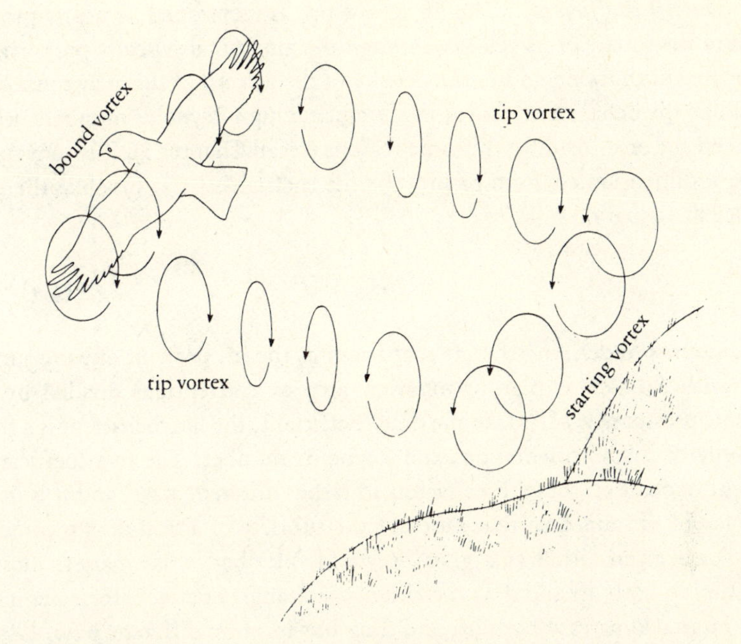

The bound vortex around the wings, the tip vortices, and the starting vortex for a gliding aircraft. *In toto,* the vortices form a complete vortex ring.

FIGURE 11.7

as for drag minimization; the rounded leading edge of conventional wings has the basic role we saw for streamlined objects a few chapters back.

The distribution of the lifting force on the surface of a wing is of some interest. More of the lift usually comes from the reduced pressure on the top than from the excess pressure on the bottom, quite consistent with our notion of the role of the upper convex surface. The detailed distribution of lift on an airfoil can be determined in the same way as can the pressure distribution around a cylinder: a series of holes is drilled in the surface and led by internal tubes to a manometer. It turns out that the center of lift is relatively near the leading edge of the airfoil, usually being at or in front of the point of maximum thickness. It also happens that as the angle at which air meets wing (the angle of attack) increases, the center of lift shifts forward, with consequences for pitching moments and stability.

The lift coefficient. The lift of a wing, as mentioned, is nearly proportional to the square of its velocity through the air. It is also nearly proportional to the area of the wing and to the density of the air. All of this may sound very much like the behavior of drag at moderate and high Reynolds numbers, which is indeed the case. And for the same reasons (recall Chapter 5), it is convenient to use a dimensionless form of lift, the lift coefficient, C_L, just like the drag coefficient (equation 5.4):

$$L = \tfrac{1}{2} C_L \rho S U^2 \tag{11.4}$$

This amounts to defining the lift coefficient as the lift per unit of wing surface (L/S, with S taken as plan or top view area by convention) divided by the dynamic pressure, $\tfrac{1}{2}\rho U^2$. Like the drag coefficient, the lift coefficient is a function only of shape, orientation, and Reynolds number. The two coefficients, though, depend on these three factors in rather different ways, and it is in the interplay of lift and drag that much of the subtlety of airfoil design come in.

For a given airfoil at a given Reynolds number, what matters most in determining both lift and drag coefficients is its angle of attack. Increases in the angle of attack increase both lift and drag but in quite different ways. Lift increases from zero to some maximum at an angle of (typically) around 20° and then drops off again, while drag is never zero and increases continually up to very high angles of attack. In practice, the angle of attack is the immediate independent variable, and it is reasonable to plot the lift and drag coefficients against the angle of attack. Gustav Eiffel (the famous tower builder), had a better suggestion. Lift coefficient (ordinate) can be plotted against drag coefficient (abscissa) and the angle of attack treated parametrically and merely noted on the curve. Figure 11.8 gives an example of such a "polar diagram"—a fast, easy view of the characteristics of an airfoil. Two most interesting data practically jump out. A line through the origin tangent to the curve is a line whose slope gives the maximum ratio of lift to drag possible with that airfoil; as we'll see, maximizing that ratio is often a primary objective in airfoil design. Also, the point at which that tangent line touches the curve gives the angle of attack at which the maximum occurs.

Aspect ratio and induced drag. Figure 11.5 includes a definition of something called the aspect ratio of a wing; in simplest form it is the ratio of length or span to width or chord. Thus long skinny wings have high aspect ratios. The definition is a bit loose since for aerodynamic as well as mechanical reasons most practical wings taper from center to tip, but the definition will be

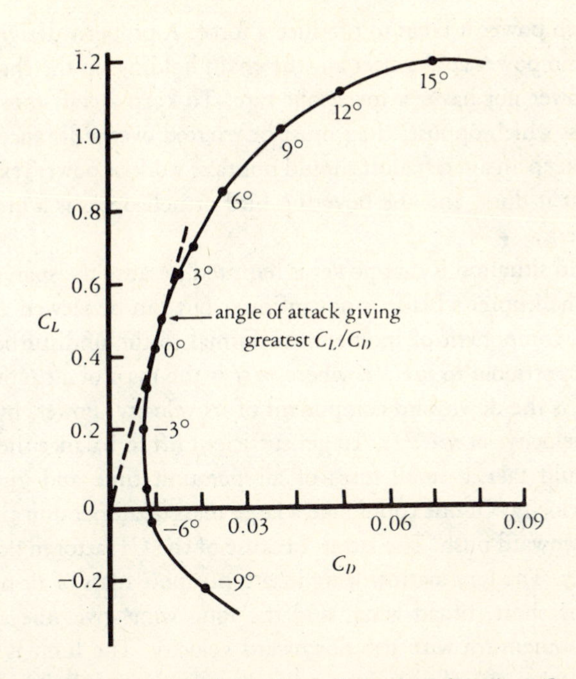

A polar diagram for an airplane wing. C_L is plotted against C_D (usually with different scales) and the angles of attack are noted on the curve. Any line which passes through the origin is a line of constant lift-to-drag ratio. The one tangent to the curve indicates the maximum lift-to-drag ratio obtainable: it is tangent at the angle of attack that gives the maximum ratio.

FIGURE 11.8

good enough. Aspect ratio is far more important than one might guess, and its consequences merit a paragraph or so of explanation.

It is quite possible to measure the lift and drag of something approximating an infinitely long wing: one just puts big end plates on a model or runs the model from one side wall to the other in a wind tunnel. If real wings, with aspect ratios much lower than infinity are compared to an infinite (or "two-dimensional") airfoil, the real ones are inevitably the poorer—they produce less lift and suffer more drag. It is a simple matter to blame the deterioration on the presence of tip vortices on a finite wing, and the explanation is both true and intuitively reasonable. The stubbier the wing, the more it is influenced by what happens at the tip.

But there is a more basic and important issue underlying the inferiority of finite airfoils. To ask what power is necessary to sustain an aircraft in the air is to ask how much power it takes to produce a force. A properly designed machine should take no power at all; after all, the chain holding up the chandelier consumes no power nor has it a metabolic rate. To keep a craft moving requires power: thrust which opposes drag must be exerted over a distance at a certain rate. But to keep an aircraft aloft should not take work or power, except that we all know that it does, and the hovering bird or helicopter is a profligate consumer of energy.

The odd situation is that power is required because the span of the bird's wing or the helicopter's blade is not infinite. Lift can be viewed as the rate of creation of a component of momentum normal to the undisturbed airstream and thus proportional to mU_v/t, where m/t is the mass of air moved per unit time and U_v is the downward component of its velocity. Power, by contrast, is force times velocity, or mU_v^2/t. To get sufficient lift to balance the weight of a craft one could take a small mass of air per unit time and give it a large downward velocity. Or one could take a large mass of air per unit time and give it a small downward push. The latter, because of the U^2 factor in power, will be much cheaper. The long narrow wing intercepts more mass of air per unit time than does the short, broad wing; and the long wing gives the air the same downward momentum with less downward velocity. The limit is the infinite wing, which takes an infinite m/t and imparts to it a negligible U_v. Thus its power consumption approaches zero. The finite wing must consume power, and the lower its aspect ratio, the more power it must consume.

How does a finite wing signal its requirement for power? The only way it can do so is by having more drag than an infinite wing. The extra drag is termed the "induced drag" and is the drag incurred as a consequence of producing lift with a less-than-infinite wing. Another way of looking at the induced drag is to recognize that in giving air a significant downward velocity the air stream crossing a wing is deflected from the horizontal (Figure 11.9). The lift vector, normal to the local air stream, is no longer normal to the overall free stream or (in level flight) to the horizon. It can be treated as a resultant and resolved into two components: a new vector normal to the free stream but a little less than the old one, and a drag vector parallel to the free stream. The latter, the induced drag, adds to the other sorts of drag—skin friction and pressure drag. Induced drag times velocity is the price in power paid for staying aloft with a wing of less than infinite span or aspect ratio; thus long, skinny wings are—on energetic grounds—better than short, broad ones. Put in terms of a polar diagram, reduction of the aspect ratio pushes the wing's curve toward the right and toward the abscissa by producing less lift, more drag, and a lower maximum lift-to-drag ratio.

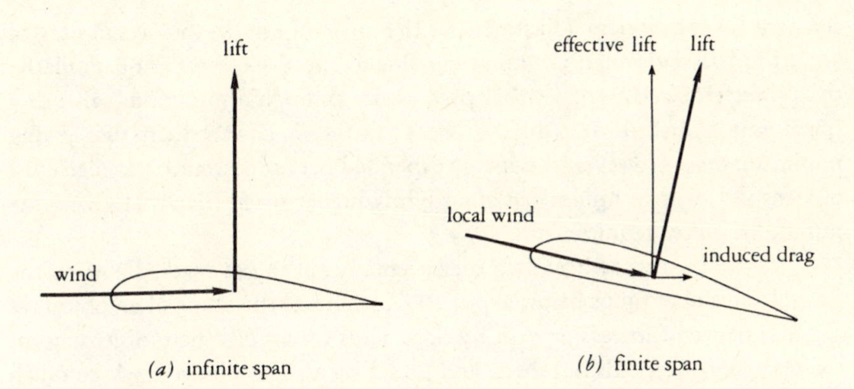

(*a*) infinite span (*b*) finite span

The origin of induced drag in a wing of finite span. Since the wing gives the air a significant downward push in (*b*), the net local wind direction crossing the wing is no longer the same as that of the free stream as it had been in (*a*). Thus the lift vector is tipped back and resolves into a lesser effective lift and the induced drag.

FIGURE 11.9

The induced drag and the resulting induced power vary in an odd but reasonable way with the velocity of the airfoil through the fluid. As speed increases, the airfoil comes in contact with more fluid per unit time. Since its requirement for lift is nearly constant (equal to the weight of the craft), the airfoil need give the fluid passing it less of a downward deflection at higher speeds. Therefore, as speed increases, the local airstream is more nearly horizontal, the lift vector more nearly vertical, and the induced drag is less. The induced drag, then, is greatest when speed is least; it is very nearly inversely proportional to the square of the speed of the airfoil relative to the undisturbed fluid. This dramatic increase in induced drag at low speeds is one of the main reasons why, in practice, flying slowly is so expensive.

Profile drag. The remaining drag of an airfoil is termed the "profile drag;" it represents the combined pressure drag and skin friction discussed in Chapter 5. (Profile drag is sometimes either called parasite drag or included, with the drag of non-lift-producing components, in the parasite drag. See the footnote on page 115). By contrast with induced drag, profile drag is present even when an airfoil produces no lift; like induced drag, it increases sharply with increases in angle of attack. Like parasite drag (but unlike induced drag) it increases nearly in proportion to the square of speed. As a result, if lift is kept constant by appropriate adjustment of the angle of attack, the overall drag of an airfoil passes through a minimum value as speed increases. And similarly, the power required to offset that drag passes through a minimum value at a speed just a little lower (due to the U factor in power) than that giving minimum drag. However, the energy expended per unit distance travelled (cost of transport) will be minimized at a slightly higher speed than that giving the minimum power requirement.

As we've seen earlier, drag coefficients begin to rise markedly when the Reynolds number drops below about 1000, whatever the shape of an object, as a consequence of increasing skin friction. Thus for airfoils the profile drag increases at low Reynolds numbers, and under such circumstances may be much larger than the induced drag.

Stall. Lift is nearly proportional to the angle of attack, but only up to a point. At angles above some critical value lift typically drops, sometimes abruptly. In physical terms, the flow has begun to separate on the upper surface of the airfoil, the circulation is much reduced, and the airstream is not so effectively deflected downward. The phenomenon may be clearly seen in the polar plot of Figure 11.8. Stall is a Bad Thing, but under ordinary conditions the stall point occurs at an angle of attack well above that which gives the best lift-to-drag ratio. During low speed operation, as in take-off and landing, aircraft use high angles of attack so that the higher lift coefficient can compensate for the lower velocities. Under such special but not extraordinary circumstances, devices which postpone stall are enormously useful; and a very wide range of schemes have been investigated. Mises (1945) has a particularly useful discussion of stall and high lift devices.

The most frequently mentioned candidate for an anti-stall device in a living flier is the "alula" of bird wings, a group of feathers attached to the anatomical thumb on the leading edge of the wing. Nachtigall and Kempf (1971) found that when the angle of attack is high (30° to 50°), as just prior to landing, the alulae are commonly erected and give lift increases of up to 25%. Their photographs of flow show the expected mechanism: air is persuaded not to separate so readily as it flows across the top of a wing at high angles of attack. The alulae seem to have little effect on lift or drag at lower angles of attack.

Wing area. While lift can be controlled through adjustment of the angle of attack, use of the latter as a primary control parameter would necessitate operation quite often at sub-optimal lift-to-drag ratios. Varying wing area with changes in flying speed might reduce greatly the range over which the angle of attack must be altered. To a limited extent, large commercial aircraft do so with various wing extensions which emerge during take-off and landing. But such mechanical arrangements are complex and heavy. Birds, though, make dramatic alterations of wing area as speed changes; Pennycuick (1968) has carefully documented the phenomenon in gliding pigeons. And in flapping flight, the configuration of a bird's wing is changed continuously during the stroke cycle (Bilo, 1971; 1972).

Wing area, of course, is chosen in the design of an aircraft. Slow craft have large wings and fast ones have small wings in order to produce an appropriate amount of lift without requiring operation at unreasonable or uneconomical angles of attack. A common measure of the relative area of a wing is the so-called "wing loading"—the weight of the craft divided by the area of the wings. Successful people-pedalled planes operating at 5 or 6 m s^{-1} have the lowest values among human aircraft, about 20 N m^{-2}. A 747 jet transport operating at 250 m s^{-1} has a wing loading of about 2000 N m^{-2}. The situation for flying animals in considerably more complicated since wings are not always just fixed, lift-generating devices; they move up and down and produce thrust, and the small fliers beat their wings much more frequently than do the large ones. An Andean condor has a wing loading of about 100 N m^{-2} (McGahan, 1973); a wren of 25 N m^{-2} and a bumblebee, 50 N m^{-2} (Greenewalt, 1962); a fruit fly has a wing loading of 3.5 N m^{-2}.

Of course there is a basic allometric problem in wing loading. If force reflects mass and mass reflects volume, then the lift needed will be proportioned to the cube of a typical linear dimension. For an airfoil, though, lift is proportional to area and thus to the square of a linear dimension. Constant wing loading therefore requires that shape change with size: a large flier will have disproportionately large wings. Among animals, larger ones do in fact have relatively larger wings than do smaller ones, but the differences are not sufficiently great to keep wing loading constant, as indicated by the figures just mentioned (Alexander, 1971). In compensation, larger creatures typically fly a bit faster. This allometric problem implies that angels must be either of unusually low mass, supersonic, or buoyed up by contemplation of their divine mission.

Shape and Reynolds number. Conventional aircraft airfoils are designed or have been selected (by a process resembling natural selection) for operation at Reynolds numbers in the millions. Birds, bats, and insects operate at values, based on chord length, not exceeding half a million—about 50,000 for typical birds, about 5000 for large insects. For conventional airfoils, everything gets worse as the Reynolds number is lowered. Profile drag, as mentioned earlier, increases. The maximum obtainable lift coefficient decreases, as do the maximum possible lift-to-drag ratio and the angle at which stall occurs (Goldstein, 1938). For an airfoil whose best L/D (infinite aspect ratio) was 80 at a Reynolds number of 6,500,000, the best L/D was only 47 at 310,000. An airfoil for which the stall angle was 18° at Re = 3,300,000 had a stall angle of 12° at 330,000 and 9° at 43,000. Some deterioration of performance at low Reynolds number is unavoidable, but much of these commonly quoted data simply reflect the use of inappropriate airfoils.

Some effort has gone into the design of airfoils for model airplanes (Schmitz, 1960). A moderately cambered flat plate (thickness of about 5% of chord) proves far better than a conventional airfoil: at Re = 42,000, the flat plate has a maximum lift coefficient of 1.05; the conventional airfoil of about 0.7. The cambered plate has a best lift-to-drag ratio of about 11.0; the conventional airfoil, 4.5. Only in minimum drag coefficient are they similar. Tested at a Reynolds number of 168,000, however, the conventional airfoil is better in all respects. Another (alleged) source of information on airfoils for model craft is Rabel (1965).

At high Reynolds numbers, even minor surface irregularities may have a great effect on the performance of an airfoil. But airfoils such as insect wings, selected for operation at low Reynolds numbers, are nothing if not irregular. While exceedingly thin, insect wings are usually somewhat corrugated; not only do veins protrude, but the veins are not coplanar, presumably giving some increase in the stiffness of the wings. These irregularities may have some aerodynamic consequences. Newman et al. (1977) made models of a cross section of a dragonfly wing (Re = 10,000) with a highly pleated forward half; they obtained some truly strange polar plots and regarded eddies between the folds as being of great importance. While separation occurred near the leading edge at low angles of attack, the flow reattached further downstream up to much larger angles. On the other hand, Rees (1975) made a folded sheet metal model of cross sections of a hover fly (syrphid) wing and compared this with a smooth model representing an envelope around the corrugations. His somewhat tidier data show little difference between the two, and the polar plots are nicely consistent with my data (Vogel, 1967b) on fruit fly wings. Apparently, at Reynolds numbers of 450 and 900, the flow behaves as if the folds were filled. This matter of the function of the profiles of insect wings would certainly repay further investigation; probably real wings should be compared to model wings of varying degrees of abstraction.

more about biological airfoils

Now that the basic operation of airfoils is clear, let's turn our full attention to the lift-producing airfoils of organisms. These are a diverse group of structures. For one thing, they function over a Reynolds number range of four or five orders of magnitude; the equivalent range for all human-carrying, subsonic aircraft is less than three orders. For another, they serve a diversity of functions, with the production of a net circulation about a structure practically the only common feature. Even the generation of a force is in at least one case irrelevant. And finally, they have evolved from a wide variety of precursers in a large number of animal and plant groups. It is highly likely that many lift-producing airfoils have escaped identification as such due to an insufficiently general appreciation of the varied roles these devices can serve.

A lift-producing sand dollar. First, a brief look at what must be the least obvious use of an airfoil to have emerged to date. There is a sand dollar, *Dendraster excentricus,* on the beaches and tidal channels of the Pacific coast of North America which, when exposed to currents, routinely assumes a most un-sand-dollar-like posture (Figure 11.10a). It buries its anterior edge in the sand and erects its posterior in the flowing water above instead of lying, half-buried, like a bump on the bottom. Moreover, these sand dollars are neither randomly nor regularly spaced out on the sand but cluster at densities of 200 to 1000 per square meter.

The existence and peculiarity of this behavior was well known but its functional significance was enigmatic until O'Neill (1978) demonstrated that each sand dollar functioned as a "lifting body." A sand dollar, flat or slightly concave on its oral (lower) surface and convex on its aboral (upper) surface, develops substantial circulation about itself. Since one end is in the sand, it has only one tip vortex and an effective aspect ratio about double that based on its measurements. But the lift is directed horizontally, and the lift as a force proves to be of little interest. Instead, the circulation and cambered profile bring food particles closer to its feeding appendages on the oral surface. Drag is also not of direct consequence, at least at velocities below those which might dislodge the creature. Aggregation of individuals turns out to be highly valuable in mutual improvement of feeding currents; usually the oral surface of one individual faces the aboral surface of the nearest neighbor (Figure 11.10b) much like the wings of a multi-winged aircraft. The sand dollars seem to regulate their density to maintain optimal gaps between the individuals, adjusting their spacing

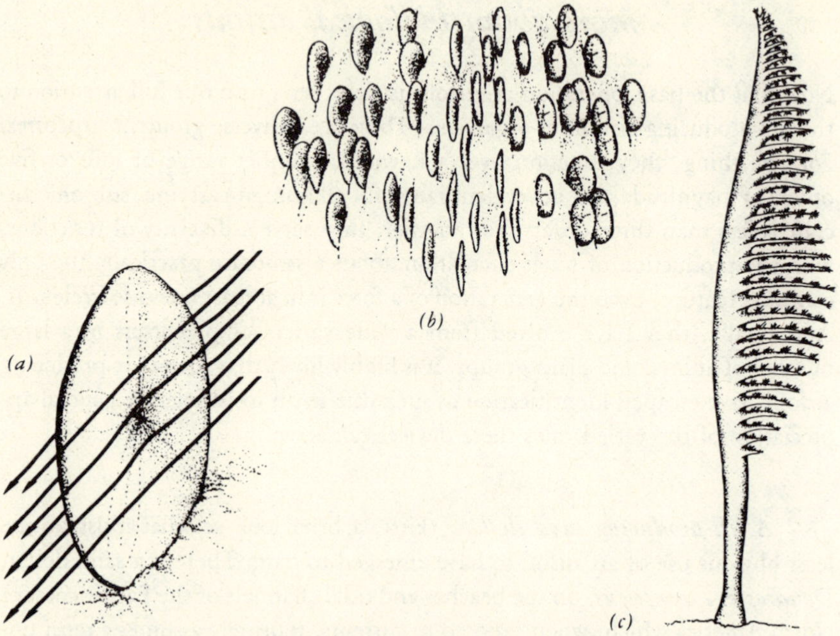

(a)

(b)

(c)

Flight isn't the only use for lift. (*a*) At least one sand dollar stands on its anterior edge and operates as a colonial multi-winged craft to bring feeding currents closer to the oral surface (*b*). A sea pen (*c*) may get lift from its leaflets to partly offset its tendency to bend over in currents.

FIGURE 11.10

from 1 to 10 cm in proportion to the square of the average surge velocity. As hydrofoils of symmetrical leading and trailing edges which get lift at near-zero angles of attack, they are well suited to take advantage of the bidirectional surge of waves. And populations from sheltered locations are slightly more cambered and have slightly higher lift coefficients than those from more exposed localities.

This species of sand dollar cannot be the only sessile animal to exploit a lift-producing airfoil or hydrofoil. Best (personal communication) has some evidence that sea pens (Anthozoa: Pennatulaceae) can generate lift. The central rachis of a sea pen stands erect, with leaflets extending downstream and with the rachis rotating to keep this orientation in changing tidal currents (Figure 11.9c). It seems likely that lift on the individual leaflets helps to keep the sea pen erect in water currents.

TABLE 11.1
Characteristics of animal airfoils

	Re	*AR*	$C_D$min	$C_L$max	@ α	C_L/C_Dmax	@ α
Anas platyrhynchos (mallard duck) wing (1)	60,000	9.0	0.12	0.9	28°	2.7	16°
Passer domesticus (house sparrow) wing (1)	23,000	4.5	0.17	1.1	30°	2.7	20°
Petaurus breviceps (marsupial flying squirrel) entire (2, 3)	100,000	1.0	0.10	0.6	34°	2.0	34°
Schistocerca gregaria (desert locust) hindwing (4)	4000	5.6	0.06	1.13	25°	8.2	7°
Tipula oleracea (crane fly) wing (5)	1500	6.9	0.10	0.82	35°	3.7	13°
Melalontha vulgaris (cockchafer beetle) elytra (6)	1100	4.2	0.17	0.83	25°	1.9	15°
Drosophila virilis (fruit fly) wing (7)	200	5.5	0.33	0.87	30°	1.8	15°

References: (1) Nachtigall and Kempf, 1971; (2) Nachtigall et al., 1974; (3) Nachtigall, 1979a; (4) Jensen, 1956; (5) Nachtigall, 1977b; (6) Nachtigall, 1964; (7) Vogel, 1967b.

Bird wings. Table 11.1 gives the salient aerodynamic characteristics of a variety of animal airfoils used in flight. The first two entries are birds, a duck and a sparrow, with data from mounted wings in a wind tunnel. By the standards of either large insects or small airplanes, these are very bad wings, with best lift-to-drag ratios of only 2.7. The lift coefficients seem low and the drag coefficients very high. What the data really illustrate are the pitfalls of working with these mechanically abnormal wings in wind tunnels. I once rationalized some analogous results by noting that for a machine designed by the evolutionary process, the only appropriate operating conditions are its natural conditions. Measurements under other circumstances are more likely to show substandard than superior performances. With insect wings we've been fairly fortunate in the decent congruence of results on living and decorporate wings; bird wings are much less cooperative.

Just how good bird wings really are is evident from determinations of lift and drag coefficients made from living, gliding birds in nature or wind tunnel; how such data can be derived will be discussed in the next chapter. For a pigeon, Pennycuick (1968) measured a maximum lift-to-drag ratio of about 6, with drag including body drag as well as wing drag. At low speeds, the bird got its lift coefficient up to 1.3 at a Reynolds number of about 40,000. Bigger birds do still better. A falcon has a maximum lift coefficient of 1.6 at a Reynolds number of 80,000 and a best lift-to-drag ratio of 10.4 (Tucker and Parrott, 1970). A petrel reached a maximum lift of 1.8 at a Reynolds number of 50,000; the high value was probably facilitated by use of its alula and by spreading its primary feathers as additional high lift devices (Pennycuick, 1960). Its best lift-to-drag ratio was about 8.5. A large vulture reached a lift-to-drag ratio of 15.5 at a Reynolds number of 140,000 (Pennycuick, 1971), a condor performed in nature at 13.7 at 360,000 (McGahan, 1973), and an albatross was estimated to achieve an L/D of 18 (Pennycuick, 1960). In fact, models built around specific cross-sectional profiles of pigeon wings perform very much better in wind tunnels than do whole, intact wings (Nachtigall, 1979c).

Beyond changes in shape and the unknown role of surface flexibility, there are other tricks a bird can play to increase the apparent quality of its airfoils. As a result of the pair of tip vortices and the deflection of the airstream, there is a region of downwash just behind a bird, so a bird will suffer extra induced drag if it flies just behind another. Conversely, there is a region of upwash lateral to and behind a bird; and, by proper positioning, another bird can reduce its induced drag by flying in this region. A flock which extends laterally, not vertically, can derive mutual benefit (Higdon and Corrsin, 1978). A V-formation is particularly good, and Lissaman and Shollenberger (1970) calculated that a formation of 25 birds could get an increase of 70% in distance travelled for a given expenditure of energy compared to each bird flying alone. An analogous scheme may be applicable to fish; Weihs (1975) has worked out the optimal spacing pattern for a school of fish.

If an airfoil moves just above a solid surface, its performance is improved by a phenomenon termed "ground effect", mainly as a result of a decrease in induced drag. Withers and Timko (1977) have pointed out that skimmers, flying close to the surface of calm water, can achieve about a 20% reduction in their requisite power (assuming an average wing-to-water distance of 7 cm). A skimmer flies with its lower mandible partly submerged, but the mandible is well streamlined, and its hydrodynamic drag is negligible compared to the aerodynamic forces on the bird.

Yet another trick may be employed by the Wilson's storm petrel, which has been described as walking on water. Withers (1979) suggests that it operates somewhat like a kite, which stays aloft as long as it is tethered to the ground. If the petrel faces into the wind and is blown backward by it over otherwise stationary water, it can maintain sufficient airspeed to stay aloft by dangling its high-drag feet in the water as sea anchors.

A gliding mammal. Table 11.1 also gives the characteristics of a gliding phalanger, *Petaurus breviceps,* an engaging marsupial of extraordinarily close resemblance to the North American flying squirrel, *Glaucomys volans;* the full polar plot for *Petaurus* is given in Figure 11.11. These data were obtained on whole, performing animals and are probably quite reliable. The interesting point about this airfoil is its low aspect ratio of unity, rather a bad state of affairs if these animals are viewed as purely aerodynamic devices. The low aspect

FIGURE 11.11

Polar diagram for the flying phalanger, *Petaurus.* Maximum lift is obtained at an angle of attack of 34°, which is also very close to the angle giving maximum lift-to-drag ratio (from data of Nachtigall, 1979a).

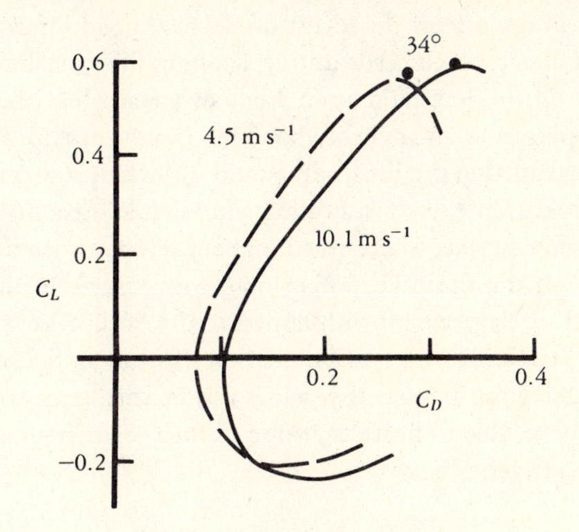

ratio limits the maximum lift and increases the induced drag. The polar curve is pushed to the right; and the line of maximum lift-to-drag ratio, the tangent to the curve, intersects essentially at the stall point.

Under such circumstances, anything which either postpones stall to higher angles of attack or reduces the suddenness of separation and stall should be of substantial benefit. Nachtigall (1979b) tested a set of flat plates of *Petaurus*-like camber and aspect ratio with and without fur coats. Best performance was obtained with *Petaurus* fur covering the upper surface and lying in its normal orientation. Application of fur increased the stall angle from 30° to 33° and the maximum C_L from 0.72 to 0.76 and reduced the suddenness of separation. Other kinds of real and synthetic furs were not as effective. The maximum lift-to-drag ratio was a little worse with fur, but that reflects an increase in drag, not a decrease in lift. The former may be no detriment to performance since in flight these animals gradually increase their angles of attack to very high values, around 60°, to limit the speed of impact by using high drag for braking.

Insect wings. The characteristics of three insect wings of different sizes —those of a locust, a crane fly, and a fruit fly—are listed in Table 11.1; their complete polar plots make up Figure 11.12. These insect wings are basically beating rather than gliding wings, so their operating situation is exceedingly complex. Both velocity and angle of attack vary from base to tip, and only in the case of the locust hindwing were the data obtained under reasonably realistic conditions. Table 11.1 also gives data for the elytra, or wing covers, of a beetle; these are held out as fixed wings (tipped well above a horizontal plane), with the hindwings acting as pusher-propellers behind them.

And only in the case of the locust do we have good information on the actual angles of attack which occur during flapping flight. A locust hindwing achieves its best lift-to-drag ratio at an angle of attack of 7°, but during the downstroke it operates at an angle of about 13° (Jensen, 1956). However, it is not clear that maximizing the lift-to-drag ratio is the best strategy at all points in a complex stroke. More likely, a beating wing should use a high angle of attack during the downstroke, where drag is not entirely detrimental, and a lower angle of attack on the upstroke, where drag with respect to the local wind amounts, in part, to negative lift with respect to the earth. Also, operation at a higher angle of attack than that giving the best lift-to-drag figure will provide additional lift; the latter implies that wings can be smaller; smaller wings are cheaper and may be able to beat faster due to their lower moment of inertia, providing more lift yet.

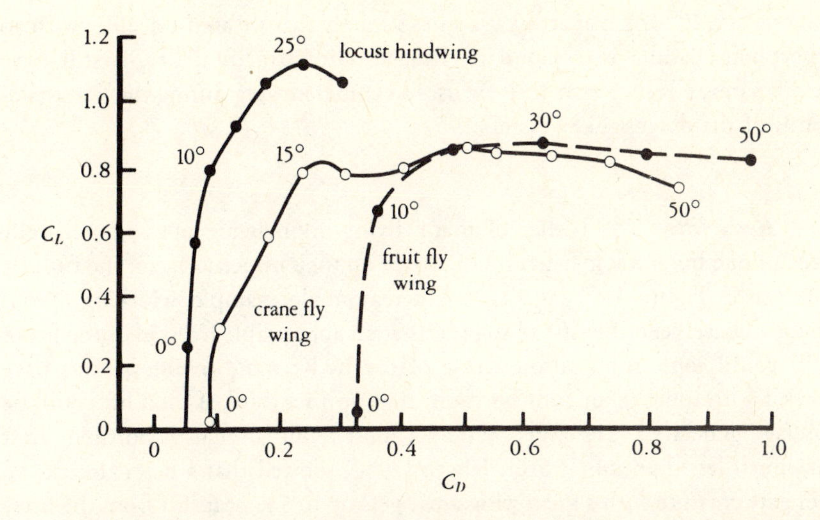

Polar diagrams for the wings of three insects—a locust hindwing (Jensen, 1956), a crane fly (Nachtigall, 1977b), and a fruit fly (Vogel, 1967b). Further data is given in Table 11.1.

FIGURE 11.12

In comparing the three insect wings, we see the pernicious effect of low Reynolds number on performance. Drag gets much worse relative to lift as the Reynolds number decreases, and the drag becomes almost all profile drag rather than being about half induced drag. Thus, making longer and more slender wings is of little help. The shift of the polar curves to the right may come from an increase in the profile rather than induced drag in *Petaurus,* but the effect is the same. The best operating point is pushed up to higher angles of attack, in fact to the stall point for an insect the size of a fruit fly. The benefits of stall-resistant design get more important, and there is some evidence for anti-stall devices in small insects. A fruit fly wing doesn't stall in the usual sense; its lift just levels off above a certain angle of attack. This levelling occurs because with so much drag, there is simply very little momentum flux behind the wing to be deflected downward. But a flat or cambered plate at the same Reynolds number does stall, and the lift coefficient drops at angles above about 25°. It is not entirely clear what is different about the real wing, although some indirect evidence points to the tiny hairs (microtrichia) on the surface of the wings (Vogel, 1967b). The fact that camber has a marked effect even at Reynolds numbers around 200—a cambered wing gets 50% more lift than a flat one at a 20° angle of attack—argues strongly that we aren't dealing with an amorphous paddle in a cloud of attached air. Structural details still have aerodynamic effect. So the fruit fly uses a cambered wing during the latter two-thirds of the downstroke (Vogel, 1967a).

Body lift. The bodies of many flying animals are not symmetrically streamlined but are somewhat more convex on their upper surfaces; the fruit fly fuselage in Figure 7.3 (page 123) is a reasonable example. Hocking (1953) seems to have been the first to suggest that an appreciable fraction of the lift of a fly could come from airfoil action of the body; more recent workers have viewed with some skepticism his claim that up to a third of total lift could be fuselage generated. The aspect ratio for most flying animals is horrible, after all, much less than unity. Still, Jensen (1956) showed that a desert locust, an elongate creature with a great wing area, gets up to 5% of its lift from the body and has a lift-to-drag ratio for the body of 0.8. A fly has a more promising shape than a locust for obtaining lift from the body, so Hocking's suggestion shouldn't be dismissed. And Nachtigall (1964) found that the body of a cockchafer beetle could produce substantial lift, but with a maximum lift-to-drag ratio of only 0.4 and that with the front end pitched up about 40°. Whether such angles do occur in forward flight is unknown, as is the effect of the wind generated by the beating wings as it passes the body.

More interesting, perhaps, is the "swooping" flight of small birds, in which the wings are folded and held close to the body. In zebra finches, according to Csicàky (1977), as much lift as drag is produced when the body is inclined with a head-up pitch of 20°. Body lift should be mainly significant in descending body-gliding. If a bird keeps the initial heading which gave it a negligible body angle at the start of a body-glide, its increasingly downward trajectory would increase the body angle and lift as the glide or swoop progressed. Rayner (1977) has recently analyzed the phenomenon from a theoretical viewpoint.

The limits of circulation. The practical limiting Reynolds number for airfoils which produce lift by creating circulation cannot be very much lower than that of fruit flies. The only data I know of at much lower Reynolds numbers are those of Thom and Swart (1940) for cambered airfoils at $Re = 10$ and $Re = 1$. At a Reynolds number of 10, the best lift-to-drag ratio was only 0.43: far worse than the fruit fly's 1.8; it occurred at an angle of attack of 45°.

At a Reynolds number of one, the best lift-to-drag ratio was 0.18, again at about 45°. Despite very high angles of attack, the deflection of the wake of the airfoil was never more than about 11°. We have entered a world in which fluid is highly resistant to being set into circulation by any fixed (non-Flettner) airfoil, so thick is the attached cloud of fluid on any object. As far as I know, the use of a Flettner type of rotating airfoil has never been investigated at Reynolds numbers anywhere near this low.

Mechanisms other than circulation which might generate lift are still available at Reynolds numbers in the range of unity, but some possibilities are unpromising. One might envision a scheme whereby a flat plate was moved upwards while oriented parallel to the flow and downwards while oriented perpendicular, with lift produced as a consequence of the difference in drag between the two strokes. The main difficulty with this idea is that at low Reynolds numbers the difference in drag is small—about 1:1.5 at $Re = 1$ (Thom and Swart, 1940). A widely used textbook of physiology suggested alternative strokes at different speeds, a fast downstroke following a slow upstroke, with lift produced as a result of the dependence of drag on the square of velocity. But at such low Reynolds numbers drag is well known to be dependent on the first rather than the second power of velocity, so the scheme produces no net force at all.

What ought to work are alternative strokes of an appendage whose area varies between power and recovery strokes. We might call this the Mary Poppins umbrella-folding flight system, using as a model a loose umbrella thrust rapidly up and down. The system is known in nature; Nachtigall (1974) gives a good explanation of how the arrangement of flexible hair fringes on the legs of swimming beetles produce a propulsive force. Presumably many tiny crustaceans use the same system, at least in part. In at least three orders of small insects there has been a convergence toward wings of structure similar to that of beetle legs (Figure 11.13), as first noted by Thompson (1942), who felt that this structure allowed the insects to "row" through the highly viscous air. More recently, Kuethe (1975) has suggested that the most common of these bristle-winged groups, the thrips (*Thysanoptera*) really use their wings in a manner analogous to the waving flagella of spermatozoa—waves of bending may be propagated laterally to generate a force. In either case, it seems that a sheet of bristles, by virtue of a thick boundary layer, functions much like a solid plate at low Reynolds numbers. Some actual experimental work on these common but aerodynamically peculiar thrips would certainly be worthwhile.

We'll resume the discussion of what can be done to effect propulsion at low Reynolds numbers in Chapter 13.

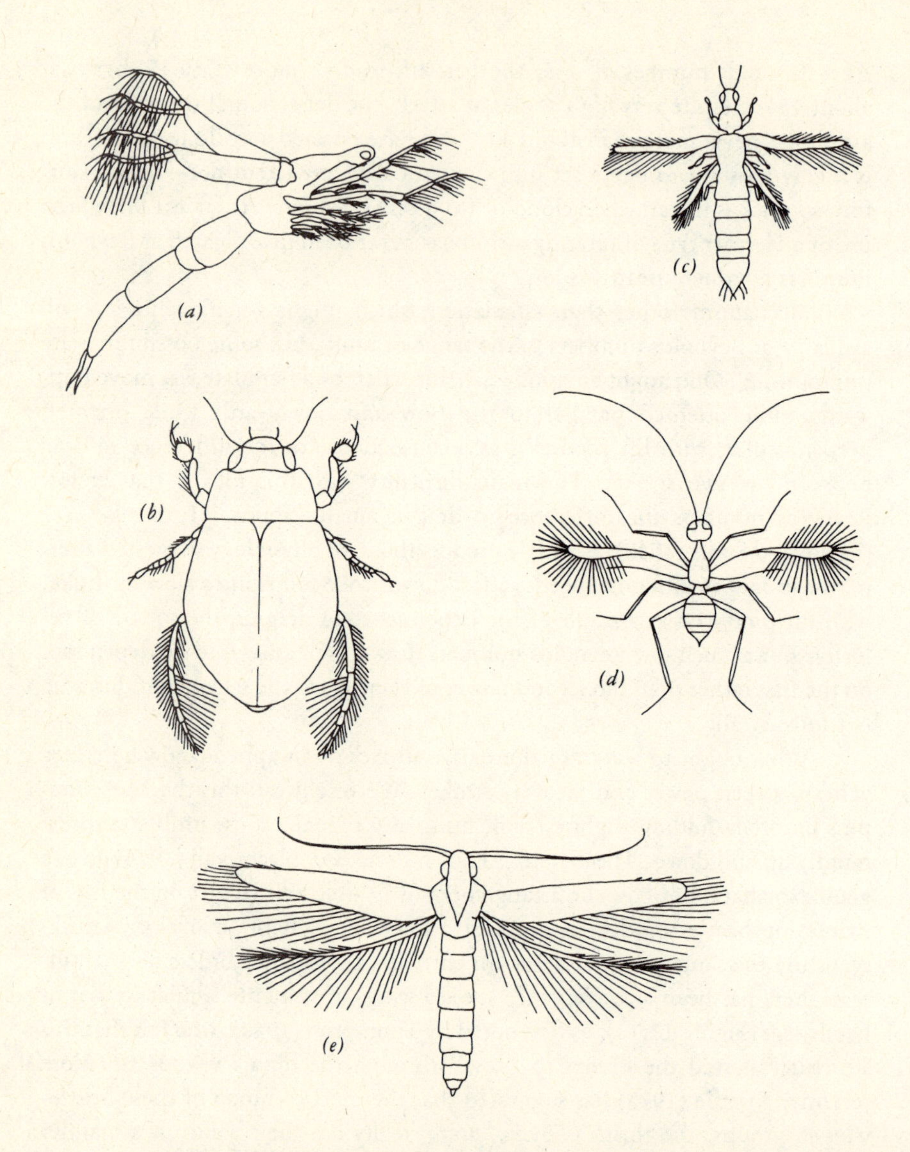

Animals with propulsive appendages which lack a continuous membrane: (*a*) a rather aberrant cladoceran crustacean; (*b*) a diving beetle; (*c*) a thrips; (*d*) a mymarid wasp; (*e*) a tiny moth. All but the beetle are quite small.

FIGURE 11.13

CHAPTER 12

gliding and thrust production

AN AIRFOIL IN A HORIZONTAL AIRSTREAM produces an upward force, but, unfortunately, it experiences drag as well. So unless acted on by some external force, it soon loses airspeed, loses lift, and falls down. This chapter is about the ways in which a lift-producing flier can arrange either to descend slowly and steadily or to avoid descent altogether. The subject is complicated, so the discussion will necessarily be somewhat superficial. In particular, I'll give little attention to the actual kinematics of flapping flight, where what is known defies my attempts at facile summarization. Similarly, I'll not discuss fish locomotion, another situation in which circulation generates lift which is used in propulsion. For details on the former, the reader should see Jensen (1956), Nachtigall (1966) and Bilo (1971, 1972); for a look at the latter, Lighthill (1969), Webb (1975), Alexander (1977) and Hoar and Randall (1978) should be useful.

gliding

Certainly the most obvious steady-state sort of flight is what is called "simple gliding." It's most easily defined as a situation in which an airfoil moves through the air, losing altitude just rapidly enough to maintain both a steady speed and a vertical force the same as its weight. The air is assumed to be still. In simple gliding no forces are unbalanced, so the aerodynamic resultant force, the outcome of lift and drag, must balance precisely the weight of the craft.

From the statement that weight and aerodynamic resultant are equal and opposite, it is an easy matter to derive the angle at which the craft descends. With the aid of Figure 12.1 we can see that both the resultant force (R) and the weight (W) are obviously vertical. The oncoming wind (U), equal and opposite to the path of the craft, and the drag (D) are both obviously perpendicular to the lift (L). Therefore, the angle between the resultant and the lift must be the same as the angle between the horizontal and the wind, by elementary geometry. The former is the angle whose tangent is the ratio of drag to lift. The latter is, of course, the angle of descent, called the *glide angle*. Thus the glide angle is set by the lift-to-drag ratio:

$$\cot \theta = \frac{L}{D} = \frac{C_L}{C_D} \qquad (12.1)$$

FIGURE 12.1

The relationship between the glide angle, θ, and the lift and drag forces for an airfoil.

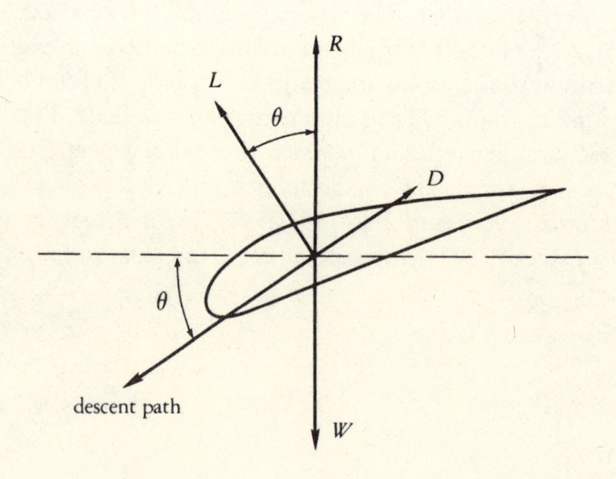

It is a pleasure to discover something as simple and familiar as the lift-to-drag ratio taking on such a central role! To minimize the glide angle, one needs merely to maximize the lift-to-drag ratio. This, by the way, is the basis for the indirect determination of the ratios cited in the last chapter: glide angle and flying speed can be measured on freely gliding birds even if lift and drag cannot.

If the air is otherwise still, then minimizing the glide angle maximizes the distance a simple glider will go before reaching the ground. An albatross with a lift-to-drag ratio of 18, released 1 km above ground, could glide a horizontal distance of 18 km. A phalanger (*Petaurus*) with a best lift-to-drag ratio of 0.6 would go at most 0.6 km. Long, skinny wings and the consequent low induced drag are clearly advantageous, and whether bird or person-carrying craft, good gliders typically have wings of high aspect ratio. A high-performance sailplane (Schemp-Hirth SHK) with an aspect ratio of 20 can go 39 km horizontally for each kilometer of descent (Tucker and Parrott, 1970).

This business of distance traveled versus descent is initially counter-intuitive. For moderate and high Reynolds numbers, speed affects both lift and drag in much the same manner, so the lift-to-drag ratio is not particularly speed-dependent. Thus the glide angle is also largely independent of speed; a very heavy glider descends along nearly the same path as a light one! Weight, though, must be balanced by lift, and the latter is roughly proportional to the square of speed. So, while the heavier glider may fly the same path, it must of necessity travel faster; and it covers its horizontal distance in less time. In general, a rapid descent is disadvantageous, and most gliders, living or not, seem to avoid excessive weight. Sometimes, however, staying aloft poses hazards; and military gliders are heavily loaded affairs, balancing the pitfalls of high-speed impact against the insecurity of a long, slow flight.

The relationship between glide angle and lift-to-drag ratio also explains why gliding animals are fairly large. Small size means relatively more profile drag and, as we've seen, a lower maximum L/D. Therefore, small size almost invariably implies a higher glide angle. An albatross may descend at about 3°; a falcon can do no better than 5.5°, a pigeon about 9.5°. Insects are far worse: even if the fuselage contributed no drag at all, a crane fly would descend at 15° and a fruit fly at nearly 30°. Simple gliding, based on the production of lift, just isn't very good at low Reynolds number. Perhaps that is the rationale for the use of "swooping" with closed wings in small birds (Csicaky, 1977). In the size range of small insects, gliding is largely replaced by "ballooning" or "parachuting": maximizing drag, ignoring lift, and covering distance by maximizing time aloft in the wind while traveling horizontally at the speed of the wind. This is the world of small spiders, some insect larvae, and a great many wind-dispersed seeds.

A convenient way of viewing the performance of a glider is with a so-called "glide polar," a curve on a graph with flying speed on the abscissa and sinking speed on the ordinate, as in Figure 12.2. On such a graph, a line from the origin tangent to a craft's curve is a line of least slope or best glide angle; the tangent point determines the flying speed (and sinking speed) at which that best glide angle occurs. Increasing the weight of the creature or craft would shift its curve to the right and downward without much change in the slope of the tangent line. The lines for individual craft are not straight, mostly as a result of the way drag varies with speed. At low speeds induced drag is relatively high and at high speeds profile drag is dominant, with a minimum drag coefficient in between.

FIGURE 12.2

Glide polars for a variety of craft—SHK sailplane, vulture, and falcon (Tucker and Parrott, 1970). Also shown are single data points for some other gliders: (1) Rüppell's griffon vulture (Pennycuick, 1971); (2) Andean condor (McGahan, 1973); (3) desert locust (Jensen, 1956); (4) monarch butterfly (Gibo and Pallett, 1979); (5) flying phalanger (Nachtigall, 1979a).

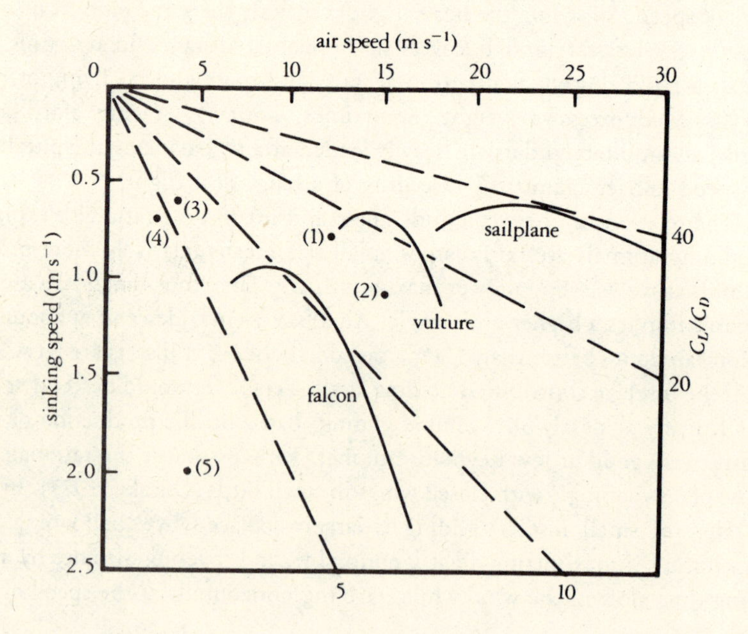

The glide polar gives a useful comparative view of several other parameters. Time aloft may be more important than maximum still-air glide distance. A horizontal line tangent to a curve is tangent at the flight speed for which time aloft is maximized; its intersection with the ordinate gives the sinking speed from which the time aloft may (knowing altitude) be figured. The greatest time aloft is achieved at a lower speed than the least glide angle, and a slow glider can do as well as a faster one with a better lift-to-drag ratio. Thus if time aloft is the important factor, birds don't look quite so bad in a comparison with sailplanes. Even insects, if large, aren't disastrous. A desert locust, despite a lift-to-drag ratio of 8.2 for its hindwing (and much less for the entire insect) descends at 0.6 m s^{-1}—in the same range as a vulture or sailplane (Jensen, 1956). A monarch butterfly has an overall lift-to-drag ratio of 3.6, and therefore a glide angle of 15.5°; but it travels at only 2.6 m s^{-1}, and consequently descends at only 0.68 m s^{-1}—again in the same range (Gibo and Pallett, 1979).

Another parameter which emerges from the glide polar is the minimum glide speed. It is the left end of the curve, the point at which the maximum wing area together with the maximum lift coefficient provide just enough lift to balance the weight.

Just where an animal might choose to operate on its glide polar depends in part on what the wind is doing. If the animal wishes to maximize distance traveled and is assisted by a tail wind, a lower flight speed than that which would maximum L/D is the best tactic. If it faces a head wind less than its flying speed, it is best off picking a higher flight speed than that which gives the maximum L/D. If it faces a head wind greater than its flying speed, it should land as quickly as possible (Tucker and Parrott, 1970).

Simple gliding may be used for seed dispersal; the most famous example is the seed-leaf (really the fruit) of the Javanese cucumber, *Zanonia macrocarpa* (Figure 12.3). It is 12 to 15 cm in span, about 50 cm^2 in area, and weighs, according to Bishop (1961), about 50 g. (The weight of 0.175 g given by Hertel (1966) is certainly in error; no figure in Hertel's book should be accepted without confirmation.) The *Zanonia* seed-leaf formed the model for a successful series of gliders around the turn of the century. It seems that after Otto Lilienthal was killed in a birdlike glider (1896), attention was focused on the need for stability and on the obvious instability of birds. The instability of a bird is almost certainly concomitant with its maneuverability and highly sophisticated controls (Smith, 1952). The seed-leaf was, of course, intrinsically stable. But the model did not lend itself easily to the addition of an engine, and the design was returned to the botanical back burner.

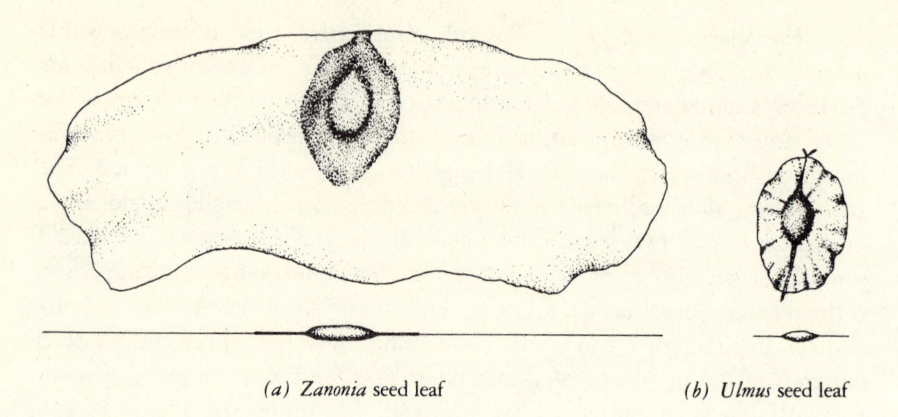

The gliding seed leaves of (*a*) the Javanese cucumber, *Zanonia macrocarpa,* and (*b*) the elm, *Ulmus rubra,* in plan view and (somewhat idealized) side view.

FIGURE 12.3

The rarity of the *Zanonia*-type gliding seed has been explained by McCutchen (1977) as due to its inferiority compared to autogyrating or autorotating designs in the presence of any turbulence. Such simple gliders should be useful mainly in the still air of the interior of a rain forest. In fact, smaller and less elegant plain winged seeds with some central concentration of mass are not all that uncommon: examples are the fruits of the hoptree, *Ptelea trifoliata* and of the elms (*Ulmus*) (Figure 12.3). These seem to be simple gliders, if a bit less stable than *Zanonia.* Burrows (1975) has pointed out that they ought to be subject to the same "fugoid" oscillations as some aircraft—an undulating flight path caused by a continuous interchange of kinetic and potential energy. In a fugoid, a steep glide increases speed and thus lift, bringing the craft onto a course of less and less rapid descent until it loses so much speed and lift that it again turns downward. But experimental work on such fruits is sorely lacking.

Gliding, of a sort, is practiced by many animals other than birds, bats, and insects. Flying squirrels and phalangers have already been mentioned. There is also a flying lemur (*Cynocephalus*), and a flying snake (*Chrysopelea*). Gliding quality varies, and these creatures may be closer to parachutes than gliders, but all have some area-increasing adaptations and appropriate behavioral responses (Oliver, 1951). The arboreal lizard, *Anolis carolinensis,* is reported to have a glide angle of 68° and the tree frog, *Hyla venulosa* of 57° (Oliver, 1951); neither shows particularly strong anatomical adaptations. Probably the most spectacular reptilian glider is the flying lizard of the Philippines, *Draco volans,* which has a broad patagium supported by rib extensions. Hairston (1957) estimated its glide angle at 11°, certainly well within the range of the flying squirrels.

soaring

Atmospheric motions are, of course, far more varied and complex than just steady winds. For organisms, the speeds of air movements are of the same order as or higher than their own flight speeds. The situation must enforce respect for local air movement but also permits all sorts of schemes for capitalizing on atmospheric motion. Air in motion with respect to the ground provides a source of energy; and flight, per unit time, is the most energy intensive form of animal locomotion. In simple gliding, an organism heads earthward, the only real question being its time, speed, and place of arrival. In soaring, as commonly defined, altitude is maintained or even gained by using the energy of the wind.

For the purposes of explanation, at least, it is useful to distinguish between two general forms of soaring. In the simpler, "static soaring," a craft or creature takes advantage of a region in which air is moving upward. Somewhat trickier is "dynamic soaring," which involves no upward air movement but only a temporal or spatial gradient in wind velocity from which an animal must extract the power necessary to stay aloft. Pennycuick (1972) gives a good account of the various sorts of soaring; I'll just briefly mention the possibilities in order to emphasize their diversity.

The simplest sort of static soaring is "slope soaring," or flying in air which is moving upward along the side of a hill. The updraft may extend above the crest of the hill and may have a fairly complex structure, but it's possible to continuously sink with respect to the local air without sinking with respect to the earth (Figure 12.4a). Slope soaring is the usual sort practiced by hang-gliders. Sometimes a version of slope soaring is possible in the lee of a range of mountains, in the large standing vortices created by wind over the peaks; it also appears possible in the lee of ships.

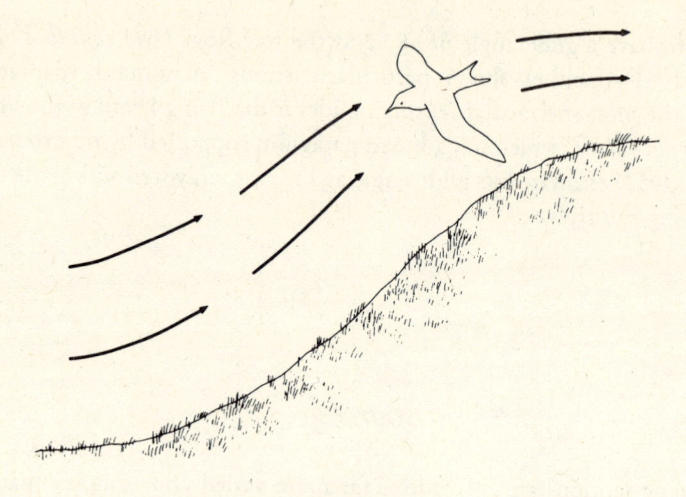

(*a*) slope soaring

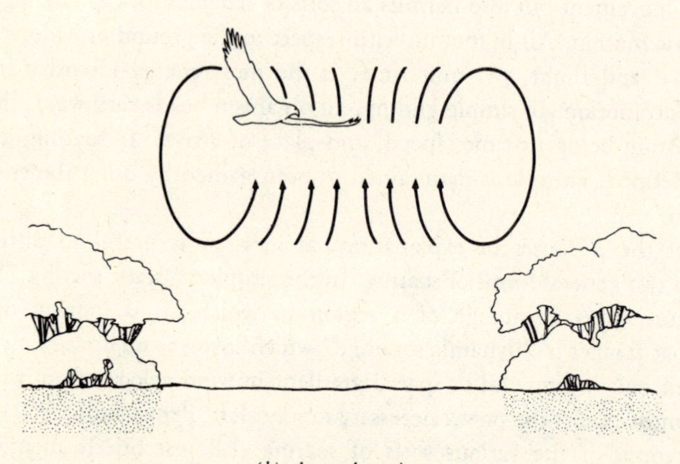

(*b*) thermal soaring

(*a*) Slope soaring: if the air moves up the hill, the bird descends with respect to the local air but does not descend with respect to the earth. (*b*) Thermal soaring: a bird (here rather larger than life) can, by circling within a rising vortex ring, descend with respect to the local air but remain within and ascend with the ring.

FIGURE 12.4

A second kind of static soaring is "thermal soaring." A "thermal" is a horizontal vortex of rising air rather like a smoke ring. The toroidal vortex has air rising in the middle of the ring and descending peripherally as the whole structure rises. Thus if an animal can circle in the inner portion of the vortex ring it can continously sink with respect to the local air but not fall out of the invisible elevator (Figure 12.4b). In this area of central North Carolina, plowed fields surrounded by forests seem to heat up and generate a continuous set of thermals. On summer afternoons these vortices are marked by circling turkey vultures. My very casual observations suggest that the interstate highway system has been a bonanza for the buzzard business: the pavement generates thermals and the traffic provides fresh food. A vulture just has to glide along the highway, ascending on a thermal every once in a while. Even without highways, cross-country thermal soaring is quite common. White storks, according to Pennycuick, use the scheme for long-distance migration. Desert locusts clearly employ thermal soaring on suitable occasions, with pauses of up to 150 seconds in the beating of their wings (Roffey, 1963). And monarch butterflies could not make their long migrations without using both thermal and slope soaring (Gibo and Pallett, 1979).

Of the possible types of dynamic soaring, the best understood is that done in the wind gradient near the surface of the ocean. Cone (1962) provides an analysis of such soaring; the possibility was recognized a hundred years ago by Lord Rayleigh (1883). This "gradient soaring," as done by an albatross, has two alternating phases: a downwind, downward glide and an upwind, upward glide (Figure 12.5). A bird well above the earth or ocean glides downward,

FIGURE 12.5

Dynamic soaring in a wind gradient. The bird alternately ascends and descends, extracting energy from the gradient by quite a different scheme than that discussed in Chapter 4. The bird reverses its heading at the two marked points.

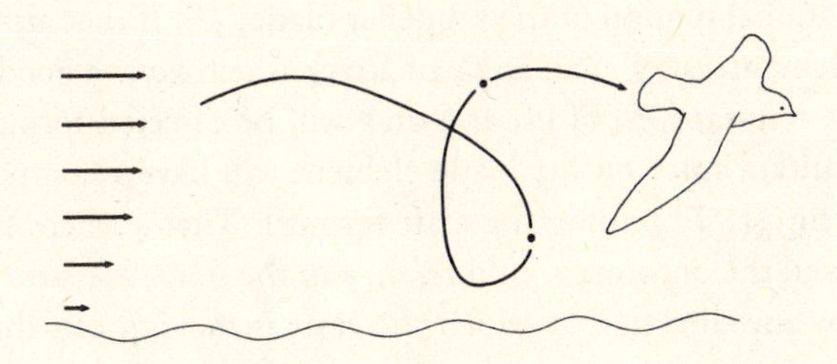

gaining speed as it loses altitude and encounters slower, lower air. Nearly at the lowest point, it turns into the wind, dives further, and then begins to ascend, facing upwind. As it ascends, it slows with respect to the earth, but speeds up with respect to the surrounding air as it encounters faster oncoming air at higher altitudes. As the gradient becomes less pronounced and the air's speed more uniform, the bird turns downwind and begins its downward glide again. The overall motion is a series of vertical loops progressing downwind.

Other sorts of soaring are also possible: a bird ought to be able to use a temporarily unsteady wind to permit soaring by flying through gusts. Or it ought to be able to fly in and out of breeze fronts and other such irregularities. A large amount of energy is dissipated in the lowest part of the atmosphere, and only appropriate machinery and behavior are necessary to extract it.

thrust from flapping

The notion of an oncoming wind which was not horizontal proved central in explaining how simple gliding works. Still, that oncoming flow was never far from either the horizontal or the line of progression of the craft. Generating thrust requires that an airfoil encounter a flow quite far from the line of progression of the whole craft. Such a situation can be made to happen if (*and only if*) the airfoil moves with respect to the line of progression of the craft: it must rotate, flap, wiggle, undulate or otherwise meander about. But then, in a fine tour de force, thrust emerges from a structure which produces lift and suffers drag.

The origin of thrust. Before turning to reciprocating airfoils, let us consider a simpler case, that of an element of a rotating propeller, turning about an axis which happens to coincide with the overall movement of the craft (Figure 12.6). The oncoming wind seen by the blade element, U_w, results from the combination of the wind due to the craft's forward movement, U_f, and that due to the rotational motion of the propeller blade, U_r. If that airfoil, the propeller blade element, is set at an angle of attack which gives a good lift-to-drag ratio, then the resultant, R, of lift and drag will be directed forward. In other words, the resultant force on the blade element will have a component which we'll call the thrust, T, pulling the craft forward. That's where thrust comes from. Only when the oncoming wind *relative to the blade element* comes from above or below the direction in which the craft is moving can thrust be pro-

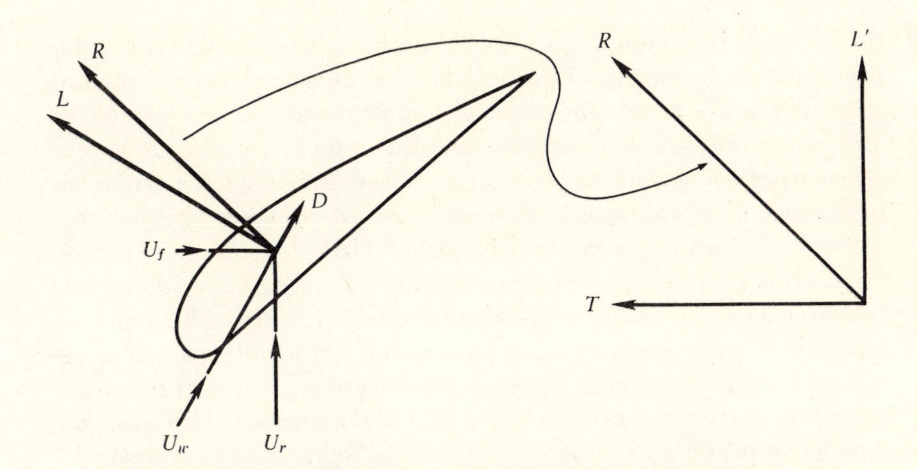

The origin of thrust. If the net oncoming wind strikes an airfoil from a sufficiently non-horizontal direction, the aerodynamic resultant may be tipped forward and may thus have a thrust (negative drag) component. The thrust component appears explicitly when the resultant is reanalyzed, as at the right.

FIGURE 12.6

duced, just as a conventional sailboat can't make headway directly into the wind.

The other component of the resultant force, that normal to the thrust, can be viewed in several ways. If the blade element is moving downward, then it is a kind of lift. But for a rotating propeller, downward movement of one blade element is always balanced by upward movement of one or more elements of the one or more other blades. In an aerodynamic sense, therefore, this other force is not of great consequence for a rotating propeller. Still, it has to be counteracted in order to keep the propeller turning. Thus the force (L' here) multiplied by its distance from the axis of the propeller's rotation gives the torque which the engine must supply to keep that element of the blade moving. If the analysis is extended to all elements of all blades, the sums give the overall thrust which (at steady speed) must just balance the drag of wings, fuselage, and all else except the propeller; and they give the torque (and power) required of the engine.

Practical propellers are twisted along their length from hub to tip. The reason should be obvious: near the hub, the direction of the oncoming wind is mainly a matter of the wind due to the craft's progression, while nearer the tip the direction of the oncoming wind will be quite different, with a large component due to the rotation of the propeller. To maintain an optimal angle of attack all along its length, the propeller must be twisted. At very low speeds, little twist is necessary, with more twist needed as speed is increased if the propeller rotates at a constant rate. Aircraft of the non-living sort make a crude approximation of variable twist by allowing rotation of the entire rigid propeller around its long axis, so-called "variable pitch." Flying organisms with beating wings actually change the degree of lengthwise twist. Such adjustment, though, is absent in beating wings which operate at low Reynolds numbers; a fruit fly wing is operated without lengthwise twist. Apparently, at low Reynolds numbers the lift-to-drag ratio is no longer sufficiently sensitive to small changes in angle of attack for twist to matter much. For a beating wing, of course, the direction of twist must be reversed after each half-cycle if twist is employed.

Advance ratio. It should be clear that no thrust can be generated if the relative wind striking a rotating propeller blade is too close to the axis of progression. This has the effect of setting an upper limit on the speed of the craft, a limit which can be raised by using a longer propeller or a faster rate of rotation. Critical, then, is the ratio of the forward speed of the craft and the speed of the blades of its propellers. The usual index of this quantity is the *advance ratio*, J:

$$J = \frac{U_f}{nd} \tag{12.2}$$

where U_f is the craft's forward speed, d is the diameter of the propeller, and n its rate of revolution. For ordinary airplanes, advance ratios are generally under 4.0. For a propeller without provision for pitch adjustment, the angle of attack of its blade elements will vary with the advance ratio and its operating efficiency will depend very strongly on flying speed. Mises (1945) gives a very good introduction to the blade element view of a propeller as well as to advance ratio and its relationship to propeller efficiency.

For the reciprocating propellers of flying animals, formula (12.2) is clearly inadequate. We must replace d by twice the wing length, l, and replace the rotation rate by the wingbeat frequency times the amplitude of the wingbeat (the stroke angle, ϕ) as a fraction of 180°:

$$J = \frac{U_f(90°)}{nl\phi} \tag{12.3}$$

Using this formula, the advance ratio for a desert locust in fast forward flight is about 1.7 (data from Jensen, 1956); for a fruit fly it is about 0.77 (Vogel, 1966). The latter reflects the lower maximum lift-to-drag ratio of a wing at low Reynolds number: with a lot of drag, it takes a lot of wing beating to get a decent forward component on the aerodynamic resultant force. The wingbeat frequencies of small insects are, in fact, very high. Fruit flies beat their wings at around 200 times per second, mosquitoes at 300 to 600 (Greenewalt, 1962), midges up to about 1000. The all-time record for the rate of reciprocation of any animal appendage is held by a midge with its wings almost completely amputated to reduce the moment of inertia: 2218 beats per second (Sotavalta, 1953; not Guinness).

The advance ratio gives a handy way of making a crude estimate of the top flying speed of an insect. One merely assumes some value of J depending on the size of the insect and a moderate amplitude (say 120°). Length is easily measured, and data on frequency have been collected by Greenewalt (1962), so U_f can be calculated. The results may be crude, but they're better than some of the grotesque guesses in the entomological literature!

Flapping flight. In birds, bats, and insects, the flapping wings combine the functions which airplanes divide between fixed wings and propellers. Thus animal flight involves a kind of cross between a fixed-wing craft and a helicopter. In practice, the wing stroke isn't a simple reciprocating up and down analog of a vertical propeller but instead usually takes the form of an inclined ellipse or figure eight (Figure 12.7). The downstroke moves forward as well as downward; it produces mainly lift but also some thrust. The upstroke goes backward as well as upward, and it produces mostly thrust but also some lift (Jensen, 1956).

FIGURE 12.7

A somewhat idealized path of an insect's wing as it might be projected onto the surface of a cylinder. The solid curve uses the insect's body as a frame of reference; the dotted lines refer to the earth as the insect flies along. Notice that the wings nearly meet at the top but not at the bottom of the stroke.

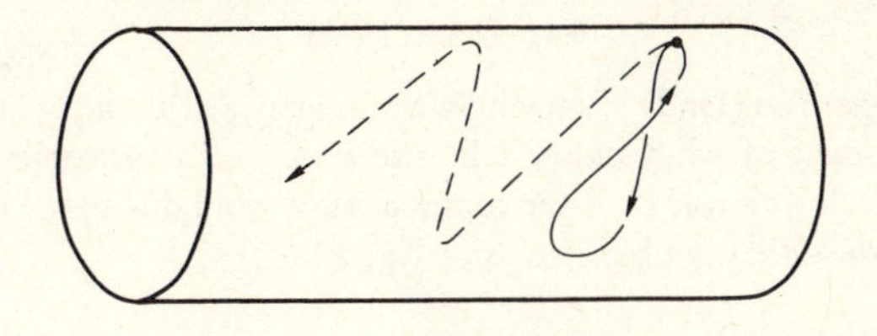

In very low-speed flight or in hovering, the stroke plane is almost fully horizontal in most flapping fliers: the wings go back and forth rather than up and down, although it is customary to refer to upstroke and downstroke as before. For the reasons discussed in the section on induced drag in Chapter 11, hovering requires the greatest expenditure of energy per unit time of any form of flapping flight. Indeed, with a limit of about 250 W kg^{-1} for the power output of muscle, hovering for more than very brief periods is not possible for creatures much larger than hummingbirds (Weis-Fogh and Alexander, 1977). Concomitantly, the amplitude of the wingstroke is greatest in hovering, and the assumptions of any theory which attempts to account for the aerodynamics of flapping flight are put to their harshest test. Two good papers on the aerodynamics of hovering are those of Weis-Fogh (1973) and Ellington (1978). It might be noted that the shift of stroke plane is mainly accomplished in some hoverers (hummingbirds, fruit flies) by tilting the entire body so the head is uppermost, but in others (desert locusts, hover flies) by altering the plane of the stroke with respect to the rest of the body.

four kinds of moving airfoils

Propellers aren't the only sort of moving airfoils which deal with thrust and power; four arrangements are possible, and at least three out of the four have clear biological examples. Figure 12.8 considers a blade element of each of these devices; they can be categorized as follows:

1. A propeller has, ideally, a horizontal shaft and inserts power into a horizontal airstream.

2. A windmill also has a horizontal shaft but extracts power from the horizontal airstream.

3. A helicopter rotor has a vertical shaft and inserts power into the airstream, taking air from above and thrusting it out below the plane of the rotor.

4. An autogyro, descending, has a vertical shaft and extracts power from the airstream; it takes air from below and retards its passage upward through the plane of the rotor.

FIGURE 12.8

(*Facing page*) Four kinds of moving airfoils, analyzed as in Fig. 12.6. U_i is the component of wind induced by the action of a helicopter blade in pushing air downward; U_d is the component of wind due to the descent of an autogyro. Other symbols are as in Fig. 12.6.

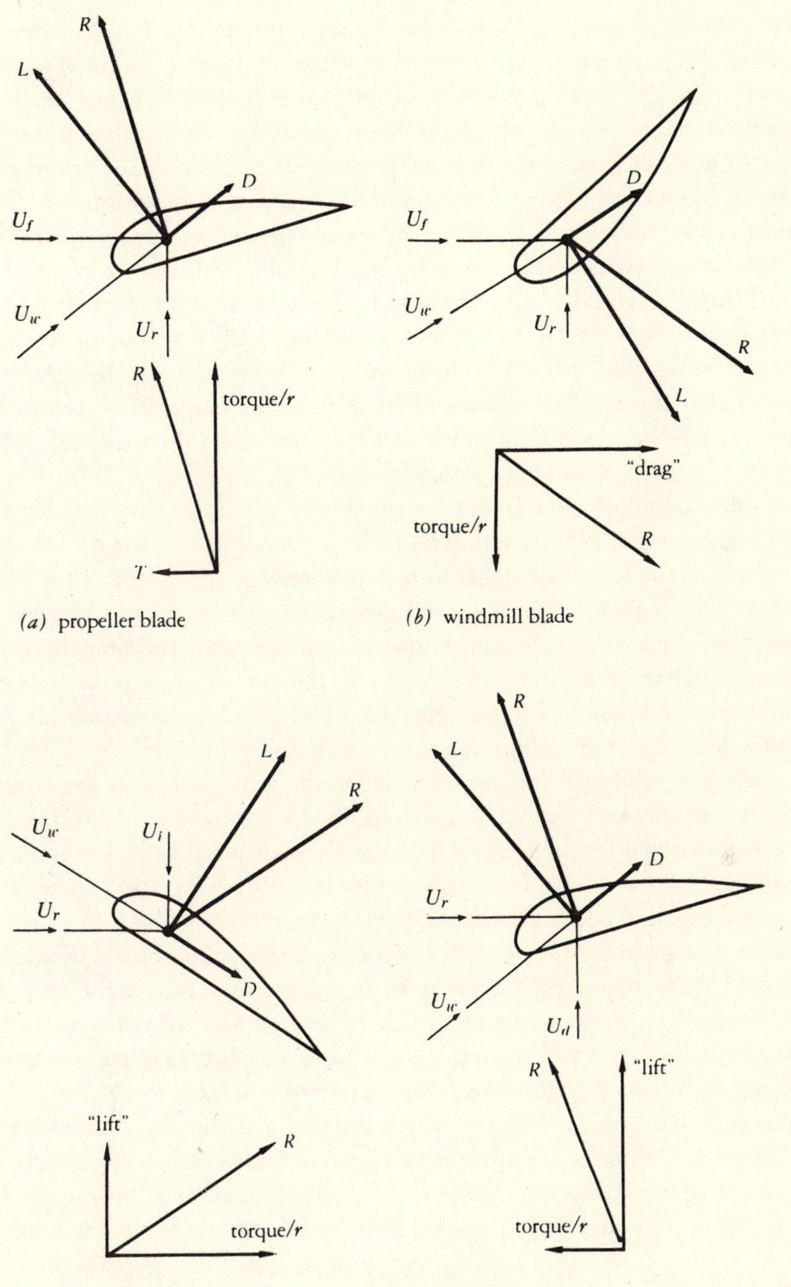
R
L
D
U_f
U_w
U_r
R
torque/r
T
(a) propeller blade
D
U_f
U_w
U_r
R
L
"drag"
torque/r
R
(b) windmill blade
L
R
U_w
U_i
U_r
D
"lift"
R
torque/r
(c) helicopter blade
R
L
D
U_r
U_w
U_d
R
"lift"
torque/r
(d) autogyro blade

We've already mentioned the propeller (Figure 12.8a), and its operation is essentially the same whether it rotates as an airplane propeller, oscillates as a beating wing, or reciprocates as the dorso-ventrally elongate tail fin of a scombroid fish. The flow comes from ahead, and the aerodynamic resultant generates a force which the engine must counteract. Notice that to keep a positive angle of attack, an airfoil which is not symmetrical top to bottom must have its convex side facing upstream and its concave surface downstream. Only when one considers the direction of the oncoming wind as seen by the moving blade does the arrangement appear logical. The rule, by the way, is general and quite useful for all axial fans and blowers in home or lab: *blades should be concave downstream.* And you can't reverse the direction a fan blows by reversing the fan hub on its shaft; either the motor must be reversed as well or the twist and camber of each blade must be reversed. I once took an ordinary ventilating fan and reversed the hub on the shaft; the wake speed dropped from 80 to 20 cm s^{-1}; the direction was, naturally, unchanged.

A windmill (Figure 12.8b), on the other hand, must have the "lower" flat or concave side of its airfoils facing into the wind, just as (and for the same reason that) the sails of a sailboat work best when concave upwind. Thus a propeller makes a poor windmill or generator unless the blades and hub are reversed on the shaft. If the hub is reversed on the shaft, the propeller-now-turned-windmill will, of course, revolve in the opposite direction. In windmills, the wind produces a resultant force which is composed of a torque (per unit radius), now the useful component which drives whatever the windmill is connected to, and of a drag force parallel to the wind. In a sense the latter is aerodynamically irrelevant although it tends to blow the windmill downwind—a serious matter if the supports for the rotor are inadequate. Wheel-and-axle arrangements are unknown in biology outside of the bacteria, and I know of no close biological analog of a windmill. Some leaves seem to translate in a circle in a wind, restrained by a petiole which can't rotate freely, but the significance (if any) of these movements appears to be unknown. One might look for reciprocating rather than rotating airfoils, by analogy with the action of beating wings as propellers. Probably, aquatic organisms are better candidates than terrestrial ones, and probably the power extracted will have to be invested in something like fluid movement which requires a minimum of transducing equipment: it's clear that muscle cannot be used backwards as a mechanical-to-chemical-energy transducer. There are "biological windmills" in the general sense of devices that extract energy from velocity gradients in fluid media

(Vogel, 1978a), but they don't seem to include any devices, even quixotic ones, which do so by direct application of lift-producing airfoils. Birds performing dynamic soaring do extract energy from velocity gradients, but again the mechanism is a long way from that of a windmill.

Helicopter rotors (Figure 12.8c) are the technological analogs of hovering fliers, the main difference being their rotation instead of reciprocation. The wind now comes from above, and the airfoil must be concave downward and downwind with the leading edge elevated. Rotation produces an upward resultant; the latter resolves, in turn, into a vertical force (lift with respect to the earth) and a torque which, as with a propeller, the engine must counteract. Since there is typically no downward component of the wind except that which the rotor causes, helicopter rotors should be as long as mechanically practical to intercept a large volume of air per unit time and thus keep the induced-power requirement to a minimum. This advantage of long blades is the main reason for the impracticality of aircraft in which the propellers used for fast forward flight are shifted toward a vertical axis for short take-offs.

Autogyros (Figure 12.8d) may be less familiar to the reader than the previous three devices, but they are completely competent flying machines and are commercially available at relatively modest cost (Bensen Aircraft Corp., Raleigh, N.C.). They look superficially like helicopters, but the overhead rotor is not powered, another propeller is necessary for level flight, and they cannot ascend vertically in still air. An autogyro, powered, has the plane of its passive rotor tilted a little so that approaching air strikes the underside of its plane of rotation (a helicopter uses the opposite tilt). Air passes through the rotor, and in being retarded, it generates lift and a drag with respect to the craft. The propeller counteracts the drag. Unpowered, an autogyro descends slowly with rotor passively turning; in effect it is a set of airfoils (the blades) gliding earthward in a tight helical path. As in any of these devices, air must strike the concave side of the blades, so the blades have to be concave downward (as in helicopters). But in contrast to helicopter blades, the blades of the autogyro must have their leading edges depressed rather than elevated.

If a helicopter loses power, it can revert to autogyration, but the pitch of the blades must be rapidly shifted to the leading-edge-downward orientation. Otherwise the rotor will turn backwards and the airfoil profiles will be inappropriate. If pitch is adjusted, though, both camber and lengthwise twist are proper, unlike the shift from propeller to windmill.

CHAPTER 12

momentum flux and actuator discs

It is sometimes useful to view a thrust-producing or power-extracting rotor as an infinitely thin disc through which fluid passes and, in passing, gains or loses momentum at a certain rate. This idealized mechanism for changing momentum flux is termed an "actuator disc;" its area is the area swept by blades or beating wings. A stream of fluid, in passing a thrust-producing rotor, is accelerated, and therefore, by the principle of continuity, the stream contracts. According to the momentum theory of propeller action, the velocity of fluid passing through the actuator disc is midway between that upstream and that downstream. Mises (1945) succinctly presents the conclusions of momentum theory for the efficiency of propeller or helicopter rotor operation.

The main use of this view of thrust transduction for us is as justification for making wake-traverse measurements in a manner analogous to that described for drag in Chapter 4. If an airstream of uniform velocity, U_0, passes through an actuator disc such as the beating wings of a tethered insect, the thrust is given by an analog of equation (4.18):

$$dT = \rho U_0 U_1 dS_1 - \rho U_1^2 \, dS_1 \qquad (12.4)$$

The power extracted or inserted is thrust times velocity, so:

$$dP = \rho U_0 U_1^2 \, dS_1 - \rho U_1^3 \, dS_1 \qquad (12.5)$$

Equations (12.4) and (12.5) must be integrated over the wake as described in Chapter 4.

Using this approach, I once mapped the velocities in the wake of a tethered fruit fly beating its wings in otherwise still air ($U_0 = 0$). The resulting power output of 16 microwatts was the same as that obtained years earlier by C.M. Williams, who made a fly move a load up an incline by revolving a "fly wheel." Williams, for measuring the horsepower of a fruit fly, was the target of sarcastic comment in the popular press (Chadwick, 1953).

the flight of samaras

An unpowered autogyro is a kind of glider and, like any glider, it can have no unbalanced forces under steady-state conditions. Initially, the aerodynamic resultant is tipped forward, pointing in the direction of rotation, and produces a torque which accelerates the blades, as indicated in Figure 12.8d. As the blades speed up, the oncoming wind becomes more nearly horizontal, and the resultant tips back until it is vertical and equal to the weight of the craft.

Autogyros of this sort are quite common in nature, having evolved independently in many families of plants (Figure 12.9). As mentioned in the previous chapter, some are Flettner-rotating autogyros, but more are simple lift-producing airfoils. The latter range in size from about 1 cm to 18 cm in

FIGURE 12.9

The autogyrating descent of a maple samara (*a*); it describes a low cone (*b*) as it rotates about the seed as an apex.

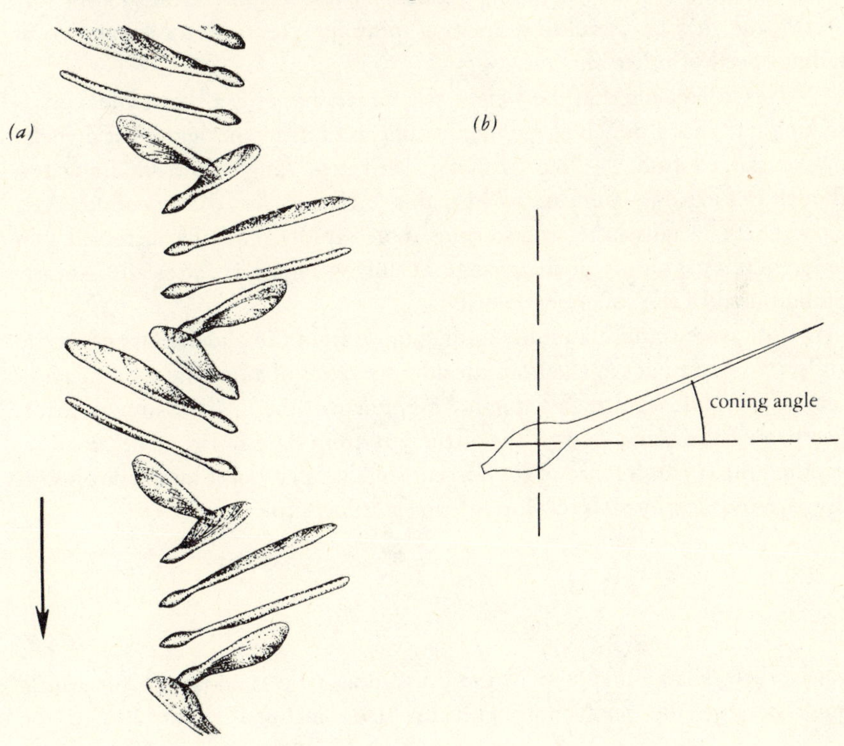

length, have from one to three blades, and span several orders of magnitude in mass. The best known of these samaras are those of the maples (*Aceraceae*), but they also occur in some tropical *Juglandaceae* and *Leguminoseae* as well as in most of the *Pinaceae* and many of the *Cupressaceae* among the conifers. The definitive study of the aerodynamics of samaras is that of R.A. Norberg (1973); some further information is given by Hertel (1966) and McCutchen (1977).

For an autogyro, a low sinking speed is obviously desirable, which implies that a low rather than a high advance ratio is preferred. A fairly large, one-bladed samara of the Norway maple, *Acer platanoides,* with a functional blade length of 3.7 cm, falls at 90 cm s^{-1} and revolves 13 times per second; its advance ratio (equation 12.2) is 0.9. The smaller samara of the Norway spruce, *Picea abies,* has a blade length of 1.1 cm, a falling speed of 64 cm s^{-1}, revolves 20 times per second, and thus has an advance ratio of 1.5 (data from Norberg, 1973). I measured an advance ratio of 1.3 for red maple (*Acer rubrum*) samaras, whose functional blade length is about 1.7 cm. The advance ratio should increase as the Reynolds number decreases because the poorer lift-to-drag ratios at low Reynolds numbers mean relatively less force available to spin a samara and more drag opposing the spin. These advance ratios do not vary greatly and may be useful in estimating spinning rates from measurements of falling speed of other samaras.

It is interesting that the weight of a samara has only a slight effect on its sinking rate. For a bunch of red maple samaras of different degrees of dryness, masses ranged from 11.2 to 57.2 mg, a fivefold range. The sinking rates, though, varied only from 62 to 115 cm s^{-1}; that is, less than twofold. As a heavier samara falls faster it also spins more rapidly, and the increased spin reduces the magnitude of the change in sinking rate. Lift, after all, is nearly proportional to the square of velocity.

Just as we could obtain lift-to-drag ratios from the glide angles of simple gliders, so we can get L/D's from the advance ratios of autogyrating samaras—they're just gliders going in a spiral. An estimate of the L/D for a highly effective part of the blade, four-fifths of the way from base to tip, is obtained by dividing the circumference of the descent spiral at that point by the downward distance traveled in each revolution, which reduces to:

$$\frac{L}{D} = \frac{0.8\pi}{J} \qquad (12.6)$$

For Norberg's large maple seed, the lift-to-drag ratio is about 1.5 (or a little more because the seed spins with its blade inclined about 20° to the horizontal); the Reynolds number, based on chord length, is 3000. For his small spruce samara, the lift-to-drag ratio is 1.7 at a Reynolds number of 600. These figures are a little worse than those for insect wings (Table 11.1); selection may not be as single-mindedly concerned with lift-to-drag ratios in passively falling samaras.

non-steady-state effects in flapping flight

If an airfoil starts suddenly from rest in a moving fluid, the circulation about the airfoil does not fully develop for the first two or three chord-lengths of travel; apparently, the starting and bound vortices interact destructively until the distance between them increases somewhat. Thus the lift on an impulsively moved airfoil will not appear right away. In flapping flight, this "Wagner effect" should have a distinctly detrimental effect on performance (Weis-Fogh, 1975a): wings may start and stop several hundred times each second and travel no more than a few chord lengths during each stroke. Consequently, the steady-state lift coefficients measured on isolated wings in wind tunnels probably overestimate the lift actually available in flapping flight.

At the same time, it is not a difficult matter to estimate the average steady-state lift coefficients required to support a hovering creature's weight. One needs to know only a few data—wing length, shape, area, wingbeat frequency, and stroke amplitude; Weis-Fogh (1973) has derived the appropriate formulas and applied them to a variety of insects, birds, and a bat. The results of his calculations are most interesting. Even with no Wagner effect, insects the size of fruit flies and smaller as well as butterflies should not quite be able to hover with their observed wingbeat unless unreasonably high lift coefficients are assumed. A fruit fly would need a lift coefficient of 1.0; the steady-state wind tunnel results suggest that it gets 20% less. A smaller insect, the tiny wasp, *Encarsia formosa,* with a wing length of 0.7 mm and a Reynolds number of about 20, needs a lift coefficient of 1.7 (Ellington, 1975). Movies show that *Encarsia* beats its wings in ordinary fashion, so it must fly with lift based on circulation. But it is hard to see how such a high lift coefficient can be had even under steady-state assumptions at such low Reynolds number.

Weis-Fogh himself (1973, 1975a, 1975b) has suggested ways in which flapping can actually be turned to advantage. In these very small insects, *Drosophila* and *Encarsia,* the wings come very close together at the top of the stroke. (The original evidence included some pictures I published without

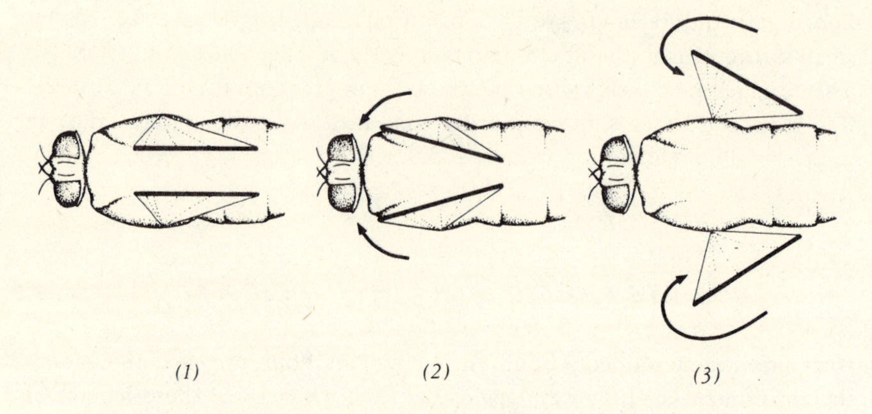

Weis-Fogh's "clap and fling" mechanism for the initiation of circulation through the interaction of a pair of wings. The wings are shown cut chordwise about halfway along their spans. At the start of the downstroke (*1*), the two wings part first in the front (*2*); air passing around the leading edges to fill the gap begins the circulation (*3*).

FIGURE 12.10

noticing anything significant.) As the wings separate (Figure 12.10), the leading edges part first. The space between them must be filled with air, and that air must come around the leading edges: it is, in fact, a circulation, begun and well-established at the very top of the stroke. The mechanism has been termed the "clap and fling." This circulation, incidentally, is created quite independently of the classical scheme for generating circulation described in the previous chapter. Also, it does not depend on viscosity and thus should be largely independent of the Reynolds number.

In few if any animals do the wings meet at the bottom of the stroke, so the trick is not repeated. But another non-steady phenomenon, the "flip" may occur. The wing is rotated very rapidly from the angle of attack of the latter part of the downstroke to that of the upstroke. This rapid rotation appears to throw earthward the bound vortex of the downstroke and to initiate the opposite circulation of the upstroke in a manner perhaps analogous to the Flettner rotation of a flat ash or tulip poplar seed.

Birds larger than hummingbirds can hover, if at all, only for brief periods; but even during these periods their lift must balance their weight. Furthermore, rapid take-off must require momentarily high values of lift. U.M. Norberg (1975) estimated that a flycatcher (*Ficedula hypoleuca*) needed a lift coefficient of 5.3 to hover—a very high value which takes some special explanation. She is dubious about whether the clap-and-fling is of much importance in the flycatcher. The lift coefficient needed by the bat *Plecotus auritus* to hover is also too high to be consistent with steady-state aerodynamics (U.M. Norberg, 1976) as is the case for a dragonfly, *Aeschna juncea,* (R.A. Norberg, 1975); in neither is the wing stroke appropriate for a clap-and-fling. On the other hand, pigeons quite audibly clap their wings together when rapidly ascending.

So where are we? It's become clear in the past decade that non-steady-state phenomena can't be blithely ignored. It's also fairly clear that the clap-and-fling mechanism, although the best understood, isn't the only operative non-steady phenomenon. The subject has become controversial as well as complex, with various proposals appearing, mainly in the *Journal of Experimental Biology*. Probably the most promising development is work on a more sophisticated analytical approach to flapping flight. In this scheme, the inevitable time-averages of momentum theory and steady-state assumptions of blade element theory are replaced by consideration of the location, strength, and histories of the various vortices produced in flapping (Ellington, 1978; Rayner, 1979).

CHAPTER 13

flow at very low Reynolds numbers

IMAGINE LIVING IN A WORLD in which the following scenario is entirely plausible. You pour cream into your morning coffee; the cream remains in a blob in mid-coffee. So you stir the cup three times counterclockwise, spreading out the cream. Then, after distilling the daily disasters from the newspaper, you resume stirring but absentmindedly do so clockwise. After three circuits of the spoon, the blob of cream reappears, separated from the coffee, and you mutter maledictions about the nuisance of mixing liquids. The mental outlook improves not a bit when the drain in the sink plugs up—bailing out the contents is unavoidable. No small bucket is handy, but fortunately a colander works almost as well to transfer the water to an adjacent tub. Mildly mollified, you go off in your motor boat, pointing the propeller, a spiral cylinder in the front, out into the main channel. It just takes far too much fuel to move along the shore. In mid-passage, the motor quits, and the boat immediately stops. Having neglected to fasten your seat belt, you tumble over the bow but at least have the presence of mind to get quickly back into the boat. So only a little water needs to be scraped off. A sticky mess, nonetheless.

The principal peculiarity of the world described is the exceedingly high viscosity of its liquids. It is the world, as Howard Berg puts it, of a person swimming in asphalt on a summer afternoon. It bears close resemblance to the world of flows of very low Reynolds numbers, the world of what is called "creeping motion," the everyday world of every microscopic organism which lives in a fluid medium. Flow at very low Reynolds numbers may seem bizarre to us, but the range of flow phenomena with which we commonly contend would undoubtedly seem even stranger to someone whose whole experience was at Reynolds numbers well below unity. How could we make credible, for instance, the idea of turbulance?

At very low Reynolds numbers, flows are typically reversible: a curious temporal symmetry sets in, and flow may move matter around but doesn't have to particularly disorder what it moves. Concomitantly, mixing is exceedingly difficult: spreading one fluid through another miscible one need not immediately mix the two. Inertia is negligible compared to drag: when propulsion ceases, motion almost immediately ceases. Berg (Purcell, 1977) has calculated that if a bacterium suddenly stopped rotating its flagellum, it would coast to a stop in a distance equal to a tenth the diameter of a hydrogen atom. Separation behind bluff bodies is unknown: separation results from inertia, the tendency of fluid to continue to move downstream rather than curve around the rear of an obstacle. But where inertia is negligible, fluid goes around curves and corners with magnificent indifference. Velocity gradients are what the fluid abhors. Streamlining is a fine way to increase drag; the extra surface exposed in the process incurs extra skin friction with little drop in pressure drag. Shape matters, but to a lesser extent and in a different way than what we're used to. Propulsion in fluid media is possible, but not by imparting local rearward momentum to the fluid any more than we impart rearward momentum to part of a floor when we walk on it. Nor can one create circulation about an airfoil. Boundary layers are thick because velocity gradients are gentle, and the formal concept of a boundary layer has little or no utility. The drag of an object moving through a fluid may be much increased by walls around the fluid a hundred or more object-diameters away. Vortices exist only in the upper end of this realm; they're very regular, mostly core, and dissipate rapidly if their voracious appetite for energy isn't constantly satisfied. Turbulence, of course, is unimaginable. Galileo must recant: an object of density greater than that of the medium falls, but a large object falls much faster than a small one. Terminal velocity is reached almost immediately after release; after that, the balance of drag and weight is all that sets speed.

This queer and counterintuitive range has only limited technological interest. It's the range of sedimenting particles, settling fog droplets, and slow percolation through porous beds. But its biological importance cannot be overemphasized. The vast majority of organisms are tiny, and they live in this world of low Reynolds numbers.

The reversibility alluded to is dramatically illustrated in a film loop (FM-115, from Encyclopaedia Britannica Educational Corp.); you can do it almost as well yourself, by injecting a bit of colored glycerin beneath the surface of a beaker of clear glycerine and stirring with a rod, first in one direction and then in the reverse.

The dependence of the tactics of living on size and Reynolds number is perhaps best illustrated by some calculations given in a charming essay by Purcell (1977). Consider a bacterium about one micrometer long, swimming through water at about 30 $\mu m\ s^{-1}$. The Reynolds number is about 3×10^{-5}. The bacterium can swim, but should it? Diffusion brings its food to it; to increase its food supply by 10% it would have to move at 700 $\mu m\ s^{-1}$. Nor should it thrash about since stirring is out of the question. The only reason to swim is to seek a more concentrated patch of food—greener pastures. It is as if we've encountered a cow who eats some grass and then, rather than walk around, waits for the grass to grow again.

Flows at low Reynolds numbers may seem peculiar, but they are orderly (Purcell calls them "majestic") and far more amenable to theoretical treatment than the flows we've previously considered. The most sacred Navier-Stokes equations, mentioned several times already, contain inertial and viscous terms. A very low Reynolds number means a preponderance of viscous forces, so it is reasonable to ignore the inertial terms. With this simplification, explicit solutions to many problems are possible, and the advent of large computers has permitted approximate solutions to others. In general, curves describing drag and other phenomena are regular and lack any of the bumps, bulges, and bluffs we've had to contend with earlier. So we're able to rely more on equations and less on semi-empirical graphs. The relevant fluid mechanics is well-treated in detail by Happel and Brenner (1965), with clarity and brevity by White (1974), and with a more biological perspective by Hutchinson (1967).

drag

At high and moderate Reynolds numbers, inertial effects were all-important, and we found that drag depended on the product of projecting area and dynamic pressure. A dash of Reynolds number dependence accounted for the vagaries of flow which resulted from the residual action of viscosity in the boundary layer. At Reynolds numbers below about 1.0, drag depends mainly on viscous forces and ought to be largely independent of fluid density. The customary drag coefficient, C_D, as defined by equation (5.4), is often still used, perhaps to maintain consistency with the high Reynolds number branch of the field. There's nothing wrong with using C_D; it's just the result of a definition and has no automatic phenomenological implications; but it may be misleading in practice. Thus the drag coefficient for a sphere is $24/Re$ up to about $Re = 1$; as White (1974) notes, this implies a Reynolds number effect where none really exists.

At low Reynolds numbers, from either dimensional reasoning or the definition of viscosity (equation 2.3), drag should be proportional to viscosity, to velocity, and to some linear dimension. Drag should also be dependent on the shape of an object. As at high *Re*'s, we can separate the effects of shape from those of size, speed, and viscosity by introducing a coefficient which depends only on shape and orientation, the *viscous drag coefficient.* To be completely unambiguous about the source of the size index in the definition, let us define two slightly different coefficients, one (C_{DVS}) based on the square root of wetted surface area and the other (C_{DVV}) based on the cube root of volume. Skin friction may be more closely related to surface, but the problem of carrying guts and gonads is rather volumetric. Thus:

$$D = C_{DVS}\,\mu S^{1/2} U \qquad (13.1)$$

$$D = C_{DVV}\,\mu V^{1/3} U \qquad (13.2)$$

It might be noted that the relationship between these and the usual drag coefficient can easily be obtained by recognizing that all refer to the same measurable drag, so, for example:

$$\tfrac{1}{2} C_D \rho S U^2 = C_{DVS}\,\mu S^{1/2} U \qquad (13.3)$$

At low Reynolds numbers, then, all we really need for determining drag are (1) data on one or the other viscous drag coefficient for the shape and orientation of interest, together with (2) information on size, speed, and the viscosity of the medium. The coefficient should not vary with viscosity, velocity, size, or Reynolds number.

Spheres. The first, simplest, and the most useful formula for drag at low Reynolds numbers is Stokes' law for the drag of a sphere:

$$D = 6\pi\mu a U \qquad (13.4)$$

The viscous drag coefficients are $C_{DVS} = 5.317$ and $C_{DVV} = 11.69$. Stokes' law is trustworthy up to Reynolds numbers of about 1.0. Above that, various theoretical treatments have been attempted, but the most useful formula I've seen is a curve-fit equation given by White (1974):

$$C_D = \frac{24}{Re} + \frac{6}{1 + Re^{1/2}} + 0.4 \qquad (13.5)$$

The characteristic length in the *Re* is the diameter of the sphere and the surface in C_D is the projected area. The equation is a useful approximation of the line in Figure 5.4 up to a Reynolds number of 2×10^5, that is, to the great drag crisis where the boundary layer becomes turbulent.

It is worth noting that of the drag predicted by Stokes' law, two-thirds comes from skin friction and one-third from a fore-and-aft pressure drop. The negligibility of inertia does not imply that flows are not associated with pressure differences; it still takes a difference in pressure to persuade a fluid to move from one place to another. Indeed, it takes relatively more pressure at low Reynolds numbers because of the greater retarding effects of viscosity, as one can see in the Hagen–Poiseuille equation (10.2).

Circular disks. Exact solutions are also available for the drag of circular disks. For a disk which faces the oncoming flow, one for which the flow is parallel to the axis of revolution,

$$D = 16\mu a U \qquad C_{DVS} = 6.383 \qquad (13.6)$$

For a disk with its faces parallel to the flow or the axis of revolution normal to the flow,

$$D = 10.67\mu a U \qquad C_{DVS} = 4.255 \qquad (13.7)$$

Several points ought to be mentioned. These are "worst" and "best" orientations with respect to drag for an object which appears about as anisotropic as anything we're likely to devise. The difference, though, is only 1.5-fold, about what we noted for nearly flat plates at a Reynolds number of 1 (Table 5.1, page 78). The two orientations do not converge further in drag with further decrease in the Reynolds number. Also, the comparison shows that while shape has a non-negligible effect on drag, the effect is scarcely overwhelming, a point made earlier in connection with flat plates. Referred to surface area, the disk normal to flow has a higher drag than a sphere, rather like the situation at intermediate, subcritical Reynolds numbers. Streamlining may not have its usual meaning, but evidently, elongating a body fore and aft can still lower the drag per unit of surface. One can't really refer to volume since, ideally, the disk lacks volume entirely.

Cylinders. As we'll see later, the drag of cylinders takes on great importance in explaining propulsion by flagella and cilia. As it turns out, the drag of cylinders is computed by considering them as very long prolate spheroids; in this way theoretical equations have been derived for drag with the long axis of a cylinder either parallel to flow or normal to flow. According to Cox (1970), the drag of a cylinder of length l and radius a in the parallel orientation is:

$$D = \frac{2\pi\mu Ul}{\ln(l/a) - 0.807} \qquad (13.8)$$

For a cylinder whose length is fifty times its diameter, the viscous drag coefficient, C_{DVS}, is 6.598 while C_{DVV} is 24.33. The drag with the axis of revolution normal to flow is:

$$D = \frac{4\pi\mu Ul}{\ln(l/a) + 0.193} \qquad (13.9)$$

For the cylinder of length fifty times the diameter, the coefficients are $C_{DVS} =$ 10.448 and $C_{DVV} = 38.53$. One occasionally runs across analogous equations in which the numerical constants in the denominators are -0.5 and $+0.5$, respectively; but, as Brennen and Winet (1977) point out, these are solutions derived for zero Reynolds number, and inertia always makes a small but quite real contribution.

It is occasionally claimed that a cylinder normal to flow has just twice the drag of one parallel to flow. This twofold difference would be reached only for a cylinder which is either infinitely long or of zero thickness. For a length-to-diameter ratio of half a million it is still only 1.86. Still, these figures are above the ratio of 1.5 for the orientations giving most and least drag for disks: oddly enough, cylinders are more anisotropic than disks.

For a circular cylinder normal to the flow, White (1974) offers the following empirical formula, useful in the intermediate range of Reynolds numbers from unity to 1×10^5; it may be compared to Figure 5.3, (page 72). The drag coefficient is the conventional one, with area taken as projecting area, and the Reynolds number is based on the diameter of the cylinder:

$$C_D = 1 + 10.0\, Re^{-2/3} \tag{13.10}$$

Spheroids. Many biological objects may be reasonably approximated by either oblate (pancake or discus) or prolate (cigar or football) spheroids, as mentioned in Chapter 6 in connection with drag at higher Reynolds numbers. To take the extreme cases, such spheroids move with their axes of revolution either parallel or normal to the free stream; with two classes of spheroids and two orientations, we have four situations of interest. These fill in gaps between spheres, disks and cylinders. Loosely speaking, disks are the ultimate result of squashing spheres into oblate spheroids, while cylinders are the final result of stretching spheres into prolate spheroids. Happel and Brenner (1965) give appropriate, if mildly messy, equations for these cases; surface areas of spheroids can be calculated from formulas for areas and eccentricity in Beyer (1978).

Table 13.1 presents the viscous drag coefficients for a variety of spheroids as well as for the sphere, disk, and cylinder already considered. A number of items are noteworthy. First, the drag coefficients are not particularly shape-dependent. Shape matters, but far less than at high Reynolds numbers. Orientation matters also, but again not very much, and no alteration of orientation can make even a twofold difference in drag. Relative to surface area, a rather long prolate spheroid traveling lengthwise or a flat disk traveling edgewise incurs the least drag; a circular cylinder moving crosswise incurs the most drag. Relative to volume, a prolate spheroid twice as long as wide and moving lengthwise gives the least drag. Both it and a 1:2-oblate spheroid moving edgewise have less drag relative to their volumes than does a sphere: not much, but enough to embarrass glib claims that spheres have least drag. The lowest drag oblate spheroid has a length-to-diameter ratio of 1:1.5 and moves edgewise (Zaret and Kerfoot, 1980).

Fluid spheres. If a sphere of fluid rises or falls through a fluid medium, the passage of the medium will induce a toroidal motion within the sphere (Figure 13.1). The result will be lower drag on the sphere; it is as if the no-slip condition were partially relaxed. The phenomenon depends on the relative viscosities of sphere and medium: for a droplet of water in air the effect is negligible while for a droplet of air in water there is, in effect, perfect slip.

TABLE 13.1

Calculated drag of simple bodies at very low Reynolds numbers

		flow parallel to axis of rotation			*flow normal to axis of rotatation*		
	l/d	C_{DVS}	C_{DVV}	U/U_s	C_{DVS}	C_{DVV}	U/U_s
Circular disk	1:50	6.383	36.60	0.319	4.255	24.40	0.479
Oblate spheroid	1:4	6.174	16.10	0.726	4.857	12.67	0.923
Oblate spheroid	1:3	6.012	14.82	0.789	4.950	12.14	0.963
Oblate spheroid	1:2	5.794	13.34	0.877	5.074	11.68	1.001
Sphere	1:1	5.317	11.69	1.000	5.317	11.69	1.000
Prolate spheroid	2:1	4.896	11.17	1.047	5.608	12.80	0.914
Prolate spheroid	3:1	4.762	11.39	1.027	5.858	14.01	0.835
Prolate spheroid	4:1	4.734	11.77	0.993	6.093	15.15	0.772
Circ. cylinder	50:1	6.598	24.33	0.480	10.448	38.53	0.304

l/d: length along axis of rotation over maximum width

C_{DVS}: viscous drag coefficient based on $S^{1/2}$, (equation 13.1)

C_{DVV}: viscous drag coefficient based on $V^{1/3}$, (equation 13.2)

U/U_s: terminal velocity relative to that of sphere of equal volume and weight. Axis of rotation is what a biologist would call the axis of radial symmetry.

Happel and Brenner (1965) give the following formula:

$$D = 6\pi\mu_{ext}\, aU \frac{1 + (2/3)(\mu_{ext}/\mu_{int})}{1 + (\mu_{ext}/\mu_{int})} \tag{13.11}$$

where μ_{ext} is the viscosity of the medium and μ_{int} that of the sphere. For perfect slip (gas in liquid), the drag is two-thirds of that given by equation (13.4) for a rigid sphere:

$$D = 4\pi\mu_{ext}\, aU \tag{13.12}$$

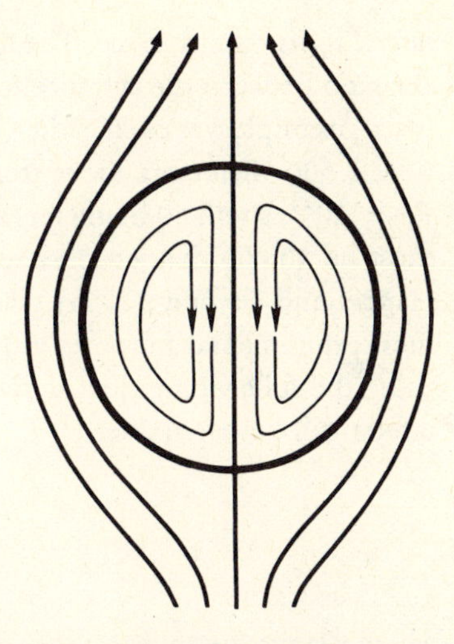

FIGURE 13.1

Internal circulation of a fluid droplet in an external flow. Note the toroidal vortex ring induced within the droplet; it's similar to an ascending thermal (Fig. 12.4b) and to secondary flow near a bend in a pipe (Fig. 10.8).

A converse possibility should not be ignored: actively moving a membrane around a sphere or spheroid, extruding it at the front and absorbing it behind, is a reasonable way of developing thrust at low Reynolds numbers. It is a scheme which should work in *any* sufficiently deformable medium.

Orientation. At intermediate and high Reynolds numbers, a body which is symmetrical about each of three mutually perpendicular axes will either tumble or will take up an orientation with the maximum cross-sectional area normal to the direction of motion. Typically this will maximize its drag, for better or worse. A proper arrow requires feathers to maintain its orientation. A *Pilobolus* sporangium, according to the late Robert Paige, tumbles in flight. But for symmetrical objects moving through fluids in the range of Stokes' law, any orientation is stable; a particle, released into an unbounded medium, retains its original orientation if it is symmetrical and has uniform density. It is no difficult matter, though, to equip an object with protrusions which will ensure that the object will seek a preferred orientation following any rotational displacement (McNown and Malaika, 1950; Hutchinson, 1967).

Wall effects. Organisms frequently move near solid surfaces or air–water interfaces, and every measurement in a wind tunnel or flow tank involves the imposition of a wall somewhere near the test object. Just how close a wall can be without significantly affecting the pattern of flow and the forces on an object is a question of considerable interest. The problem is most acute at low Reynolds numbers, where a body influences the flow for a great distance lateral to its outer surface. The lower the Reynolds number, the larger must be the ratio between the distance to a wall and the diameter of an object. A "gee whiz" example was cited earlier. At a Reynolds number of 10^{-4}, the presence of a wall 500 diameters away from a cylinder doubles the effective drag; at $Re = 10^{-3}$, a wall 50 diameters away dominates the determination of drag and must be 10,000 diameters away before its influence is trivial (White, 1946). Happel and Brenner (1965) give a general treatment of wall effects, but for most experimental purposes a quick index is adequate. Robert Zaret suggests using the following guide, derived from White (1946), for telling which wall effects will be negligible:

$$\frac{y}{l} > \frac{20}{Re} \qquad (13.13)$$

where y is the distance to the nearest wall, and l is a characteristic length of the object. The formula is usable only at Reynolds numbers below unity and is a fairly stringent criterion which could be shaved a bit if necessary.

The same wall effects should influence the calibration of hot wire and hot probe flowmeters when these are used near walls, so appropriate caution should be exercised (White, 1946).

Unless an object moves axially in a cylinder or midway between plates, another effect of a wall is to introduce some asymmetry in the drag of even a symmetrical object. A sphere falling near a wall will begin to rotate in the same direction as if it were rolling along the wall. The rotation may, in turn, generate other forces. Under some circumstances, constriction of streamlines between moving object and wall will produce "lift," attracting the body toward the wall.

terminal velocity

If an object whose density differs from that of the medium is released, it accelerates either upward or downward. At low Reynolds numbers it almost immediately stops accelerating, having reached a constant, so-called "terminal" velocity. Not that terminal velocities don't exist at high Reynolds number: they are just less commonly of biological relevance and they take quite a distance to achieve. As in gliding, the absence of acceleration implies a balance of forces: weight minus buoyancy (net body force) against resistance (drag force).

Consider a solid spherical object of density ρ falling in an ambient medium of density ρ_0. The gravitational force is given by the gravitational constant times the difference between the mass of sphere and mass of displaced fluid, $(\rho - \rho_0)(4/3)(\pi a^3)g$. The drag is given by Stokes' law (equation 13.4): $6\pi\mu a U$. If we equate the two and solve for velocity, we get the very well known equation,

$$U = \frac{2a^2 g(\rho - \rho_0)}{9\mu} \qquad (13.14)$$

This works for the same conditions as Stokes' law, that is, up to a Reynolds number of about 1.0. Above that, a somewhat more complex formula can be derived from equation (13.5). Notice that terminal velocity is proportional to the square of the radius: a bigger object falls faster, not because it has less drag (which it doesn't—it has more), but because it has more volume and hence a larger net weight for a given density. But for a given investment of mass, a larger sphere will fall more slowly than a smaller one, at least in a medium such as air whose density is much lower than that of any solid sphere.

For objects with negligible displacement such as disks or for any objects moving in a medium whose density is much less than their own, the expression for terminal velocity can be simplified to:

$$U = \frac{mg}{(K)\mu a} = \frac{mg}{C_{DVS}\,\mu S^{1/2}} = \frac{mg}{C_{DVV}\,\mu V^{1/3}} \qquad (13.15)$$

where K is the numerical constant of equations such as (13.4), (13.6), or (13.7). Here the argument is more obvious that for a given mass, the larger object falls more slowly: velocity is inversely proportional to radius. To use equations (13.15), one simply selects an appropriate viscous drag coefficient and combines it with the needed square root of surface or cube root of volume. Selection of a coefficient is simplified by the fact that the latter is not especially shape-dependent, so Table 13.1 should give a reasonable selection of values. The table also includes terminal velocities relative to those of spheres for the shapes considered.

Stokes' radius. Free descent or ascent at terminal velocity is an ordinary event, particularly among dispersing seeds, spores, some baby spiders, gypsy moth caterpillars, and planktonic organisms. The shapes of such small objects are extremely diverse, and for some comparisons an index of effective size may prove useful. Most of these objects are either quite small or have elaborate filamentous outgrowths of very small diameter. In either case, the Reynolds number is fairly low. A reasonable index is the radius of the sphere of the same mass which would ascend or descend at the same rate; we can designate this as a_s. It has been called the "nominal radius" by Kunkel (1948); I think the name "Stokes' radius" is less ambiguous. Either way, the index can be obtained by measuring the mass and velocity of an object and then solving the following equation for a_s:

$$mg - \frac{4}{3}\rho_o g \pi a_s^3 = 6\pi\mu a_s U \qquad (13.16)$$

The equation is, of course, the now-familiar one which equates drag with the difference between weight and buoyancy. The density of the object has been eliminated so one need not prejudge the object's effective size. The resulting Stokes' radius can be viewed as a measure of the effectiveness of any fluff or protrusions in acting as a drag-generating pseudosolid body.

passive sinking and dispersal

One way to achieve dispersal if ambient currents of air or water are available is to get up into the flow, fall out as slowly as possible, and be passively carried by the fluid motion. With respect to the local fluid, the dispersing objects may be descending slowly, but with respect to the earth below they may be moving rapidly along quite a different path. The issue of getting up was briefly touched on when boundary layers were considered. Even if boundary layers did not exist, however, the distance a slowly sinking object might travel should increase with an increase in the height at which it begins its journey. The two important parameters are the speed of the flow and the length of time the object can remain aloft. The former is often an environmental fact of life, although attached seeds and spores commonly wait for a wind before release. The latter, for a given release height and ignoring convection, is inversely proportional to the terminal velocity of free fall, given by equations (13.14) and (13.15). A fair idea of the tactics of minimizing sinking rate will be apparent from inspection of the equations and the data in Table 13.1.

Seeds and spores. A reasonable number of measurements of sinking rates are available in the literature, although few include enough morphological data to permit comparison with a Stokes radius. Sheldon and Burrows (1973) measured the sinking rates of a large number of wind-dispersed Compositae, but their results are expressed in a form from which the primary data on weights and diameters cannot be reconstructed. Sinking rates of these parachuting seeds ranged from 19.2 to 219.8 cm s^{-1} in still air. A few years ago, to demonstrate the use of a vertical wind tunnel, I measured masses and terminal velocities of some milkweed seeds (*Asclepias* sp.). These were large, even by milkweed standards (Wilbur, 1976, has a nice paper on milkweed seeds strategies), and these seeds are, in any case, well above the range to which Stokes' law is applicable. The Stokes radius, calculated either by equation (13.16) or by assuming a C_D of 0.45, proves to be about 1.1 cm—well below the measured 2.7 cm radius—at a descent speed of 30 cm s^{-1}. So the fluff doesn't behave quite like a solid sphere. My earlier comments on these results (Vogel, 1969) are best treated as fluffed lines of a misguided youth.

Wind-dispersed seeds are, of course, very common and greatly varied in shape, size, and mechanisms for using wind. Those of the "parachuting" sort, that is, drag-maximizing rather than lift-producing seeds, are usually fairly small. Harper et al. (1970) mention that such seeds are particularly prevalent among plants of open habitats and that seeds from these places are smaller, on the average, than those from woodland margins which, in turn, are smaller than those of forests. Part, but only part, of the explanation must be the availability of wind in open habitats. In addition, open habitats commonly represent early successional stages and are settled by plants of the r-strategist sort which produce large numbers of small, highly vagile seeds. For a temporary inhabitant of a patchy environment, wind dispersal makes good sense. Forests, by contrast, are later stages or climax communities filled with K-strategists which tend to produce larger seeds for more assured propagation in a "filled" habitat. If heavier seeds are to be wind dispersed, their higher Reynolds numbers should favor lift-producing schemes such as autogyros and gliders. Thus, for example, a drag-producing circular disk would need an area of 24 cm^2 to descend at the same rate as one of Norberg's (1973) maple samaras of 5.6 cm^2. But, for autogyrators at least, a brief but fast free fall precedes steady spinning; the fast fall may be a minor matter for a seed released from a tree but would use up most of the height of release in a lower plant.

Wind dispersal is probably most effective with small propagules; these have low terminal velocities even without the benefit of bristles or fluff. Ingold (1953) cites, as a record of sorts, 0.05 cm s^{-1} for the rate of fall of the 4.2-μm diameter spore of *Lycoperdon pyriforme,* but he considers 1 cm s^{-1} as a more typical rate for fungal spores. Gregory (1973) provides a collection of data on size, density, and terminal velocities of spores, although he didn't feel that, given even very slow winds, these low sinking rates were significant. But this view was roundly criticized by Schrodter (1960) who quite rightly pointed out that seeds and spores are always sinking relative to the local air, and, on the average, as much air descends as ascends. Schrodter also noted that fungal spores are usually spheroids rather than spheres; this might reduce their sinking rates relative to the mass they transport. Pollen grains are comparable to spores in size and must present a similar situation with respect to wind dispersal.

The role of convection currents and other atmospheric irregularities in the dispersal of all these wind-borne propagules is clearly significant. But this isn't a book on micrometeorology, so I shall merely refer the reader to Burrows (1973) for a theoretical treatment of the relevant processes and to Salisbury (1976) and Van der Pijl (1972) for qualitative discussions.

Phytoplankton. The situation for planktonic organisms is far different from that of biological objects sinking through air; indeed, dispersal may be only a minor aspect of the buoyancy business. Of particular interest are the phytoplankton, many of which are not competent swimmers but all of which function photosynthetically only near the surface of the water. Local, wind-driven currents are nearly ubiquitous, and apparently it is possible for individuals of a species to remain negatively buoyant provided their sinking rates are sufficiently low. Almost everything of which a plankter is made is denser than either fresh water (1.00×10^3 kg m^{-3}) or sea water (1.025×10^3 kg m^{-3}): the siliceous walls of diatoms have a density of 2.6×10^3 kg m^{-3}; calcium carbonate may approach 3.0×10^3 kg m^{-3}; and even cellulose is 1.5×10^3 kg m^{-3}. Some forms, such as the dinoflagellate, *Noctiluca,* have a large vacuole in which they concentrate ammonia and exclude potassium and divalent anions to reduce density. But the scheme can achieve positive buoyancy only in sea water. Fat is apparently not ordinarily important as a flotation material, and gas vacuoles are common in fresh-water, but not marine, forms.

Smayda (1970), from whom the comments above are derived, notes the conspicuous absence of spherical marine phytoplankton. He has a collection of data which, he asserts, shows that sinking rate is inversely related to surface-to-volume ratio, and hence the more irregular species sink more slowly. But surface-to-volume ratio incorporates effects of both shape and size: smaller size would give a higher ratio and, for a given shape and density, a slower sinking rate. Out of curiosity, I've reanalyzed the data for diatoms (17 species), eliminating size from the surface-to-volume ratio by using a dimensionless surface-to-volume ratio, $S^{1/2}/V^{1/3}$, and eliminating size from the sinking rates by referring each to the sinking rate of a sphere of the same size. A graph using these new variables is pure scatter or very nearly so. Thus most of the variation in the original data is easily attributable to the size variation rather than to the shape variation of the diatoms. The results also imply that shape is much less important in setting sinking rates than is the density of the diatom relative to the medium, the main unmeasured residual variable.

A notable exception to the rule that larger means faster sinking is a condition analogous to that which makes fluffy parachutes useful in airborne seeds. The increase in size must make enough of a reduction in effective density to offset the effect of size *per se.* Hutchinson (1967) points out that a mucus coating can reduce sinking rate if the density difference between organism and mucus is more than twice the density difference between mucus and medium. The criterion is not devastating; both desmids and diatoms, among others, are often coated with thick mucilage. The disadvantage of this scheme is presumably the longer diffusion path in and out of the organism.

Sinking rates of phytoplankton are indeed low, ranging from about 0.1 to about 7 meters per day (Smayda, 1970). One meter per day is typical, which is about three orders or magnitude below the rates typical of fungal spores in air. One might ask why the organisms don't simply achieve neutral buoyancy; I think the answer has two components. First, hydrodynamic manipulation of shape and orientation can only reduce sinking rate; it can't stop the sinking altogether. The only real solution is density alteration. Second, though, is the really awkward factor. Neutral buoyancy may not be desirable at all (Lund, 1959; Smayda, 1970). A phytoplankter which doesn't move with respect to the water may deplete the local nutrients (Munk and Riley, 1952). These organisms are much larger than the bacteria for whom a wait-for-diffusion strategy is appropriate. Slow sinking may be quite a good thing and adjustment of sinking rate a useful variable in compensation for changes in environmental quality. The reported increases in sinking rates of cultured diatoms as nutrients are depleted (Smayda, 1974; Titman, 1975) may be a deliberate, selected response, a result of some specific physiological mechanism rather than any passive physical or chemical change.

For small, self-propelled animals, sinking is probably not too much of an issue. Plesset, et al. (1975) give figures for the ciliate protozoan, *Tetrahymena pyriformis,* from which sinking rates can be compared to swimming rates. *Tetrahymena* is 50 μm long and 30 μm wide and close to a prolate spheroid in shape; they report its density in fresh water as 1.076×10^3 kg m^{-3}. Even with the unlikely assumption that it always sinks parallel to its long axis, its calculated sinking rate of 56 $\mu m\ s^{-1}$ is still eight times less than its swimming speed of 450 $\mu m\ s^{-1}$.

propulsion at low Reynolds numbers

Circulation, as mentioned in Chapter 12, isn't worth much at Reynolds numbers of about ten or lower, so other schemes must be used for generating thrust. One which will work is an alternation of high and low drag strokes in opposite directions, what we earlier called the "Mary-Poppins-umbrella system." The drag of a recovery stroke can be kept below that of a power stroke in at least two ways. The first is to change the effective area of the moving appendage between the strokes, the second is to change the orientation of the appendage so that it either recovers traveling parallel to the flow or through the slower fluid near the body. Changing the speed of the strokes accomplishes little; these are reversible flows with a time-independent character so net progress cannot be made by using a fast power stroke and a slow recovery stroke.

Reciprocating appendages. Generally, when the area of an appendage is varied, the orientation is changed as well. Nachtigall (1974) has described very clearly how such a combined system works in the swimming appendages of aquatic beetles. The outer segments of the propulsive legs are little more than tubes to which bristles attach. The bristles are spread by the thrusting stroke of the leg until they lie in a plane; if the leg is flattened, they lie in the plane of flattening. Two-thirds to three-quarters of the thrust is attributable to the bristles. Even at these not-so-low Reynolds numbers, they produce up to 54% as much thrust as would an equally broad solid surface. During the recovery stroke, the hairs rotate back against each other and against the leg segments. In addition, the leg joints are flexed during the recovery stroke. At least one beetle has flattened blades in place of cylindrical bristles; spread during the thrust stroke, these overlap like drawn Venetian blinds; they give about 90% as much thrust as a solid surface (Figure 13.2). Drawing an appendage forward near the body may involve a steep velocity gradient, but it minimally counteracts the work of the power stroke, whatever its intrinsic cost.

Tiny crustaceans use much the same system of folding bristles and appendages, although the details vary with the body form of these diverse creatures (Strickler, 1977; Zaret and Kerfoot, 1980). They are certainly accomplished

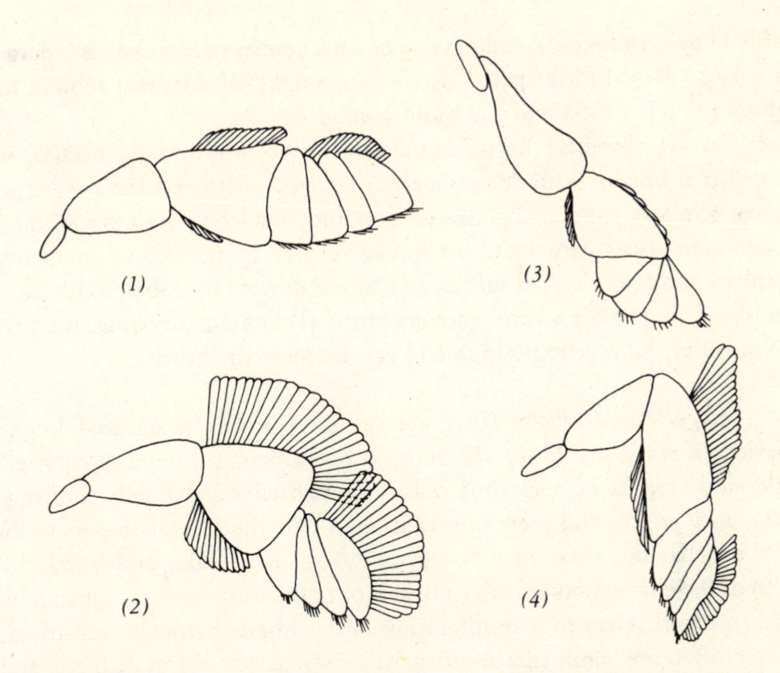

The rowing stroke of the hind leg of the beetle, *Gyrinus*. The oarlets fold back during the recovery stroke (*3*) and (*4*).

FIGURE 13.2

swimmers: the highest relative rate of movement of any aquatic animal, 200 body lengths per second, was recorded for a copepod crustacean; in such an escape reaction the Reynolds number of a millimeter-long animal may reach 500 (Strickler, 1975a). It must be admitted, though, that speeds this high are reached with the aid of a flapping abdomen.

Swimming around at low Reynolds number may present perils and opportunities for the particular perceptual world of these animals. A body moving through a fluid disturbs the fluid at a considerable distance lateral to itself, and the disturbance ought to be detectable to another body, perhaps to a predator or prey. In effect, to swim is to scream, "Here I am" (Strickler and Twombly, 1975). Conversely, the forces on a moving animal might tell it something about who and what may be lurking in its immediate surroundings. But perhaps active locomotion generates such complex signals that it "jams" the receipt of information about the outside world (Zaret, personal communication). Some copepods, at least, normally move intermittently, gliding and sinking just a little between strokes. A sinking rate of a millimeter per second, as reported by Strickler (1975b) for a species of *Cyclops,* might be sufficient motion for an animal to tell a bit about the world around it.

In the moderate Reynolds number world of swimming beetles, we've noted that bristles form dense sheets and may be flattened and overlapped to make a plane surface. By contrast, in the smaller and slower world of a cladoceran crustacean, hairs are spread further apart; velocity gradients are gentler, and fewer bristles suffice. In a model devised by Robert Zaret, the propulsive action of the second antenna of the cladoceran, *Bosmina,* is rather insensitive to the precise position and gap between the bristles.

Flagella and cilia. These are the most widely recognized locomotor devices of small organisms. There is no fundamental difference between the cilia and flagella of eucaryotic cells, although cilia are usually shorter (5 to 10 μm in length) and more numerous on a cell. A cell rarely has more than a few flagella, but these may be up to 150 μm long (Jahn and Votta, 1972). Often, flagellar action refers to a helical or undulatory wave propagation while ciliary action refers to a reciprocating beat without particular regard to any other difference. Both cilia and flagella are about 200 nm in diameter and are musclelike devices with contractile machinery inside (Holwill, 1977). Bacterial flagella, however, are something quite different; they're composed of a single filament rather than the conventional nine radial doublets and two central singlets, and they're only 20 nm (rather than 200) in diameter and lack any enzymatic activity (Berg, 1975). The arrangements and actions of these organelles are quite diverse. An organism or cell may be covered with cilia or have distinct bands of them; it may have one or more flagella on the anterior which pull, one or more on the posterior which push, or even a flagellum extending laterally which moves the creature in a helical path. The motion of a flagellum may take the form of planar, undulating waves either from base to tip or tip to base; a

FIGURE 13.3

The stroke of a cilium, extended during the power stroke and flexed down near the surface for recovery. Note that the cilium moves normal to its long axis during much of the power stroke and moves parallel to its long axis during much of its recovery.

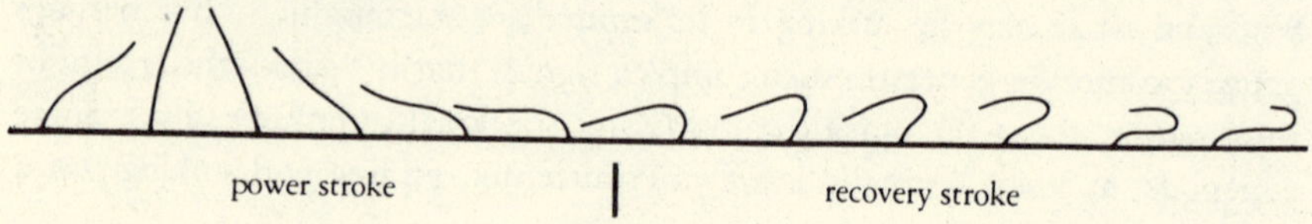

power stroke

recovery stroke

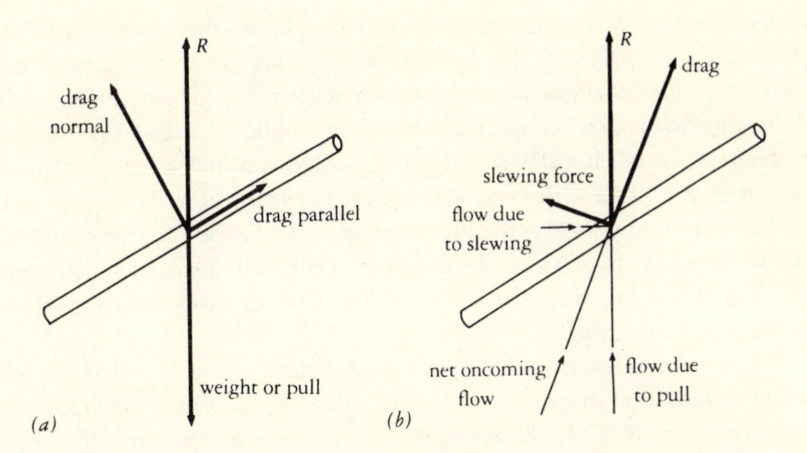

An obliquely oriented cylinder, pulled downward (*a*), slews forward (to the left) (*b*). The slewing results from the difference in resistance to motion along and transverse to its long axis (*a*); this difference gives rise to a kind of lift, the slewing force, when the resultant is reanalyzed (*b*). The overall effect is like a poor glide (Fig. 12.1.).

FIGURE 13.4

flagellum may propagate helical waves from base to tip; or it may produce spiral waves. Ciliary coats or bands beat more like our description of the swimming appendages of beetles or microcrustaceans—an extended thrusting stroke is followed by a flexed recovery stroke in which the cylinder moves lengthwise close to the surface of the organism (Figure 13.3). Groups of cilia are usually coordinated, and a host of patterns of coordinated beating have been described and named. Bacterial flagella are rigid helical structures which are driven in a circular path around the long axis of the helix by proper rotary motors to which they attach; they are the only truly rotational (wheel-and-axle) devices known in organisms (Berg and Anderson, 1973).

Does a common physical mechanism for propulsion underlie all this diversity? The basic scheme is best introduced by considering what would happen if a section of a circular cylinder is pulled obliquely through a viscous medium (Figure 13.4). We've previously noted that the drag of a cylinder oriented normal to flow is greater (1.6 to 1.8 times) than the drag of the same cylinder parallel to the flow. Sir G.I. Taylor was, I believe, the first to point out (1951) that this effect can be used for propulsion. If a cylinder is pulled obliquely, the difference in drag on the two extreme orientations generates a resultant force which is not in the direction opposite the movement of the cylinder. On a long tether, the cylinder will slew off to one side, much like a water skier with skis askew or a wedge which persists in going lengthwise even if hit by a glancing blow. If a cylinder is dropped obliquely, it slews off toward the direction in which its lower end points, looking just like a very poor glider; you can try it yourself with a fine wire dropped in a tank of water. According to Happel and Brenner (1965) the best orientation for slewing for the cylinder is with its long axis about 55° to the direction of the pull; it will then slew about 20° off the direction of the pulling force. The analysis is much like the one we did for a gliding airfoil.

If we then expand our view to a long, helical cylinder, a little thought should make it clear that rotating the helix will make the whole thing move forward. The game is much like screwing into a cork except that some slip occurs; the helical elements do not quite advance axially but, in the case just described, $55° - 20° = 35°$ to the long axis of each element. It makes little or no difference whether the helix truly rotates, as in a bacterium, or whether it translates in a circle, as does a eucaryotic flagellum.

Nor does it make much difference if the undulation is in a plane. Pieces of cylinder are still oriented obliquely and moved laterally. In planar undulation, though, not all of the flagellum can contribute, since some of its length is "wasted" at the upper and lower ends of each wave. On the other hand, a planar undulation doesn't tend to throw the rest of the organism into rotation in the opposite direction.

These devices are all (nearly all, really) cylindrical—none are finlike or paddlelike, which might seem better. But remember that circulation has no part in the scheme, and the difference in drag is central. A flat plate normal to flow has only 1.5 times the drag of one parallel to flow; a cylinder (depending on its length) has a ratio of 1.6 or better. For a disk, the maximum angle of slewing will be less that 12° instead of nearly 20°, so moving a paddle like a fishtail won't work as well as oscillating a cylinder.

Cilia, of course, typically beat rather than undulate. But the drag difference is still crucial—they just alternate normal and parallel exposure instead of combining the two in an oblique presentation. Neither ciliary nor flagellar propulsion can really be described as efficient. A ciliary tip velocity of 4 mm s^{-1} produces an average flow past the cilium of only about 1 mm s^{-1} (Sleigh, 1978). Still, propulsion seems to take only a small part of the metabolism of these animals, and a competitor must pay quite the same price. Ciliary propulsion may have fringe benefits. Because water is actively moved very close to the surface of the animal, the influence of the animal on the flow is far more limited further away than if it were passively sinking at the same speed. So propulsion using many cilia may be "quieter" than reciprocating a few large appendages or even sinking and might give less signal to either predators or prey. Furthermore, it may augment diffusive exchange by reducing the amount of semi-stagnant water around an animal.

Usually animals using ciliated surfaces for locomotion are both larger and faster than those using a few flagella. Where sizes overlap, the ciliated forms go about ten times faster than flagellated ones. But, as Sleigh and Blake (1977) point out, propulsive efficiency drops with increasing size for either and no large swimmer in a hurry uses either scheme: muscles give another order-of-magnitude increase in speed.

Besides the references cited here, a voluminous literature treats cilia and flagella, including complex hydrodynamic analyses and treatments of wall effects, body oscillations, coordination, bioconvection, and so forth. The best sources seem to be the first volume of the pair edited by Wu et al. (1975) and the review by Brennen and Winet (1977). As a final point, let me emphasize that the swimming of an elongate fish may look like that of a sperm or flagellate, but the two are based on very different physical principles. The former is built around circulation and the induction of rearward momentum in the fluid; the latter about neither but rather on a difference in drag in two extreme orientations.

filtration

Filter feeding usually involves a crucial component of flow at low Reynolds numbers: edible particles suitable for capture are rarely over a millimeter in diameter, and sieving devices likewise are small. Flow speeds at the filter are most often slow, partly, one may guess, because fine filters may not be mechanically robust and partly because a large filtration surface will, by continuity, usually result in low speeds. The only high Reynolds number filter-feeder may be the baleen whale, and little is known about the details of its operation.

The pre-eminent problems of filter feeding from the viewpoint of fluid mechanics are those of ensuring contact between edible particles and filtration surface and of keeping the resistance of the filter low enough so flow goes through it. To a large extent, these are antithetical requirements: a fine-meshed filter will have a high resistance to flow. The viscous character of flow at low Reynolds number can only make matters worse: if you can bail with a sieve, you can't easily filter with it. The problem can be illustrated by considering how much fluid goes through and how much around a filter. I once took a large pinnate antenna of a male saturnid moth and looked with a microanemometer at how much of the air that its path would intercept would actually go through it—the figure was only about 20 to 25% (the air seemed to go more readily from front to back!). By contrast, 50% of the light impinging on the antenna went through. Calculations of the efficiency of antennae as pheromone detectors should incorporate a correction for viscosity.

It is remarkable just how low the Reynolds number of a filter-feeding device can get. Both the tiny microvilli which make up the collar of sponge choanocytes and the intervillar gaps are 0.1 to 0.2 μm across, and water passes through them at about 3 $\mu m\ s^{-1}$ (Reiswig, 1975a); the corresponding Reynolds number is 4×10^{-7}! The arrangement is capable of filtering bacteria and other material in the 0.1- to 1.0-μm size range (Reiswig, 1971b; 1975b). The geometry of a choanocyte is not particularly complex (Figure 3.4, page 31), and work on scale models in high viscosity fluid might be illuminating.

The physical basis for filtration by organisms has received surprisingly little attention despite the vast literature on the biology of filter feeding. (On the latter, see Jørgensen, 1966 and 1975, and Wallace and Merritt, 1980). The only paper on physical mechanisms seems to be that of Rubenstein and Koehl (1977). They stress that simple "sieving" of particles too large to pass through a filter is only one of at least five possible capture mechanisms and that each of the others permits a filter to capture particles smaller than the openings in its mesh (Figure 13.5). Particles can also be captured by "direct interception" where a steamline passes within a particle radius of the surface. Or they can

FIGURE 13.5

Five possible mechanisms of filter feeding, according to Rubenstein and Koehl (1977); (*a*) sieving; (*b*) direct interception; (*c*) inertial impaction; (*d*) motile particle deposition; (*e*) gravitational deposition.

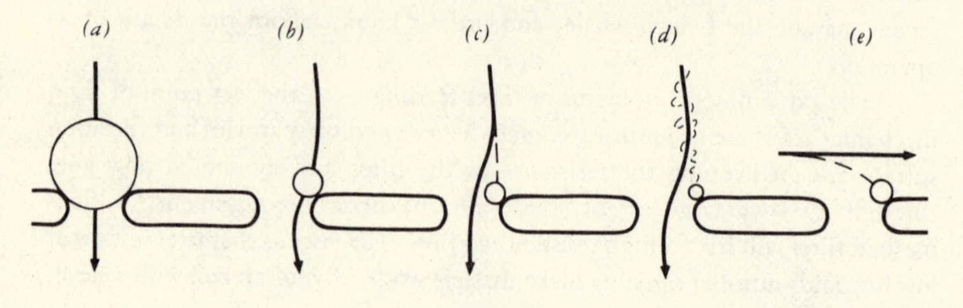

strike the solid surface by "inertial impaction," where the streamlines may bend away from the surface but the inertia of the particle keeps it close to its earlier course. Or they may hit and stick as a result of their Brownian movement —"motile particle deposition." Or they may experience "gravitational deposition," deviating from streamlines as a result of a density difference between particles and medium. Even electrostatic attraction may be possible. Rubenstein and Koehl provide handy dimensionless indices for deciding which phenomena ought to predominate under particular conditions of mesh size, particle size, flow speed, and fluid viscosity.

But the Rubenstein-Koehl model is a broad-brush view, badly wanting specific tests and more detailed analyses. LaBarbera (1978) persuaded brittle stars in a flow tank to capture Sephadex particles too small for direct sieving. The slight preference of the animals for larger particles was in good agreement with the predictions of the model. He also found a marked preference for charged particles; but, surprisingly, complete indifference to the sign of the charge. Besides tests on organisms, cognate work on physical models ought to be useful, and perhaps measurements on isolated natural filters might be practical as well. I think a promising system might be the silk nets of the hydropsychoid Trichopteran (caddisfly) larvae. Their nets are simple rectilinear meshworks and they come in a variety of mesh sizes, depending on species, instar, and current speed (Wallace et al., 1977). The animals are cosmopolitan and tolerant of experimental manipulation (Philipson and Moorhouse, 1974); the currents in which they feed are steady and unidirectional; and their biology is well known and has been well reviewed by Merritt and Cummins (1978) and Wallace and Merritt (1980).

Additional sticky problems are raised when the particles swim and are volitionally competent and when the filter is an active pump or grappling device. Strickler and his coworkers (see Alcaraz et al., 1980) have done some work on prey capture in copepods which addresses these problems; the more they investigate, the more complicated and sophisticated the copepods appear.

interstitial flow

After a wave travels up a gentle sandy beach, all of its water does not immediately run back down the beach. Instead, a substantial fraction sinks into the sand and makes its way downward and seaward through the interstices between the sand grains (Riedl and Machan, 1972). While the flow oscillates with each passing wave, the net flow is always seaward, driven by the gravitational pressure-head of the swash on the beach. A typical speed is about 0.5 mm s^{-1}; the resulting flow into the ocean, globally, is greater than that of all the world's rivers combined (Riedl, 1971b).

Flow through porous media is a well-explored subject of great interest to chemical and sanitary engineers. Its equivalent of the Hagen–Poiseuille law, Darcy's law, states that volume flow is directly proportional to the applied pressure and the cross-sectional area of the bed and is inversely proportional to the depth or thickness of the bed. The constant of proportionality, the "hydraulic conductivity," contains a factor inversely proportional to the viscosity of the fluid medium (see, for example, Leyton, 1975); thus Darcy's law is a low-Reynolds number relationship analogous to Stokes' law.

From our present point of view, the main drawback of the conventional literature on flow through porous media is its overly macroscopic approach and consequent neglect of what is going on between individual sand grains. Much biology happens between these grains; an extremely large and diverse community of small organisms lives there. This interstitial meiofauna includes members of most of the invertebrate phyla; it's particularly rich in such forms as turbellarian flatworms and gastrotrichs. The animals adhere to and crawl between the grains without disturbing the packing of the sand; they are exposed to flows of about one body length per second (it's hard to know whether to call these fast or slow flows); and the Reynolds numbers are fairly low, about 0.2, so viscous forces predominate (Crenshaw, 1980).

Much is known about the composition of this fauna (Coull and Bell, 1979, is a good starting place); much less is clear about the peculiar physical environment and how the animals respond to it. As Crenshaw emphasizes, the most important parameter is the shear stress on an animal, and not knowing field values of this variable has limited the utility of laboratory studies. When selecting a site, an organism probably picks the right shear stress rather than some grain size, and the choice typically requires only a short movement since, on a micro-scale, a beach is a fairly heterogeneous habitat.

Crenshaw (1980) has actually determined the shear stresses on meiofaunal creatures by stroboscopically photographing the passage of tiny latex spheres across animals contained in square capillary tubes. The stresses turn out to be very low—around 0.003 to 0.05 N m^{-2}—but the animals respond to these stresses in several ways. At high flow rates they do much more moving around; in sand they move predominantly upstream and downstream rather than across the flow. Some seem to exert control over the flow around them. In nature, the meiofauna live in what amounts to an array of parallel channels; Crenshaw modelled this by using an empty tube in parallel with one containing an animal. A turbellarian in a fast flow laid down an extensive network of mucus strands, effectively diverting the flow to the other channel. But the physical force of the flow may not have been an inhospitable factor; the mucus could have been part of a feeding scheme.

Evidence is increasing of functional as well as phyletic diversity in the interstitial meiofauna. Yet the convergence toward or selection for common sizes, shapes, and structures is most remarkable (Swedmark, 1964). Therefore this extensive commonality is likely to reflect the exigencies of the physical environment of sand grains and water flow. Yet the physical world of these animals is as removed from ours as it can be: try to imagine slithering through moving glop between boulders while blindfolded. The problem is the one with which we begin this chapter—that of generating hypotheses about adaptations to a world very different from anything of which we have perceptual experience.

CHAPTER 14

a few final phenomena

WHILE THE READER MAY BE MOST CONSCIOUS of how much we have covered so far, the writer is more aware of how much we have not. This chapter will contain a series of short introductions to a few more phenomena. About all these have in common is that they haven't been previously discussed and that they are or might be of biological importance.

unsteady flows

In virtually every instance, the flows mentioned so far have been steady ones, established some time in the past and persisting into the immediate future. Drag and lift were steady forces or, if they varied, could be treated as a series of steady-state situations. The real world, however, is full of unsteady or time-varying flows—winds gust, waves pass, tides turn, and objects accelerate. So additional complications rear up, together with several new phenomena which can't be treated simply as minor perturbations of previous material.

Vortex shedding. Back in Chapter 5 we noted that rows of vortices were shed behind bluff bodies at Reynolds numbers above about 40. These vortices were left behind in the wake, first from one side of the body and then from the other, and each vortex rotated in the opposite direction of the preceding and succeeding ones (Figure 5.5, page 74). In effect, a steady flow across a rigid body can produce an unsteady wake. Any object in, or crossing, the wake will be buffeted by these vortices; in an array of closely-spaced cylindrical objects exposed to a flow, the foremost ones may be less subject to damage than those further back.

But even a single bluff body is affected. Recall the earlier discussion of circulation and the conservation of vorticity in ideal fluids. Any circulation created in one place had to be balanced by an equal and opposite circulation somewhere else. In the present case, we have to consider the bound vortex around the bluff body together with all of the vortices the body has shed. The circulation about the body must be equal to and opposite the net direction of the combined circulation of the shed vortices. The implication is curious: every time a new vortex is shed, the circulation around the body must reverse direction! Circulation, of course, produces the transverse force which we've called lift. So the object is shaken in a direction normal to the flow as a result of shedding vortices; the frequency of shaking must be precisely one-half the frequency with which vortices are shed (Figure 14.1). The situation is an everyday one: suspended electrical wires "sing" aeolean tones in the wind, and some cases of whistling winds trace to the shedding of these von Karman trails. The shaking of an object induced by the shedding of vortices will be of most consequence when the object is either flexible or flexibly mounted and when the rate of shedding closely coincides with some natural oscillatory frequency of an object.

What determines the frequency with which vortices are shed? As with so many cases in fluid mechanics, a precise and universal formula is unavailable; but general guidance can be supplied. As mentioned earlier, both the Reynolds number and the drag coefficient can be obtained by dimensional analysis, assuming that four variables (length, velocity, density, and viscosity) are germane. If the same sort of dimensional analysis is done with the addition of a fifth variable, frequency (dimensions of T^{-1}), another dimensionless index emerges. In the present application, this variable is termed the *Strouhal number:*

$$St = \frac{nl}{U} \tag{14.1}$$

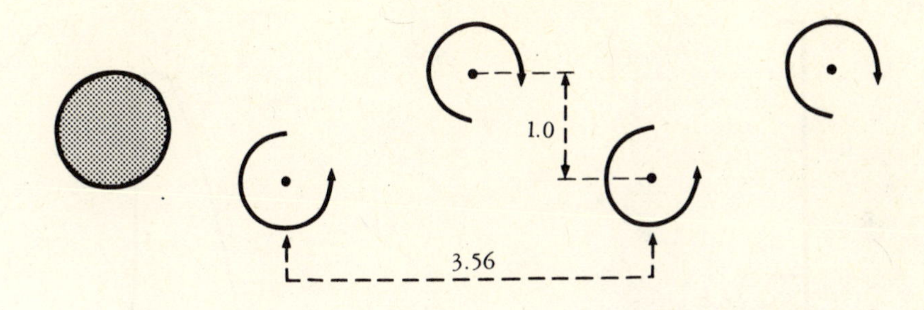

The spacing of vortices in the von Karman trail behind a bluff body: the ratio of 3.56:1 is stable.

FIGURE 14.1

where *n* is the frequency of a periodically-varying flow; *l* is a characteristic length of a solid object, usually the diameter or the distance perpendicular to both flow and the long axis of the object; and *U* is the free-stream velocity. The Strouhal number, then, is a dimensionless frequency, quite analogous to the drag coefficient, which is a dimensionless drag. And, like the drag coefficient, it is a function of shape and Reynolds number alone. Thus the Strouhal number is conveniently plotted as an ordinate against the Reynolds number, with each resulting line corresponding to a different shape. Given the Strouhal number, the size of an object, and the free-stream speed, the frequency of vortex shedding is easily figured.

Among biologically interesting shapes, data are available for cylinders and flat plates normal to flow (Figure 14.2). Fortunately, the Strouhal number varies little with Reynolds number in the range which is most likely to concern us, even though the physical character of the flow changes. For a cylinder, for example, at Reynolds numbers below 40, no vortices are shed, so the Strouhal number is effectively zero. From 40 to 150 the train of vortices (the "vortex street") is laminar, and the Strouhal number rises smoothly from 0.1 to 0.18. From 150 to 300, the vortices gradually become internally turbulent, and turbulence persists up to about 3×10^5 with a nearly constant Strouhal number of 0.20. Above that "drag crisis" the wake gets narrower, the boundary layer is turbulent, and the vortices are disorganized (Blevins, 1977). Above about 3×10^6, a discrete vortex street is re-established, oddly enough, and it persists up to at least $Re = 10^{10}$. The last measurement, by the way, is no wind tunnel determination but comes from a satellite photo of a cloud-marked vortex trail behind the mountain peak of Guadalupe Island (Simiu and Scanlon, 1978).

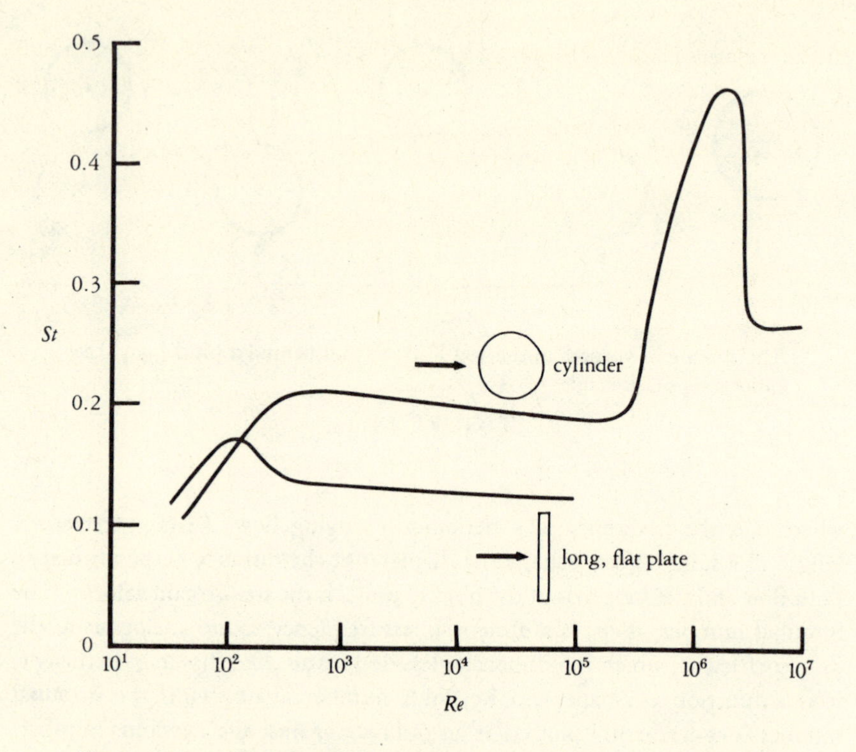

The variation of the Strouhal number with changes in the Reynolds number for a cylinder and a flat plate.

FIGURE 14.2

Another characteristic of these vortices is their relative spacing. Von Karman determined that the trails are stable if the distance between successive vortices on each side of the street is 3.56 times the distance between the center lines of the two rows of vortices, as shown in Figure 14.1. This ratio does not change with speed or the way in which the body produces the wake. If the object is free to move transversely to the free stream, the width of the vortex street will increase in proportion to the effective width of the object, now increased by the amplitude of oscillation. And the spacing between vortices will increase as well, maintaining the usual ratio. The motion of the object will, as well, increase the drag, circulation, and lift; and both the frequency and the Strouhal number will decrease (Steinman, 1955). The vortices, incidentally, do not remain stationary but move in the direction of the relative motion of the object at a rate slower than the progression of the object. Their speed is proportional to their circulation, so from measurements of the dimensions and speed of a trail of vortices, one can calculate the lift and drag of an object as described by Prandtl and Tietjens (1934).

Little in the way of biological mischief has been laid to vortices shed by this mechanism. The vortex shedding, though, is real: in a larch plantation in which trunk diameters were about 0.4 m and wind about 0.6 m s^{-1} the predominant frequency of turbulent eddies was, as expected, about 0.3 Hz (Grace, 1977). Perhaps it is of some importance that organisms do not shed vortices at frequencies at which they might oscillate. In a 20-m s^{-1} wind, a 0.2-m tree trunk will shed vortices at 20 Hz—probably too high a frequency to have any bearing on sway or wind-throw. Holbo et al. (1980) found a sharp compliance peak in 0.35-m Douglas fir at 0.3 Hz—well below the vortex shedding frequency of a high wind but well above typical gust frequencies of about 0.05 Hz. A cat's whisker of 0.3 mm diameter in a 1-m s^{-1} wind would shed vortices at 1000 Hz. But the cat's whisker is neatly tapered from thick base to fine point, so no discrete frequency should characterize the system.

One possible use of vibration induced by shedding vortices involves the spore cases of mosses (Figure 14.3). These urn-shaped structures are borne on

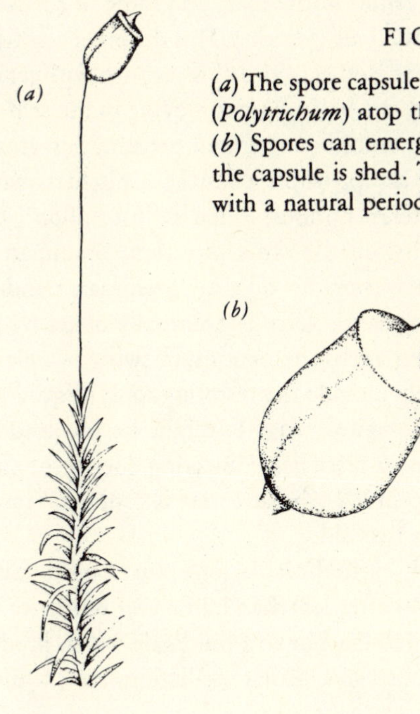

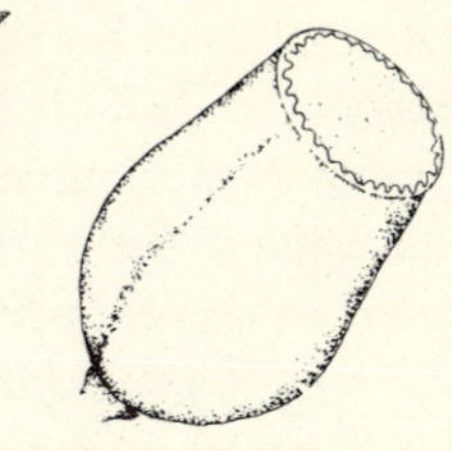

FIGURE 14.3

(*a*) The spore capsule and stalk of a moss sporophyte (*Polytrichum*) atop the leafy gametophyte.
(*b*) Spores can emerge after the lid (operculum) of the capsule is shed. The stalk is noticeably springy, with a natural period of around 0.1 s.

stiff stalks about a centimeter above the green part of moss plants. Spores are liberated into the wind after a lid drops off the spore case. Perhaps vibration of the stalk and shedding of vortices "locks in," ensuring a salt-shaker-like action when wind speed reaches some critical value. The details are uncertain, because Strouhal numbers do not seem readily available for the appropriate shapes; indeed, bodies symmetrical about an axis parallel to flow may shed toroidal vortices and induce vibration parallel rather than normal to the wind.

Vortex shedding at a frequency determined by the Strouhal number generates a purely aerodynamic or hydrodynamic forced vibration; that is, the periodic force driving the vibration exists whether or not the structure actually moves. To this mechanism must be added another of a somewhat different origin, but one which may act in concert with the former. This second scheme, involving self-excited oscillations, also releases a trail of alternating vortices; but the frequency at which they're shed is no longer directly related to the free-stream velocity.

According to Steinman (1955), the mechanism of these self-excited oscillations may be described as follows. Consider a half-cylinder with the flat face upstream (Figure 14.4a). If the half-cylinder is moving laterally as a wind strikes it, the net or relative wind will approach obliquely. The forward stagnation point will be offset from the center of the face, and more of the flow will move around the leading edge of the half-cylinder. This difference in flow will result in a net circulation around the object, and that in turn will generate a lifting force which will tend to keep the half-cylinder moving in the same direction. In other words, the flow will amplify any initial perturbation from zero speed. Lanchester, in 1907, used the principle to make a pitchless windmill which he (always enamored of foreign phrases) called a "tourbillon" (Figure 14.4b): once started, it keeps going in the same direction. Steinman (more taken with his own name) has a version he calls a "Steinman pendulum" (Figure 14.4c), in which the aerodynamic force is eventually offset by gravity but reappears in each half-cycle and keeps the pendulum swinging. He points out that a spring might alternatively provide the restoring force (Figure 14.4d); the result takes on a distinctly biological tang. It might be involved in the flutter of leaves on resilient petioles or branches. Fluttering does occur and may be adaptively significant: Schuepp (1972) showed that it could increase both heat and mass transfer as much as twofold.

The most famous case of self-excited oscillations must be the failure of the Tacoma Narrows Bridge in 1940 after less than half a year of service. It was designed for static wind loads of 2400 Pa, but in a mild gale which produced a static load of only 240 Pa it developed spectacular oscillations and, while spec-

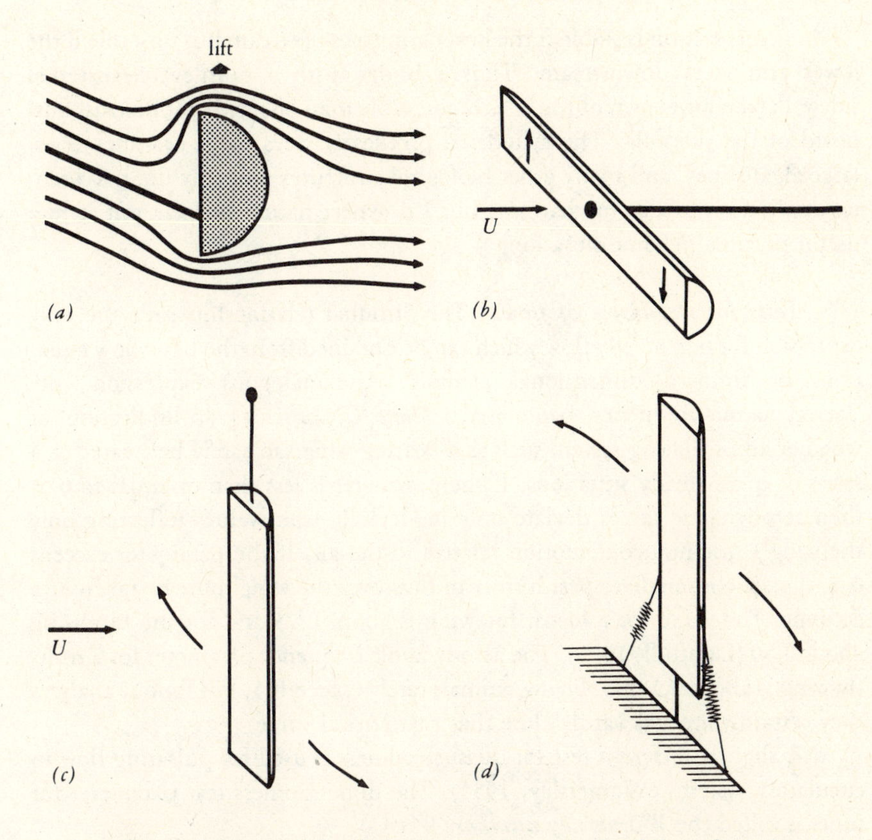

(*a*) The asymmetry of the forward stagnation point in oblique flow gives rise to a lifting force normal to the free stream; the resulting motion of the half-cylinder can maintain this asymmetry. Some self-excited oscillators: a pitchless propeller (*b*), once started, will keep rotating; a pendulum (*c*); and (*d*) a spring-restored oscillator.

FIGURE 14.4

tators watched and movies were made, broke from its cables and dropped into Tacoma Bay.

These oscillations don't depend on resonance between the frequency of speed-dependent oscillations and that of mechanical oscillations so they're worrisome to bridge builders but perhaps more versatile for organisms. The limitation is one of shape. An entire cylinder is a relatively stable profile; a half-cylinder is stable when the flat side is downstream but unstable when the flat side is upstream. A flat plate normal to flow is unstable. A beam shaped like a

"T" in cross section is stable if the lower arm faces upstream but unstable if the lower arm faces downstream. Thus a bridge with a cantilevered external sidewalk (the lower arm of the T) is more stable than one with a solid truss outboard of the sidewalk. The cylindrical profiles of trees, large sea anemones, large algal stipes, and many other biological structures probably limit accidental application of the mechanism, but I'd expect nature to have put it to a useful purpose in some situation.

Tests for steadiness of flow. The Strouhal relationship isn't the only expression for non-steady flow which can be obtained from the basic flow equations or from a dimensional analysis. An analogous expression, the "aerodynamic frequency parameter," $2\pi nc/U$, provides an indication of whether an oscillating system such as a beating wing can safely be treated as a series of quasi-steady situations. If the parameter is less than or equal to 0.5, then aerodynamic forces deviate only moderately from values reflecting only the wing's instantaneous motion relative to the air. If the parameter exceeds 0.5, then the immediate past history of flow over the wing must be taken into account. The value for a locust forewing is about 0.25 and for the hindwing about 0.50 (Lighthill, 1975). The aerodynamic frequency parameter for a fruit-fly wing is about 0.5 also. Flying animals rarely exceed 0.5, but for our analyses they certainly are awkwardly close that that critical value.

A slightly different test for quasi-steadiness is used for pulsating flow in circulatory systems (Womersley, 1955). The dimensionless test parameter for pipes is called the *Womersley number, Wo:*[†]

$$Wo = a\sqrt{\frac{2\pi n\rho}{\mu}} \qquad (14.2)$$

Here n is the frequency of a sinusoidally-applied pressure gradient. If Wo is less than unity, viscosity dominates and the profile of flow across a pipe is essentially parabolic, so the flow is said to be quasi-steady. If Wo is above unity inertial forces distort the profile and the Hagen–Poiseuille law cannot be applied in good faith. In our aorta, Wo is 10 or more, while in our capillaries it is around 0.001 (Caro et al., 1978). It might be comforting (or at least of some interest) to note that the square of the Womersley number looks like the product of the Reynolds and Strouhal numbers.

[†] Wo is called α in the original paper.

Added mass. Drag may be defined as a force which resists motion. As we've treated it, drag resists acceleration only to the extent that acceleration involves motion. For an accelerating body (as in the initial stages of free fall), the drag increases from zero at the start until it equals the accelerating force and terminal velocity is reached. Drag, however, is not the only sort of resistance which acts on an accelerating body; there is, in addition, a resistance to acceleration *per se*. The phenomenon may be introduced with an extreme example, adapted from Birkhoff (1960).

Consider a spherical bubble of air in water at atmospheric pressure much like a bubble of CO_2 at the bottom of a glass of beer. The bubble is 1 mm^3 in volume, so its mass is 1.2×10^{-9} kg. It is suddenly released and accelerated upward by the buoyant force, that is, by the difference between its weight and that of the displaced water. Since the bubble's weight is negligible in this context, the buoyant force is 9.8×10^{-6} N. The *initial* acceleration of the bubble is that buoyant force divided by the mass of the bubble—8200 m s^{-2} or 830 times the acceleration of gravity, an astonishing figure.

But for the bubble to advance upward, some water must move downward. The paths of water and gas are different: the bubble goes straight up while the water must go around, so the mechanical problem is not a trivial one to analyze. Since water is by far the denser fluid, most of the buoyant force will be invested in moving water downward rather than gas upward. For a sphere in an ideal fluid, the details have been worked out. To the mass of the bubble must be added an "added mass" (sometimes called "apparent additional mass" or, together with the mass of the object, the "virtual mass") which, for linear acceleration, is equal to one-half the mass of the displaced fluid. We can now recompute the acceleration of the bubble, taking its effective mass as 0.5×10^{-6} instead of 1.2×10^{-9}. The value we get is only twice the acceleration of gravity, a rather more credible result.

It must be emphasized, of course, that the added mass is of significance only under rather special circumstances. First, the body must be accelerating or decelerating. The added mass will reduce the rates of both acceleration and deceleration; drag, by contrast, will reduce acceleration but will increase deceleration. Second, the displaced fluid must have a mass at least comparable to that of the accelerating object. Thus, added mass is more likely to be important in water than air. (But a pendulum clock with a spherical bob will run about 10 seconds per day slower in air than in a vacuum.)

The shape of the body makes a considerable difference. It is customary to define an added mass coefficient as the ratio of the added mass to the mass of the displaced fluid. So defined, the coefficient for a sphere is 0.5, as in the example above. For a circular cylinder normal to flow the coefficient is 1.0. For a prolate spheroid twice as long as it is wide, the coefficient is 0.209; for a 3:1 spheroid it is 0.122; and for a 4:1 spheroid it is 0.082; the spheroids are progressing in the direction of their long axes (Lamb, 1932). As pointed out by Prandtl and Tietjens (1934), these figures are of somewhat uncertain applicability at low Reynolds numbers since they were computed for ideal fluids. Streamlined objects are clearly better suited to rapid acceleration in dense media as well as to minimization of ordinary drag.

Biologically, we should expect added mass to be of greatest importance in intermittent aquatic locomotion. I invoked the phenomenon for insect wings in air (Chapter 9), but such cases are unlikely to be either very common or consequential. One clearly documented case in which the added mass could not be neglected is that of the escape response of crayfish (Webb, 1979). Rapid flipping of the tail, using the large mass of abdominal muscle, moves the animal backward in a pulsating motion. As a result of the high acceleration, short duration of each acceleration phase, and density of the medium, the inertial forces associated with the added mass were ten times greater than the drag forces. Thomas Daniel (personal communication) has evidence that added mass is important in the pulsating locomotion of jellyfish.

the stability of density gradients

The assumption we've been making that fluids have uniform densities is not always realistic. Particularly when the temperature of the fluid is not uniform, density varies from place to place. The layering of a less dense part of a body of fluid on top of a more dense portion is usually stable, sometimes distressingly so. Thus an "atmospheric inversion" with hot air above cold air traps the exhalations of contemporary life in the form of haze or smog. And a lake may be severely stratified during the summer, with the surface region depleted in nutrients as material sinks through the so-called "thermocline" into the cold water beneath. Winds in the atmosphere or blowing across the surface of a lake ought to help mix things up and put matters right if they're sufficiently strong or steady, as shown in Figure 14.5. But many lakes are never wind-mixed and only "turn over" in spring and fall.

A good test is available for whether density gradients will remain stable or whether instability will set in and produce a layer of vortices. The test involves

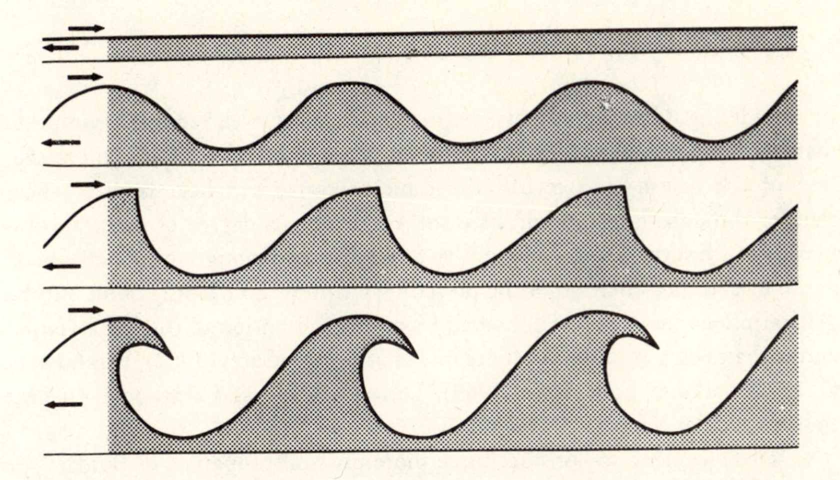

The relative stability of density gradients subjected to shear. Vortices will form in such stratefied systems if the Richardson number is sufficiently low. The consequent mixing will decrease the steepness of the gradient.

FIGURE 14.5

another dimensionless index, the *Richardson number, Ri:*

$$Ri = \frac{g(d\rho/dz)}{\bar{\rho}(dU_x/dz)^2} \qquad (14.3)$$

The expression is less complicated then it appears. The ratio $(d\rho/dz)$ is the vertical density gradient which keeps the layering stable and the ratio (dU_x/dz) is the vertical velocity gradient which promotes mixing through shear. The two can be obtained experimentally from sets of measurements of temperature and velocity versus depth. Clearly stability is favored by higher Richardson numbers; the practical test is that the Richardson number must be somewhere below 0.25 for instability to be triggered. Negative values occur, in contrast with the Reynolds and Strouhal numbers which are always positive. The density difference need not be the result of a temperature difference; it might reflect salinity gradients in an estuary, for example. Further information on the meaning and use of Richardson numbers may be obtained from Hutchinson (1957) or Mortimer (1974) for lakes or from Scorer (1978) or other meteorological texts for the atmosphere.

convective heat transfer

In considering density gradients, we've very cautiously relaxed our assumption that the temperature of a fluid is constant throughout its volume. Full relaxation of the assumption reveals the topic of convective heat transfer—heat transfer through fluid motion. This subject involves a degree of physical complexity which surpasses anything we've faced thus far, enough to fill half a book at the level of presentation of the preceding chapters. So I'll only point out the general phenomena and major variables and call attention to the more glaring and embarrassing pitfalls. For more information, see Gates (1962), Parkhurst et al. (1968), Porter and Gates (1969), Leyton (1975), and textbooks on heat transfer such as Kreith (1973).

First, we need to consider three more physical properties of fluids:

1. *Thermal conductivity, k,* is the rate of heat transfer across a sheet of material of unit area, unit thickness, and unit temperature difference between its faces. No internal movement of the material is permitted (we'll get to that later). In effect, the thermal conductivity is the reciprocal of the insulation value of the material. The dimensions of thermal conductivity are $MLT^{-3}\Theta^{-1}$, and its SI units are $\text{joules}^{1} \times \text{meters}^{-1} \times \text{seconds}^{-1} \times \text{degrees}^{-1}$. Notice that the definition is referred to volume, not mass. On this volumetric basis, water has about 22 times the thermal conductivity of air: stationary air is a superb insulator.

2. *Thermal capacity,* C_p, is the energy necessary to raise a unit mass by unit temperature; it's also called heat capacity or specific heat. The subscript, p, indicates constant pressure, because thermal capacity at constant volume is important in another context. The dimensions of thermal capacity are $L^2T^{-2}\Theta^{-1}$; its SI units are $\text{joules}^{1} \times \text{kilograms}^{-1} \times \text{degrees}^{-1}$. This property, by contrast with the previous, is referred to mass; water has a value about four times that of air. The value for water is, in fact, unusually high for common materials.

3. *Thermal expansion coefficient,* β, is the fractional expansion of volume (subtract old volume from new volume and divide by old volume) for a unit rise in temperature at constant pressure. Its dimension is Θ^{-1}, and its unit is the reciprocal degree. Water has a particularly low thermal expansion coefficient with a strange temperature dependence: maximum density occurs at 4 °C, and water expands when cooled further. Any ecologist will gladly enlighten the reader about the profound biological consequences of this odd behavior. Air is

a most ordinary substance with regard to thermal expansion: it closely follows the perfect gas law in which volume is directly proportional to absolute temperature.

Except for the thermal expansion coefficient of water, the values of none of these properties vary significantly with small changes in temperature for either air or water.

Then we have to distinguish between two quite different regimes of convective exchange. In *free convection,* movement is generated entirely by spatial irregularities in density, the latter typically caused by temperature variations within the fluid. Strictly speaking, there is no free-stream velocity. Boundary layers are peculiar (Figure 14.6): if the surface of an object is warmer than the surrounding fluid, then the fluid near the object will rise as it warms (except, of course, it the fluid is water between 0° and 4°C). The no-slip condition still holds, so the velocity of flow approaches zero linearly near the surface. But with no free-stream movement, the velocity also approaches zero, this time asymptotically, as the distance from the body increases.

In *forced convection* an external agency propels the fluid medium past the object; so the flow isn't driven by buoyancy, the free-stream flow is quite real, and boundary layers have fairly ordinary profiles. Both free and forced convection may be either laminar or turbulent; worse, mixed regimes are common; still worse, the direction of the free convective current, up or down, may not coincide with the direction of the forced current. Furthermore, the convection around a hot object in a cold fluid isn't just the reverse of that for a cold

FIGURE 14.6

Temperature and velocity profiles adjacent to a heated, vertical, flat plate in pure free convection.

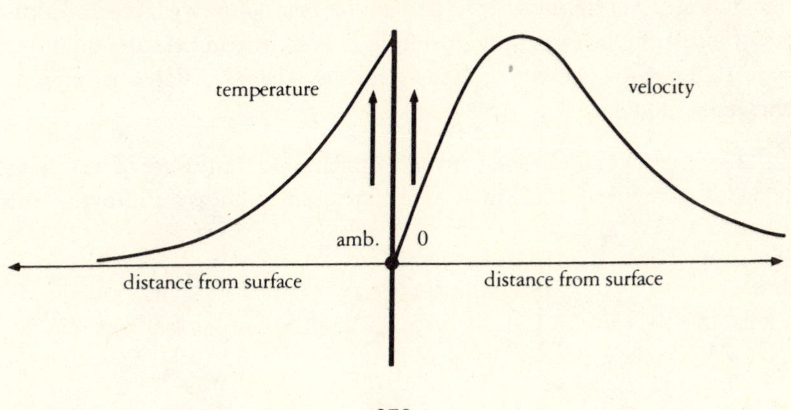

object in a hot fluid since viscosity varies with temperature. And, if convection heats or cools an object non-uniformly (as is usually the case), the conductivity of the object takes on significance.

We also must recognize that convective heat transfer is only one of several biologically important modes of heat transfer, and practical situations inevitably involve the simultaneous operation of more than one. One can't summarily neglect *radiative exchange,* heat *conduction,* or heat transfer through *change of state,* as in evaporation or melting. In addition, heat may be transferred through *mass transfer,* as when one takes in a cold drink and discharges hot urine, and through *changing the chemical composition of solutions,* although I know of no biological examples of this. Changes of state and mass transfer commonly involve flow. The innocence of previous generations of biologists about radiative exchange in particular led to some strange and amusing literature.

A set of dimensionless indices can be used to describe or to categorize situations involving convective heat transfer; I'll describe the ones the reader is most likely to encounter:

1. *The Prandtl number, Pr,* depends only on the material properties of the fluid and gives an indication of its quality as a medium for convective heat transfer. It is given as

$$Pr = C_p \mu / k \tag{14.4}$$

Physically, it is the ratio of two transport properties: the kinematic viscosity, which affects the velocity distribution, and the thermal diffusivity, which affects the temperature profile. Thus it relates the velocity distribution to the temperature distribution. A higher value indicates a poorer transfer medium: more work must be done against viscosity to move heat around. Thus a higher value indicates a better insulator. (Remember that boundary layers constitute a kind of insulating layer.) Liquid metals have Prandtl numbers around 0.01, air about 0.72, water about 6, and glycerin, 31,000. Water is typical of nonmetallic liquids.

2. *The Nusselt number, Nu,* is a dimensionless measure of heat transfer analogous to the drag coefficient as a dimensionless measure of drag. Its formula is

$$Nu = \frac{q_c l}{kS\,\Delta T} \tag{14.5}$$

Here q_c is the rate of heat transfer by convection, in joules per second; l is a characteristic length of the object immersed in the hotter or cooler fluid; ΔT, the temperature difference between the surface and the free stream; and S, the surface area of the object. Physically, it is the ratio of the temperature gradient in the fluid at the interface to the total difference in temperature between the object and the free stream. Alternatively, it can be viewed as the ratio of the length of the body to the thickness of the thermal boundary layer. Were the Nusselt number constant, heat transfer would be proportional to the temperature difference, and what the physiologists sometimes call "Newton's law of cooling" would work. But the Nusselt number is no more constant than the drag coefficient, and this so-called law is less likely to lead than to mislead. Like the drag coefficient, the Nusselt number makes a good ordinate to pair with the Reynolds number as abscissa.

3. *The Grashof number, Gr,* may be the messiest looking of these indices, but it is perhaps the simplest to envision in use. It gives an indication of the tendency for free convection to occur, and it is given as

$$Gr = \frac{\rho^2 g \beta (\Delta T) l^3}{\mu^2} \tag{14.6}$$

where β is the thermal expansion coefficient defined earlier. The Grashof number is the ratio of buoyant force times inertial force to the square of the viscous force. While the Nusselt number made a good ordinate, the Grashof number makes a good abscissa. In practice, it is a substitute for the Reynolds number in free convection situations where, with no free-stream flow, Re is zero and is thoroughly useless. Because of the squares of forces in it, the Grashof number may take on any of a more extreme range of values than can the Reynolds number. For free convection, the transition from laminar to turbulent flow takes place at Grashoff numbers between 10^8 and 10^{10}.

Two other dimensionless indices are sometimes used along with or instead of the preceding ones. The *Rayleigh number, Ra,* is merely the product of the Prandtl and Grashof numbers. And the *Peclet number, Pe,* is just the product of the Prandtl and Reynolds numbers.

In practice, much of the awkwardness and eccentricity of convection can be dealt with by plotting, for each shape and orientation, data in the form of

$$Nu = f(Gr) \quad \text{for a given } Pr \text{ for free convection,} \tag{14.7}$$

and

$$Nu = f(Re) \quad \text{for a given } Pr \text{ for forced convection.} \tag{14.8}$$

Empirically derived equations for these functions are frequencctly cited. Thus one may encounter formulas such as the following one for laminar free convection from a vertical flat plate at a Prandtl number of 0.74:

$$Nu = 0.480\ Gr^{1/4} \tag{14.9}$$

Notice how many qualifying conditions have to be attached to the formula! Even so, my advice, based on hard knocks and actual measurements, is to be skeptical about the three significant figures in the constant. Even if all the qualifying conditions are met, these formulas are basically ballpark estimates, and one significant figure is more realistic for practical problems.

How can one tell whether free or forced convection will predominate under a given set of circumstances? Another dimensionless number usually settles the matter: the ratio of the Grashof number to the square of the Reynolds number (Gr/Re^2). If this ratio is near unity, then both buoyant and forced currents are significant. If it is below 0.1, then assuming pure forced convection is safe, while if it is above 16, then one can assume pure free convection. To put the matter in a more everyday context, consider a leaf on a tree with a width of 5 cm and a surface temperature 10° above that of the air around it—reasonably normal figures. A mixed regime will occur at wind speeds between 3 and 40 cm s^{-1}, just the range in which the leaf may have a problem of excessive heating.

Convective heat transfer is as important biologically as it is complex physically. For leaves on the tops of trees, convection commonly exceeds evaporation as a mode of heat loss (Knoerr and Gay, 1965), and much of the shape of broad leaves may reflect adaptation to keeping leaves cool through convection (Vogel, 1970; Balding and Cunningham, 1976). Convection is important in both ecto- and endothermic animals; good introductions to the phenomenon in the context of temperature regulation (Schmidt-Nielsen, 1979) and as a part of energy budgets (Porter and Gates, 1969) ought to be consulted before any plunge into the primary literature. Davis and Birkebak (1975) give a useful review of the relevant concepts and data for convection through fur. Lest shoreline ecologists dismiss these matters as irrelevant, I suggest a look at Johnson (1975) and offer a quote from Vermeij (1973) on morphological adaptations: "...the tendency for sun-exposed limpets and many high intertidal and littoral-fringe neritids to possess strong shell-sculpture suggests that sculpture may play a role in maximizing convection..." Figure 9.2 (page 146) illustrates such sculpture.

Scale modeling is rarely practical in problems involving convection; there are too many dimensionless indices to be dealt with simultaneously. But modeling itself is extremely useful if a few pitfalls are avoided. First, one must be cautious in speaking of shapes which dissipate more or less heat by convection; under steady-state conditions, energy input and output rates are equal. A more efficacious shape is often one which will dissipate heat at a given rate with less forced flow and/or with a less steep temperature gradient. In practice it is often convenient to fix the temperature difference by use of a proportional heat control and then to monitor the effects of shape and orientation on heat dissipation as indicated by the energy put into the model. In addition, one must take care that the temperature distribution across the surface of the model does in fact model reality; it is not sufficient just to apply the appropriate energy load. A copper model of a leaf, heated by a lamp, is an inadequate model for any mechanically reasonable thickness of metal—too much heat is conducted laterally, and the temperature is much too uniform. As a result, the model loses too much heat at the edges, too little elsewhere, and too much overall. A metallic mouse with central heating might not be a bad model; the conductivity of the metal would mimic the thermal effect of a circulatory system.

Non-steady effects are of great importance in convective heat transfer. Winds near the surface of the earth are rarely steady, and the thermal time constants of leaves, for example, are measured in seconds. We've made some use of a pulsating wind tunnel to look at the effects of leaf thickness on energy storage and the resulting reduction in peak temperatures during wind lulls (Kincaid and Vogel, in preparation). Parkhurst and Pearman (1974) have shown that fluctuations in flow, as would be generated by turbulent eddies, considerably increase heat transfer; but much more ought to be done. When dealing with non-steady effects, it is especially important that the internal conductivity and the thermal capacity of a model be realistic. The metal mouse won't do any longer; a water-filled model with internal stirring would be better.

convective mass transport

If any fluid motion is present, diffusion, evaporation, and such molecular transport phenomena are augmented by convective mass transport. The analogy with thermal conductivity and convective heat transport is very close. Loosely speaking, mass replaces heat, concentration or density gradients replace temperature gradients, and diffusivity replaces thermal conductivity. Typically, even slight bulk movement of fluid has a great effect on the rates of transport of dissolved nutrients, respiratory gases, and water vapor, as noted earlier in Chapters 8 and 9. For dealing with convective mass transport, the dimensionless indices previously defined need only minor modification. The Prandtl number is replaced by the *Schmidt number,* given as

$$Sc = \frac{\mu}{\rho D_c} \qquad (14.10)$$

where D_c is the diffusion constant or diffusion coefficient, with dimensions of L^2T^{-1}. Viscosity and density refer, of course, to the overall fluid, while the diffusion constant refers to the diffusing molecules which may be dissolved in that particular fluid. Here air and water are very different: diffusion constants in air are ordinarily 10,000 times greater than those in water.

The Nusselt number is either redefined slightly or replaced by the *Sherwood number:*

$$Sh = \frac{q_m l}{D_c S\, \Delta[C]} \qquad (14.11)$$

Here q_m is the rate of mass transfer by convection (MT^{-1}) and $\Delta[C]$ is the concentration difference between the free stream and the solid surface absorbing or emitting the diffusing molecules (ML^{-3}). The Sherwood number may be viewed as an extra factor in Fick's law of diffusion to account for convective transport.

The Grashof number is not renamed, but the product of temperature difference and thermal expansion coefficient is replaced by the difference in density between the fluid at the interface and that in the free stream, divided by the free-stream density; the result can be used in either heat or mass transfer problems:

$$Gr = \frac{\rho^2 g(\Delta\rho/\rho)l^3}{\mu^2} \qquad (14.12)$$

If you have trouble visualizing a flow generated by a concentration difference, imagine a lump of salt suspended in a container of otherwise still water. The dissolving salt will make a locally denser region in the fluid, and free convection will occur.

Further guidance on convective mass transport may be obtained from Bird et al. (1960), Monteith (1973), or Leyton (1975). For more specific applications, one might look at Thom (1968), Caro et al. (1978) or Kingsolver (1980).

yet other dimensionless numbers

I've never seen anything claiming to be a complete list of named dimensionless numbers, but their number must certainly be large. A substantial number of forces are involved in problems of flow—viscous, inertial, gravitational, surface tensional, elastic, pressure, diffusive, thermal diffusive, and so on. Obviously, quite a few ratios of forces can be concocted! There is, in fact, a theorem which can be used to predict the number of independent dimensionless groups—the Buckingham pi theorem. It states that the number of such groups is equal to the number of separate physical parameters minus the number of fundamental dimensions. Thus four parameters (length, velocity, viscosity, and density) and three dimensions (length, mass, and time) generate one such group, usually the familiar Reynolds number. But when parameters such as the various thermal properties, surface tension, diffusivity, and so forth are considered with the addition of only the dimension of temperature, then the number of possible dimensionless indices gets very large.

Not all possible dimensionless indices are particularly useful, and not all the useful ones are really independent. Some, such as the Mach number (freestream velocity divided by the speed of sound in the fluid) should find little use in biology. Others may find more use in biology than anywhere else.

One index which has not yet been mentioned comes into play where gas-liquid interfaces are important. As promised at the start, we've ignored these interfaces and thus not worried about the world of surface waves and their associated phenomena. The index is the *Froude number, Fr,* a very simple analog of the Reynolds number. It is the ratio of inertial forces to gravitational forces:

$$Fr = \frac{U^2}{gl} \qquad (14.13)$$

A little testing of values for real fluids should convince you of the impracticality of keeping both Reynolds and Froude numbers constant as size (l) is altered.

This difficulty underlies much of the art in the hull design of surface ships. In practice, the Froude number is held constant, and models are made as large as possible to keep the Reynolds number in the right order of magnitude.

For small-scale systems, surface tension should be important; the "surface tension force" is conveniently taken as the product of surface tension (σ) and wetted perimeter. The *Weber number, We,* is the ratio of inertial force to surface tension force:

$$We = \frac{\rho U^2 l}{\sigma} \tag{14.14}$$

Perhaps we might find more useful a ratio of viscous force to surface tension force, since surface tension is most likely to be significant when the Reynolds number is low:

$$Hy = \frac{\mu U}{\sigma} \tag{14.15}$$

This index might be germane to the problem of a hydra (hence my choice of abbreviation) hanging from a water surface through a velocity gradient.

Perhaps these are enough dimensionless numbers for present purposes, so the Graetz number, the Brinkman number, the Knudsen number, the Stokes number, the Stanton number, the Cauchy number, the Eckert number, the Fourier number, the Euler number, and all the rest will forgive our lack of attention. One can, in fact, make up reasonably useful numbers for biological problems quite devoid of flow. In the last chapter, reference was made to a size-independent shape parameter, $S^{1/2}/V^{1/3}$; it is properly dimensionless and seems fraught with biological portent. Another ratio is surface tension to gravitational force, $\sigma/\rho l^2 g$, which indicates the practicality of walking on water. Water striders, if large, must be of very low density. This last ratio, with apologies to the offended, might be termed the "Jesus number."

a few final remarks

Fluid flow is not currently in the mainstream of biology. Perhaps biofluid-dynamics, to use Lighthill's conglomerate mouthful, might be part of an emerging biological or comparative biomechanics, a field for which my friend

Stephen Wainwright seems to be the chief evangelist. It is certainly no less than that; in a larger context, biofluiddynamics may be part of a renaissance of physics from its eclipse by chemistry in the study of the functioning of organisms. One can get rather attached to the field: it deals with a fascinating set of phenomena, both physical and biological. But I'm not really pushing or proselytizing for complete converts. More important is the re-education of people in conventional fields who will remain wedded to those fields, but who might come to regard flow as fascinating and relevant instead of fearsome and experimentally recalcitrant.

The preceding comment refers, principally, to morphologists, physiologists, ecologists, and paleontologists. For all these, the general message here is the same: you have to know a little about fluid mechanics to realize what your organisms must contend with and to recognize the opportunities open for them to be the cleverly adapted rascals in whom we take delight. More specific messages should emerge as well; ideally, each of you should have found parts here which speak specifically to your problems. For example, I've given much space to drag; it seems the pre-eminant phenomenon at the interface between, on one hand, flow and, on the other, morphology and solid mechanics. The interaction of shape and flow produces the forces with which mechanical design must deal. I've tried to emphasize the relationship between niche and flow: not only forces, but also filter feeding, flow-facilitated diffusion, chemosensory input, and thermal budgets are components of that relationship.

A major point underlying all of the previous chapers is that problems of flow are tractable. They are not difficult to think about, once one is familiar with fluid behavior and can envision matters from the viewpoint of the fluid. And they're not difficult to hypothesize about, given a little feeling for such matters as Reynolds numbers, boundary layers, and the like. Nor are they hard to handle experimentally—flow tanks, flow meters, and flow visualization are among the least of our technological hurdles. It isn't necessary to clutch at equations whose applicability rests only on the right variables being present, to make facile speculations about flow, or to implicate flow in the selection of shape solely on the basis of variations in shape among organisms from different habitats. I'm distressed because I don't think it is difficult to do much better; to come directly to grips with problems and not to resort to estimation, speculation, or indirect argument.

For anyone who seriously intends to pursue a flow-related investigation, this book should be considered as only a starting point. It has been an attempt to bring the fluids and the organisms into proximity, but at the cost of doing full justice to neither. Suggested sources of additional information have been indicated in the text, to the extent that I am familiar with them. In addition, the most useful sources of general guidance are marked with asterisks in the bibliography.

There are, as well, the engineers themselves—they're often better than books, although they take a little cultivation. Engineers tend to view our problems as some kind of comic relief (if one is lucky) or totally intractable (if one is not). It should be assumed that few have any idea that some really good work has been done in biological fluid mechanics. Thus under "testing insects and birds" in Pope and Harper (1966) we find the statement, "Since live models are expected to be quite small, no special tunnel techniques are needed, should tests be desired. The authors suspect that they will produce more hilarity than useful data." Put that alongside the elegant experimental work of Weis-Fogh, Jensen, Tucker, Pennycuick, Nachtigall! The bemused engineer also needs a bit of introduction to our approach to design. As a scion of an eminently practical tradition, the engineer designs things or devises rules by which efficient devices may be designed. By contrast, we start with the assumption that our organism is well-designed (with constraints, and with natural selection as our justification), and we try to figure out just why its design is a good one.

This book began with an exhortation, and I mean for it to end with the same; much polemical passion lurks beneath the pedagogical prose. Biologists seem afflicted with a great faddishness in the choice of items for investigation. One might argue the merits of the situation—that breakthroughs are made by concentrating one's troops; I'd assert that the history of science indicates the opposite. One hears that the rate of scientific progress has never been so great; one also hears that never have so many been actively pursuing science. Is it just possible that the rate of progress *per investigator* might be at a low point, with faddishness a contributing factor? Here we have an area in which people needn't trip over each other, where problems abound and investigators are scarce. Much work has been done in biological fluid mechanics: I'm certainly very much aware of the size of the relevant literature at this moment. But much of the past work has been of less than ideal quality, and it proves even easier than I had suspected to call attention to great gaping lacunae where the hand of the biologist has never set foot.

My aim in both teaching and writing has been to stimulate further investigations. I am greatly gratified whenever some student of this material takes the groundwork and carries out an investigation quite different from anything I might have conceived. I would derive equal pleasure if the written word occasionally has the same catalytic role.

APPENDIX 1

making fluids flow

THE PRIMARY TOOL IN ANY LABORATORY study of flow is some device to make the fluid and the organism (or model) move with respect to each other. Wind tunnels, whirling arms, flow tanks, towing tanks—all do no more nor less than this simple task. The biologist's requirements for flow-producing apparatus are not especially demanding in terms of size, cost, or complexity. On the other hand, the necessary apparatus is not usually in the stock of the neighborhood scientific equipment emporium, and the decisions needed to produce a suitable machine transcend mere choice from a catalog. I'll talk first about moving air and then about water, since in the present context they are strongly divergent in character. The first decision one must make is, in fact, the choice of medium; with the use of dynamically similar models (Chapter 5), the choice needn't be predetermined by the habits of the organism under scrutiny. For moving air I'll consider mainly the variety of arrangements possible and the relative advantages and disadvantages of each. For moving water, I want to offer a specific design which we've found to be especially effective.

I recommend the construction of a flow-producing device even without a specific investigation in mind. It is an engaging and satisfying task, particularly in this era of expensive apparatus and obsequious sales people. Unless our experience is grossly atypical, using such a device for demonstrations can be depended upon to generate investigations. If a choice must be made, I would opt for a flow tank rather than a wind tunnel or other air-moving system: forces are greater, ancillary equipment can be simpler, and the use of dye for flow marking is a sure crowd pleaser.

Theory rarely cares whether the test object is fixed and the fluid moves or whether the test object moves through otherwise stationary fluid. While less versatile, the latter scheme is often a quick solution to the problem of obtaining relative motion. In air, the typical device is a "whirling arm"—a horizontal rotating beam driven by a motor attached to its vertical shaft. Visual observation of a test object may not be simple, but electrical anemometers, strain gauges, and other transducers may be connected through axial twisting wires (plain) or sliding contacts (fancy). The beam must be made long enough so the rotation rate can be kept below levels at which the wake of one pass is met by the object in the next revolution. When airspeeds are high, a very long device is needed; when airspeeds are low, free convection in the room complicates matters. In water, the whirling arm is often convenient. The driving system can be mounted on a wooden superstructure over an inexpensive wading pool. Linear motion ought not to be dismissed; dragging an object through a trough of water is quite simple at biologically relevant speeds, and mounting a camera on the same carriage is only a mild complication. For very low speeds the lead screw of a lathe can be used as a sturdy mover, with some bracket from the carriage extending over an adjacent trough.

In general, however, devices which move the fluid rather than the test object are more convenient, more versatile, and much more fun. And they need not be objects of great splendor. In engineering practice, much attention is given to reducing the level of turbulence in the moving stream since tests are commonly intended to apply to machines which move through otherwise fairly still fluid. If one is concerned with locomotion, the same consideration applies. But if one is investigating attached organisms it isn't clear that a clean and nonturbulent flow is appropriate! What leaf on a tree, what sea anemone on a rock sees anything but a mass of irregular turbulent eddies? Furthermore, smooth flow is more easily obtained at low Reynolds numbers when viscosity is a potent ally than at high ones.

In building any wind tunnel or flow tank, provision for easy modification should be a major consideration. Not only do needs change with time, but a certain amount of art and luck are inescapable in the design process. The best way to proceed, I suggest, is to obtain or build a good anemometer or flow meter first. That instrument can then be used to tell when the wind tunnel or flow tank is satisfactory—when the speed range is suitable and when speeds are adequately uniform across the test area. Until satisfactory performance is obtained, modifications and emendations are routine; one starts with trivial changes of baffles and screens and ends up, if all else fails, changing taper rates of channel walls and doing similar heroics.

wind tunnels

A wind tunnel has only three basic components: a pipe, into which fits a propeller, and a motor to drive the propeller. Still, whole books have been devoted to the design of wind tunnels (for example, Pankhurst and Holder, 1952, and Pope and Harper, 1966; the first is more useful for present purposes). At least one article (Vogel, 1969) considers tunnels which are useful to biologists. But, while complications do arise, the reader should be assured that many of the factors which afflict the design of large, high-speed tunnels are quite irrelevant in building small, slow ones. In particular, mechanical support and rigidity are minor concerns, and power economy is ordinarily a secondary consideration. When thousands of horsepower are unleashed, a saving of 50 or even 10 percent is important. But for variable speed motors under about a horsepower, price is nearly independent of power, and mechanically simple designs which happen not to be power efficient can be most attractive.

Figure A1.1 shows the basic parts and arrangements common in low speed wind tunnels. Most often, the conduit is composed of three parts: the *effuser* or entrance cone, in which air is accelerated; the *working section,* where tests are carried out; and the *diffuser* or exit cone, where the air decelerates again. The section containing the propeller may be adjacent to or part of either effuser or diffuser. The use of tapered effuser and diffuser permits the air, by the principle of continuity, to be more gradually accelerated and decelerated and thus reduces the power required to achieve a given speed. If power economy is of no concern, the diffuser may often be omitted. The effuser has another role as well. In a long pipe, the flow profile is parabolic, but it more nearly approaches linearity (plug flow) after a contraction. In short, if the effuser is discarded, it shouldn't be replaced by a long length of pipe; conversely, a contraction usually needs to be inserted between a long pipe and a working section. The propeller is best located near the diffuser or downstream from the

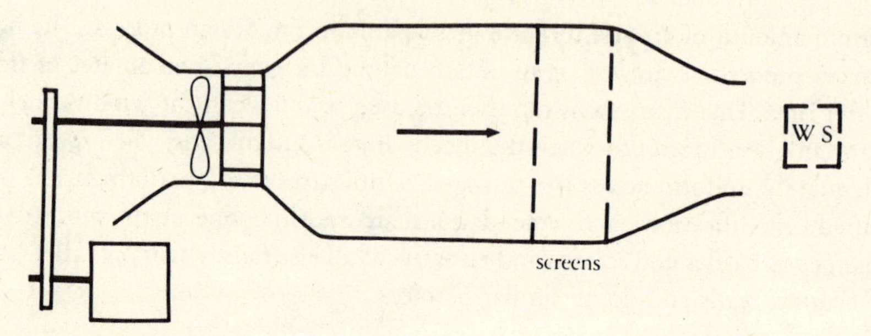

(a) open circuit, open jet

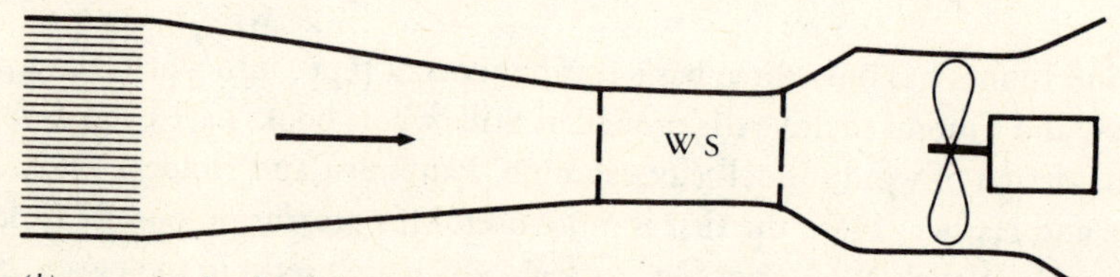

(b) open circuit, closed throat

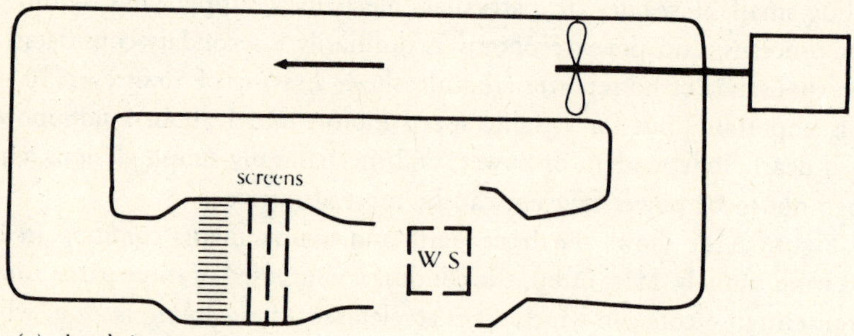

(c) closed circuit, open jet

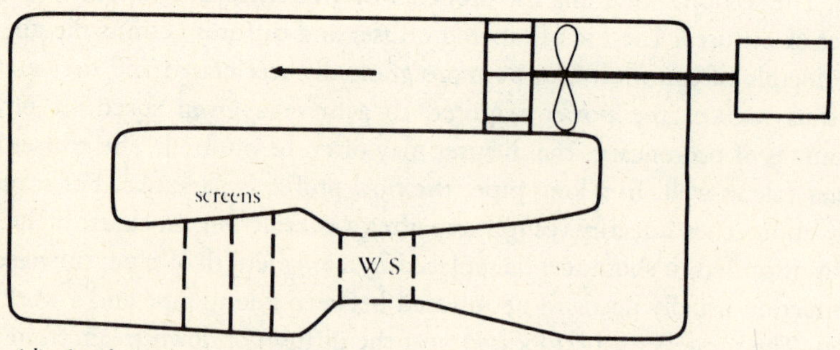

(d) closed circuit, closed throat

working section. There's trouble enough smoothing the irregularities of ambient air without having to deal in a short space with the aerodynamic pollution of a propeller. The large wind tunnel of the Duke Zoology Department (Tucker and Parrott, 1970) is similar to Figure A1.1b; visitors usually expect flying birds to face the propeller.

When outside air does not enter directly into the working section, a *settling chamber* usually precedes the final contraction. This wider channel is the preferred location for screens or "honeycomb" gridwork which reduces swirl and turbulence; more about these devices later.

Configurations. Even among tunnels used for biological investigations, the range of designs is exceedingly wide. Rather than attempting to describe a set of specific designs, I will just make a general classification of tunnels based on two characteristics. First, the presence or absence of recirculation: if the air goes around and around, one has a *closed circuit* tunnel; if room air makes a single pass through the tunnel, we speak of an *open circuit* tunnel. Second, whether or not the working section is enclosed by walls determines the distinction between *closed throat* and *open jet* tunnels. Thus four arrangements are possible, both theoretically and practically. All are useful for biologists, and each has been used at one time or another.

An open circuit, open jet tunnel (OC-OJ), as in Figure A1.1a, gives the easiest access to the working section and, for a given size of working section, is usually the smallest machine. Most of the studies of insect flight aerodynamics have used this arrangement, although with very small insects I found it to be intolerably sensitive to room air currents. The fan must, of course, be located upstream from the working section, a possible source of minor difficulty.

An open circuit, closed throat tunnel (OC-CT), Figure A1.1b, is much less sensitive to ambient air movement, and the fan can be located downstream from the working section. It is still a small machine, relative to the size of the working section. If an effuser and diffuser are not needed, it can be kept very short. I used this design for a tunnel which had a working section of 0.2 m^3 but which had to tilt 90° in an ordinary room. The design configuration seems to

FIGURE A1.1

(*Facing page*) Four arrangements for low speed wind tunnels, showing reasonable locations for motors, propellers, stators, screens, honeycombs, and working sections (WS).

me the best compromise for general biological use. The major drawback is that of access to the working section—controls must penetrate walls, floors, or doors.

A closed circuit, open jet tunnel (CC–OJ) (Figure A1.1c) is not a contradiction in terms; with an appropriate collecting cone behind the working section, almost all the air is recirculated. It gives the free access to the working section of an OC–OJ with far less sensitivity to room air movements. On the other hand, it is relatively large. I found the arrangement useful for working with flying fruit flies, but the open working section of a small version would probably not be sufficiently buffered from room air movement below about 0.4 m s^{-1}.

A closed circuit, closed throat tunnel (CC–CT) achieves full isolation of internal from external air (see Figure A1.1d). For this it has the access problems of the OC–CT and the large size of the CC–OJ. But internal temperature, humidity, and even gas composition and air pressure can be changed within the tunnel if necessary. If total enclosed volume is not excessive, the tunnel may serve a dual function by acting as a respirometer chamber as well. Still, the CC–CT configuration is best regarded as a last resort. It's far easier to use an open tunnel in some pre-existing controlled-environment chamber.

Obtaining uniform velocity. A host of schemes have been used to improve the flow in wind tunnels—to make the flow uniform in speed throughout the working section and to minimize swirl and turbulence. Contractions and settling chambers have already been mentioned. Another useful component is a screen or a set of screens, usually located upstream from the working section. These can reduce the size of vortices below the range in which they are shed and thus smooth the flow. Since a cylinder sheds alternating vortices at Reynolds numbers above about 40, the wires of the screen should be kept thin enough so their diameter-based Reynolds number is under this value. For ordinary window screen, this limits speeds to 5 m s^{-1} or less. But the criterion may be unnecessarily strict in practice: small vortices are probably preferable to the larger ones which would occur without screens.

Another useful device for smoothing flow is a "honeycomb," the name applied to any array of small longitudinal channels filling some cross section of the tunnel. These are very effective in reducing the swirling motion induced by a propeller, in collimating the flow, and in preventing "wander" of the airstream at very low speeds. Usually, honeycombs are placed in the settling length, well upstream from the working section and isolated from it by screens. In very low speed tunnels, a honeycomb may be placed quite close to a test object since sufficient time will be available for the coalescence of streams from the individual channels. Soda straws are very useful material for honeycombs. They're inexpensive in large quantities and large quantities are often needed: make sure you don't get straws which are individualy wrapped! I used 17,000 in a tunnel of 0.3 m^2 cross section. With a little effort, it is possible to use an array of straws of non-uniform length as a way of compensating for transverse velocity variations.

Padding is another useful approach to smoothing flow. The power loss driving air through a sheet or blanket will be high, but a great deal of flexibility in design may be gained if power can be wasted. Driven by a substantial pressure gradient, air will ooze through a dense barrier quite uniformly: pressures "self-equalize" in chambers whereas velocities do not. If sufficient pressure at a sufficient flow rate is available, a simple flow tank may be made with a blanket between room air and an enclosed working section and a parallel blanket between the working section and an evacuation chamber. A suitable pump such as a large vacuum cleaner is attached to the latter chamber.

Fans and motors. One can envision a wind tunnel without its own power source—perhaps a pipe atop a truck which drives at the desired speed down a highway. But a motor and fan are more convenient. The choice of both is worth careful consideration. Fans or propellers are designed to operate best at particular combinations of volume flow and pressure head. Thus a thin model airplane propeller does best at high speeds, with a large volume of air gaining a relatively low increase in pressure. However shrouded and baffled, it won't adequately push air through a dense barrier. A broad-bladed axial ventilating fan is more useful: it handles large volumes but at lower speeds and against greater resistance. A centrifugal (squirrel cage) fan will tolerate even more resistance, and a multistage centrifugal unit (as in a vacuum cleaner) will pump against still more. Power, remember, is the product of pressure drop and volume flow rate: the design of a tunnel determines how the power available may be most usefully invested. As a rule, the axial ventilating fan is the handiest; such fans are available in sizes from a few centimeters to a meter or more in diameter, and they easily adapt to ordinary belts, pulleys, and shafts. Fans with broad, overlapping blades of low pitch are usually more satisfactory than those with narrow blades and greater pitch. If an axial fan is asked to work against a substantial resistance, sets of fixed stator blades just in front of and behind the rotor usually help the fan push instead of spin air. Small fans must turn at very high rates to get even modest airspeeds, so in tunnels less than 10 cm in diameter one might place the propeller in an expanded pipe.

Power is nearly always supplied by an electric motor, mounted externally and connected by belt or shaft to the propeller. The motor ought to be a variable speed unit, and this most emphatically does not mean an AC induction motor plugged into a variable autotransformer or rheostat. The practical choice is a shunt-wound DC or universal AC–DC motor connected to a solid state (SCR) controller with internal feedback. These controllers keep motor speed relatively constant in the face of variations in load and temperature, and they maintain adequate torque at low speeds. Simplest of such combinations is a sewing machine motor and a speed control designed for small power tools. Hair dryers or leaf blowers may be of some use as drive units if they are powered by universal motors which will accept controllers. The diagnostic key to a universal motor is the pair of carbon brushes carrying current to the armature. A variable speed electric drill can also be used as a power unit and control, but one should be cautioned that many of these impromptu solutions will not work well at slow speeds for long uninterrupted periods. The best solution to providing power is a combination motor and control designed for machines or instruments; these are available from 1/20 to 500 hp, and units of up to 1 hp may be connected to 115 volt electrical outlets. Usually no gearing will be required on the motor; minor speed matching adjustments can be accomplished with a belt and a pair of pulleys.

Other notes.

1. Early in the design sequence, a choice must be made between a circular and a rectangular cross section (or some combination of the two). Doors, floors, and ports are simpler in a rectangular channel, while axial fans fit better into circular sections. If an axial fan is used in a square section, it should be shrouded to prevent back flow in the corners. If power is limited, tapered walls may help reduce the abruptness of the cross-section change at the shroud.

2. Plywood and perforated steel L-beams (''Dexion'') are handy for building rectangular sections; stove pipe and duct tape work well for circular sections. Plexiglas is the material of choice for working sections.

3. With some motor controllers the speed may be set by any parameter which can be transduced into a variable voltage—light intensity, velocity, temperature, or displacement. Thus a flying animal can ''pick'' its preferred flying speed if falling back slows the tunnel and flying forward speeds it. Or the tunnel speed can be controlled by an anemometer to compensate for changes in blockage caused by test objects; or it can be controlled by a temperature sensor to maintain the temperature of an object at some preset value.

flow tanks

The liquid analog of a wind tunnel is a flow tank; it's sometimes called a flume, or water tunnel. As with wind tunnels, devices which have been used by biologists encompass a wide range of sizes and arrangements. In most the water recirculates, but sometimes a fraction of the water is replaced in each circuit. Some have been impulsive machines in which water previously pumped up into a storage tank is suddenly allowed to rush downhill in a sluice. Some have the return circuit to one side of the working section; others have the return beneath the working section. Power is most commonly provided by a high speed centrifugal pump, but axial propellers and even paddlewheels have been used. The categorization used for wind tunnels is largely irrelevant—amost all are, of necessity, closed circuit, closed throat machines, but the extent of the exposed air–water interface varies from a single standpipe to the fourth side of every channel.

In engineering laboratories, the most common kind of flow tank consists of a horizontal trough running between two tanks. Water flows out of the upstream tank directly into the trough; it escapes over a weir or through a porous barrier from the trough and drops in a waterfall into the downstream tank. Pipes and a pump return water to the upstream tank. The arrangement can give very smooth flow which is easily controlled by a valve adjacent to the pump. The pumps themselves are convenient and widely available. But this splendid scheme suffers from several drawbacks: (1) It is profligate in power consumption due to energy losses in the waterfall and the constricted return pipes. The maximum flow obtainable with a pump which draws under 15 amperes at 115 volts is about 5 l s^{-1}; this corresponds to only 20 cm s^{-1} in a 16 × 16-cm channel. (2) Turbulence and direct heating of the water through the action of the pump are generally great and favor designs involving large volumes.

Frustrated with such limitations, we built a flow tank of somewhat less conventional design a few years ago (Vogel and LaBarbera, 1978). In particular, we wanted to move about 100 l s^{-1} and had no immediate source of the requisite five- or ten-horsepower pump. A series of these tanks has now been built, ranging from 10 to 35 cm in working section depth and diameter and from 20 to 800 liters in total volume. Largest and smallest work equally well, so we have no sense of size limitation for the design. The tanks are not difficult to construct, and they are at least an order of magnitude more efficient than the trough-cum-waterfall arrangement.

The scheme uses an approximately constant cross-sectional area for the entire channel and has no waterfall, so it eliminates the usual major avenues of power loss. Since little pressure drop need be overcome, an axial propeller can be used to move large volumes of water. The propeller turns fairly slowly, and minimal turbulence and heating result. To provide best access to the working section and least exposed air–water interface (surface waves are a nuisance), the return circuit is beneath the working section. To eliminate the need for shaft seals, the propeller shaft is vertical and emerges from the exposed water surface. The largest of these flow tanks is illustrated in Figure A1.2. I recommend the design, especially now that our initial errors can be avoided. Guidelines and suggestions for design and construction follow; they represent the accumulated lore of five tanks.

Material and proportions. Round pipe should be used for the propeller housing and return section. I will not build another tank with the lower section

FIGURE A1.2

A flow tank with a 35 × 35-cm working section. The trough is made of 3/4″ plywood; the return is 12″-diameter plastic sewer pipe; supports are wooden 2 × 4's. The window is shown as a rectangle in the center of the trough, the honeycomb as a pair of dotted lines near its upstream end. A few bearings and the sliding motor mounts have been omitted from the drawing.

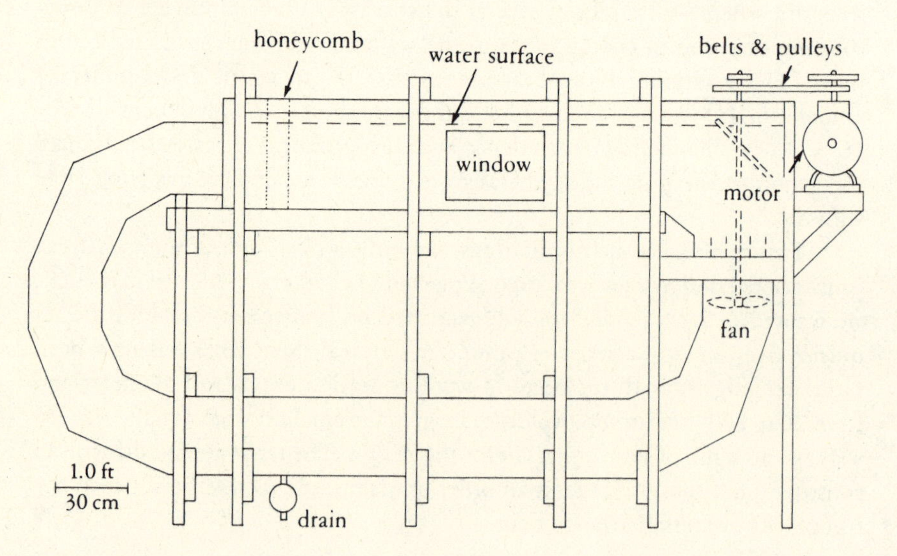

of plywood—sealing a square pipe is no simple matter. Plastic drain or sewer pipe is convenient to use and compatable with sea water, and it comes in sizes up to 15 inches in outer diameter. It might be noted that the larger sizes of pipes are heavy and the ells are costly. Pipe without internal rubber gaskets is better than that with gaskets. A rectangular channel of plywood or plexiglas serves as upper trough, with the choice depending on the size of tank and budget. Except in very small flow tanks, plywood at least 3/4 inch thick should be used to minimize flexing and simplify sealing.

To permit easy attachment and, if necessary, detachment, the trough should be slightly wider than the return pipe—3-inch pipe with a 4-inch trough; 12-inch (outer diameter) pipe with a 14-inch (inner diameter) trough. Longer troughs give smoother flow than shorter ones, although boundary layer growth (equation 8.8) may be significant in extemely long channels. A length at least five times the diameter seems minimal; around ten may be roughly optimal. Particularly high speeds may be obtained by reducing the depth of water in the trough; conversely, not making a sufficiently deep trough may limit use at low speeds. The trough, naturally, should be higher than any intended water depth by at least 20%. Splash-over may be further prevented by a fairing plate at a 45° angle at the downstream end of the trough. In practice, we let the drive shaft emerge through a slot in this plate, and water completely surrounds it.

Even fairly small flow tanks are heavy when filled, and contemplating the weight of the enclosed water is a sobering reminder of the necessity for sturdy support. No support is completely rigid, though, and small distortions upon filling should be expected. The joints between the plastic pipe and the trough should not be calked with a brittle material. We have found silicon bathtub sealant (Dow Corning or G.E.) an invaluable antidote for leaks caused by distortion since it makes permanently flexible seals. A thin coating of this sealant on glued plywood joints is also a good idea; the strong and water-resistant wood glues (such as resorcinol glue) are a bit brittle. Some sealant contains a mildew inhibitor and ought to be avoided if the tank is to be used for live organisms. Look for silicon sealant marked as safe for aquaria.

The interior of any plywood trough must be lined or painted. Any paint must be aged and soaked before organisms are introduced, and many paints still persist in "tasting bad" to animals. Our best luck so far has been with Sherwin-Williams "Tile-Clad II" epoxy enamel in white. The outside should be left unpainted so any seepage into the wood will escape and not cause rot. One might well view the tank as an inside-out boat, with external ribs and motor and an internal skin and propeller.

Collimator and baffles. Soda straws, glued together with silicon sealant, make a good upstream collimator. A pair of collimators with a space between is more effective than a single one of the same overall resistance. The total pressure drop across the collimators can easily be calculated; it should probably not exceed 1 cm of water (100 Pa) at 1 m s^{-1} or the axial fan will not work well. That 1 cm is, of course, precisely the height of water to be expected behind a collimator acting as a dam. Since the pressure drop across each straw is the same as that for all straws in parallel, and since the velocity of flow through each is the same as that through the array, a version of the Hagen–Poiseuille equation (10.5) can be applied to a single straw of the bore supplied by your local purveyor. Selection of a pressure drop then sets the lengths of the straws.

It is useful to install some baffle plates of wood or plexiglas just above the propeller. Passing the flow between parallel plates discourages formation of a vortex and drawing in of air, and it permits the tank to be used at high speeds when only partially filled. For our largest tank, four three-inch-wide plexiglas strips proved adequate. For small tanks, plastic grid of the sort used as filters for fluorescent lighting fixtures (called "egg crate") is handy as a collimator. The same grid in a large tank can be used to keep animals and solid material out of the propeller, using thin netting across the grid if necessary. Plastic window screen can do the same task, but it entails a greater pressure drop.

Propeller. The choice is usually not a critical matter. It should be as large as can be fitted into the descending arm of the return conduit. For diameters of 15 cm or more, marine propellers are useful, but the thin ones used with electric trolling motors are less satisfactory than propellers with three broad blades. For small tanks, plastic cooling fans normally used in air are adequate. Unless the fan shaft is particularly rigid, it will be necessary to use an underwater bearing, which is also a convenience for centering the fan. The stress on this bearing is slight, and a soft plastic bar across the diameter of the pipe with a hole for the shaft is sufficient.

How fast should the propeller turn? In the flow tanks we've built so far, the advance ratio (equation 12.2) lies between 0.3 and 0.7, with the velocity taken at the propeller rather than the working section. One simply has to estimate the maximum speed and assume an advance ratio (perhaps 0.5) to calculate the rate of revolution of the propeller.

Motor and drive. As with wind tunnels, variable speed motors are best. Our experience indicates that one can estimate the power required (in horsepower, with the rating conventions for small motors) by multiplying the cross-sectional area of the channel (in m^2) by the maximum speed desired (in $m\ s^{-1}$) and dividing the result by 0.2.

In almost every case, the basic speed of the motor will be higher than that needed at the propeller shaft. For small tanks, the difference is usually well within the range of a belt and a pair of pulleys; a choice of several pairs of pulleys increases the useful speed range. For large tanks, one should anticipate needing a gear box with about a 10:1 reduction in order to avoid using enormous pulleys. It is best to buy motor and gear box together, selecting the latter to bolt directly onto the former. On our most recently built tank, a 3/4-hp, 2400 RPM motor drives a 10:1 gear box. The latter connects with the fan shaft through either a four-to-eight inch or an eight-to-four inch pair of pulleys. The motor has a 20 to 1 useful speed range, so the whole drive has an 80 to 1 range. The motor and gear box are mounted on a sliding platform to permit the drive belt to be shifted without disassembly.

Other notes.

1. Don't forget to put a drain in the bottom.

2. Brass or stainless steel shaftwork and screws are necessary, especially if the tank is ever to be filled with seawater.

3. Plate glass, fixed behind molding and gasketed with silicon sealant, makes good windows for plywood tanks. The inner surface of the glass should be coplanar with the inner trough walls.

4. Surface waves may be eliminated with a floating lid for the free surface. A plexiglas lid shaped like a long, hollow barge works quite well.

5. A water-filled box (with its own drain) replacing the floor of the working section can provide a place for roots, attachments, and so forth, without reducing the effective depth of the working section.

6. It is prudent to locate a flow tank, particularly a large one, where a sudden unplanned emptying will cause neither damage nor comment.

sources of material

Industrial hardware is much less expensive than anything which might be listed in a scientific supply catalog. For large motors, gear boxes, pulleys, belts, and fans, a good source is W.W. Grainger, Inc., a whosesale supplier with offices in most large U.S. cities. For smaller motors, Minarik Electric Co., 244 E. Third Street, Los Angeles, CA 90013, should be useful. Small belts, pulleys, and shafts may be obtained from Small Parts, Inc., 6901 N.E. Third Avenue, Miami, FL 33138, or from PIC Design Corp., P.O. Box 335, Ridgefield, CT 06877.

APPENDIX 2

visualization and measurement of flow

HAVING MADE SOME FLUID FLOW, we come to the problems of watching it, measuring it, and measuring its consequences. These tasks have taxed the ingenuity of a host of investigators over the past few hundred years, and an astonishing variety of techniques have been devised. Still, every investigation seems to present its own peculiarities, so techniques inevitably require adaption as well as adoption. Several general sources are worth consulting. Pankhurst and Holder (1952), already mentioned in connection with wind tunnels, have a fine collection of lore from the days when most equipment was made by resident machinists and glass blowers. Their various manometric devices are marvelous. More recent sources include Clayton and Massey (1967), Merzkirch (1974), and Ower and Pankhurst (1977). Also worth perusing are the three volumes of papers edited by Dowdell (1974). The reader should bear in mind that most of the commonly cited or commercially available equipment is intended for large objects and fast flows, so instant applicability to biological problems will not be the rule.

APPENDIX 2

flow visualization

The basic task is making visible the path of one or more ''fluid particles''—an array of particles (pathlines) or a sequence of particles (streaklines). Criteria in any choice of techniques include such factors as (1) the degree of invasiveness, or the extent to which the use of the visualization scheme distorts the field of flow; (2) the compatibility of the scheme with the survival or behavior of the test object; (3) the compatibility of the scheme with the flow-producing device in which it is being used; (4) the ease and accuracy with which some permanent record, usually photographic, can be made; and (5) cost and convenience. Strong emphasis should be put on the last point, with the reminder that cost includes time, and time will increase with the degree to which a technique must be adapted for the use at issue.

Dye or smoke injection. Injection of some dye as a tracer, usually through a hypodermic syringe, is about the simplest possible scheme. It gives good streaklines in steady, laminar flow of liquids in flow tanks or natural streams. The technique takes well to miniaturization with the addition of a micrometer-driven syringe and drawn catheter tubing; recall Lidgard's work on feeding currents in bryozoa (Figure 9.5, page 155). Injection speeds must be low enough that the injected stream does not seriously alter the profile of the flow; ideally, an injection should be just fast enough to offset the wake of the injection apparatus. For serious photography, a motorized injector is essential.

The choice of dyes is worth some attention. Fluorescent or colloidal material has distinct advantages in obtaining adequate contrast. Our favorite is rhodamine B, which has an intense magenta color in normal and fluorescent modes; it's quite cheap if reagent-grade material is avoided. The stuff colors hands and silicon sealant, and the solvent is alcohol. Fluorescein sodium (uranine) is another useful dye—it is orange in visible light but has an intense green fluorescence even in extreme dilution. Sometimes Evans blue is appropriate because of its relatively high molecular weight and consequent slow diffusion or because it stains very few substances. Ordinary mammalian milk, both plain and colored, has been used by both biologists and engineers. Apparently, the combination of milk and commercial food coloring gives a particularly stable and non-toxic colorant.

One inevitable problem with using dyes in flow tanks is their accumulation. If the tank can be drained and refilled without great delay or expense, the problem is trivial; but when, for example, one uses artificial sea water, the matter is not minor.

In flows of air, dyes are replaced by smoke or other colloidal particles. The standard sources describe the standard arrangements, the problems of making and cooling smoke, of toxicity, of deposition, and so forth. My own inclination is toward viewing flow visualization in air as an evil to be avoided, usually by shifting to a liquid analog. Sometimes, though, large scale air movements can be simply marked; "smoke candles" (E. Vernon Hill, Inc., P.O. Box 14248, San Francisco, CA 94114) were useful for tracing the interconnections between the apertures of prairie-dog burrows (Vogel et al., 1973).

Particles. Slightly more complicated and less versatile, but ultimately more revealing is the use of suspended particles. In a time exposure, their trajectories in passing an object make streaks which can give good representations of pathlines. If illumination is provided by a repetitive stroboscope, then dotted lines result, and the distance between successive dots is proportional to the speed of flow. Ideally, particles should be very small and neutrally buoyant; otherwise their paths will deviate to some extent from true pathlines of fluid particles. An extremely wide range of material can be used for marker particles. Plastics are particularly good since they can be chosen with densities close to either fresh or sea water. I once made polymethacrylate particles slightly denser than water and then salted the fluid to bring them into suspension. (Figure 3.8 was made this way.) The longer the side group on the methacrylate, the less dense and softer the polymer; hexyl or heptyl methacrylate is about right for use in water (Vogel and Feder, 1966). Polystyrene or polyethylene should also be suitable. Latex spheres are commercially available (but expensive), and "Sephadex" microbeads have been used in several investigations. Aluminum dust works well because its high surface area largely compensates for its high density (Hersh, 1960). Dilute India ink and the punchings from computer cards have also been used.

We shouldn't ignore our own turf: biology is alive with particles which love to suspend themselves in water columns. Unialgal cultures are easy to obtain and proliferate well in either fresh water or sea water (Carolina Biological Supply Co., Burlington, NC 27215 has both cultures and concentrated media). Such cultures are useful for small-scale work. For larger particles, freshly hatched brine-shrimp (*Artemia salina*) are a reasonable choice.

A few other notes. Illumination of particles with light passing through a slit permits one to photograph a two-dimensional slice of a three-dimensional flow. The main problems with particles are provision of enough of them for the volume of a flow tank and recovering them from the tank for re-use. Again, flow marking in air is a more exasperating endeavor, and particles of any reasonable size settle out too quickly to be easily applicable to low speed flows.

Electrochemical marking. A combination of the good points of dyes and particles can be obtained by a number of electrochemical schemes. These avoid the visual pollution of the ambient flow from accumulation of dye as well as the recapture problems of particles. Offsetting the advantages is the common requirement for a medium other than pure water. Certainly the most impressive system for demonstrations is the chemiluminescent technique of Howland et al. (1966). A platinum-plated anode at between 0.2 and 1.5 volts gives a bright blue luminescence, the intensity of which indicates the local proximity of flow to its surface, in short, to the steepness of the local velocity gradient. Separation points come out dark. The solution, though, is nasty—1 N KCl, 0.01 N KOH, hydrogen peroxide, ferricyanide, and luminol—not something to which you'd blithely subject an organism. Streamers of luminescence rather than a surface glow can be produced by using a little higher potential, tetramethylammonium chloride instead of KCl, and methanol instead of water.

Perhaps a more generally useful way of producing temporary streamers (a disappearing dye, if you perfer) is Baker's (1966) thymol blue technique. A 0.01% solution of the pH indicator is dissolved in distilled water and titrated with NaOH until it just turns yellow (pH 8). A platinum wire (typically 50 μm in diameter) serves as anode and emits streamers of blue when a potential of about 5 volts is applied. If the wire is alternately insulated and bare along its length, a spatial series of points is simultaneously marked. If the current is pulsed (a physiological stimulator is suitable), particle-like sequential points are marked. The scheme works up to about 5 cm s^{-1}.

An analogous scheme which requires less chemical intervention is the production of tiny hydrogen bubbles behind a fine stainless steel cathode wire. While the bubbles are buoyant, a very fine (12 to 50 μm) and scrupulously clean wire produces such small bubbles that they scatter light well and rise slowly: the bubbles have diameters less than that of the wire. If anything, the bubbles may not be sufficiently persistent since they redissolve with a half-life of about 3 s. Again, pulsed current produces a sequence of marks, and alternately bare and insulated wire gives a spatial series. Schraub et al. (1965) give much practical advice, and Roos and Willmarth (1969) describe how the technique may be extended to very low Reynolds numbers by working in glycerin.

With any technique which uses a fine wire for generating some marker, it is necessary to keep the Reynolds number of flow around the wire to 40 or less, or vortices will be shed from the wire.

Other visualization schemes. As a general matter, flow visualization is simpler at low Reynolds numbers than at high values. But it is worth noting that many techniques are limited to high speeds and large objects. In particular, techniques based on variations in the density of the fluid caused by flow are largely inapplicable to the present kinds of problems. So interferometry, shadowgraphs, and schlieren photography will be of little use unless you are looking at heat transfer problems or can introduce pulses of heat as markers.

flow measurement

Since some of the systems we've mentioned can give decent quantitative data, the subject of visualization grades gently into that of measurement. However, if one merely needs to measure the speed and direction of flow at one or a few points, the visualization schemes are usually unnecessarily unwieldy. Instead, one commonly turns to a flowmeter or anemometer, and again, one encounters a strange and diverse technology. A few systems can be largely dismissed for more than occasional use with present problems; it happens that these are the systems with which most readers are likely to be familiar.

If you seek flowmeters in a standard scientific supply catalog, you'll most likely encounter in-line devices for measuring the average flow through a pipe. The most common sort views the height of a sphere as it is blown upward by moving gas in a gently tapered pipe. Versions exist for liquid flows and low speeds. Another common device is the rotating vane anemometer, essentially a very lightweight propeller in a shroud with some arrangement to record the revolutions of the propeller. These are usually between ten and twenty centimeters in diameter and work, in air, down to speeds of 25 cm s^{-1}. They are basically directional instruments, although misalignments of up to about 10° from the direction of flow have little effect. Equipped with a rudder, a vane anemometer can be self-orienting. Versions for use in water are not difficult to make. The meteorologists favor an omnidirectional instrument—the familiar whirling cup anemometer. These are based on the difference in drag of a hollow hemisphere or cone, depending on whether the concave or convex side faces the flow (Figure 6.5, page 92). Most have three nearly-conical cups rotating in a horizontal plane. Whirling cup anemometers perform poorly at low speeds.

The classic laboratory instruments for measuring flow speeds are Pitot-static tubes—cheap, sturdy, self-calibrating, and (compared to the previous devices) minimally invasive. Their main limitation, as we discussed in Chapter 4, is a disproportionate insensitivity at low speeds resulting from the U^2 factor in Bernoulli's equation. They shouldn't be dismissed as anachronistic; with a sensitive manometer they're useful at moderate speeds. However, it is a good practice to check the response of any Pitot-static tube against something like the movement of dye in a flow tank if the low speed limit of the device is approached; Bernoulli's principle, after all, ignores viscosity. Pitot-tubes are available from E. Vernon Hill, Inc. A related technique is the use of a "pseudomanometer" mentioned in Chapter 10. If the speed inside a small pipe is well below that of the ambient flow, a measurement of flow in the small pipe can be used to calculate a pressure drop by the Hagen–Poiseuille equation. If the two ends of the small pipe function as the apertures of a Pitot tube, then the flow of dye along the axis of the small pipe can be used as a measure of the ambient flow speed.

Yet another class of devices, favored by vascular physiologists and physicians, are electromagnetic flowmeters. These are based on Faraday's discovery that current is induced when a conductor moves through a magnetic field. If the conductor is a moving fluid such as blood or sea water and the resulting electrical currents are detected by electrodes, one has a flowmeter. These electromagnetic flowmeters are mainly applicable to flow through pipes, and they measure the average flow rate.

The fanciest bit of contemporary technology used for measuring flow is the laser-Doppler anemometer. If a particle scatters light as it passes obliquely through a laser beam, the frequency of the light is Doppler-shifted by an amount proportional to the velocity of the particle. In practice, two split beams from a laser converge at a spot, and particles which pass with the flow through that spot scatter light, some of which is picked up by a photodetector. Since the two beams come from different directions, the scattered light from each has a different frequency shift, and analysis may be limited to particles which pass both beams at once—to the convergence point. While highly accurate and non-invasive (the particles can be of the order of 1 μm in diameter), the laser-

Doppler device requires a line-of-sight path through the flow and is of limited use in confined areas. Also, the instruments are expensive and cumbersome.

A large number of flow-measuring instruments are based on the fact that cooling by forced convection is a strong function of the speed of flow. The most common device is the hot-wire anemometer, in which a very thin platinum or tungsten wire is electrically heated and convectively cooled in a moving airstream. Since the resistance of the wire varies slightly with its temperature, more rapid cooling changes the resistance. The change can either be measured or can elicit a greater heating current which in turn is measured. Hot-wire anemometers are exquisitely sensitive to very rapid fluctuations in flow and can work at relatively low speeds. For use in liquids, a variant called a hot-film flowmeter, in which a thin film of metal is deposited on an insulator, is more appropriate. Either, though, is expensive, and hot wires are notoriously fragile. Still, of the commercially available instruments, hot-wire and hot-film meters are perhaps suitable for more biological applications than any other flow-measuring devices.

thermistor flowmeters and anemometers

For looking at flows in the vicinity of organisms, the most useful sort of flow-measuring device would be a small, omnidirectional probe, rugged and simple enough for field use, sensitive to very low speeds, and (of course) inexpensive. My own preference, at this point, is for instruments built around heated-bead thermistors: they come closest to the paragon I've described. As far as I know, such instruments are not commercially available, perhaps due to the lack of easy interchangeability of the sensors. But they are not difficult to construct, and in doing so, one can adapt the design for particular requirements.

These flowmeters operate in the same manner as hot-wire anemometers: an element is electrically heated and then dissipates heat at a rate which depends on the speed of the flow past it. A change in temperature alters the resistance of the sensor, and either the temperature or the heat dissipation of the sensing element is converted to an electrical signal for meter or recorder. Thermistors are cheap, small, sturdy, and have a drastic dependence of resistance on temperature—their resistance at 50 °C is typically an order of magnitude lower than that at 10 °C.

The simplest practical thermistor flowmeter is obtained by connecting a thermistor in a Wheatstone bridge, balancing the bridge in still air or water,

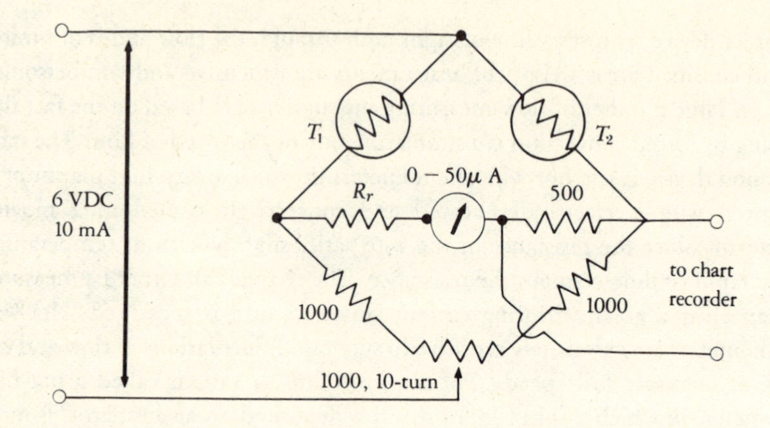

A very simple heated-bead thermistor microanemometer for use in air. T_1 and T_2 are sensing and temperature compensating thermistors, respectively, both VECO 32A402A. T_1 may be mounted, suspended by its leads, at the end of a pair of fine insulated wires run through a coarse hypodermic needle.

FIGURE A2.1

and recording or reading the imbalance when the fluid is flowing. Figure A2.1 gives such a circuit (Vogel, 1969), the one I used to map velocities behind variously abused fuselages of fruit flies. The sensor is about an eighth of a millimeter in diameter, and the response time is much less than 0.1 s. It is usable at speeds between about 0.02 and 5.0 m s^{-1}. The sensitivity or range may be altered by changing the resistor in series with the meter: 0, 4000, and 8000 ohms are reasonable values. The response is nonlinear; that is, changes in speed give greater changes in output at low speeds than at high. For many applications the nonlinearity is a convenience since it ensures a fairly constant precision relative to the indicated speed. The second thermistor serves as a rough temperature compensator; it is mounted in some wind shield exposed to air at the same temperature as the sensor. The power supply should be well regulated, so a voltage regulator must be used if the circuit is powered by batteries. Since the instrument works only in air, the leads of the thermistors may be left exposed. If necessary, the sensor probe can also be used as a temperature indicator. It is then just connected to another bridge circuit, one with a current low enough so the thermistor does not appreciably self-heat. Thermistors must always be used with substantial series resistance: because their resistance drops with increasing temperature, they will go into "thermal runaway" without some current limitation.

Although attempts to use an analogous circuit for measuring water flows were less successful, it did work for a limited range of speeds (Vogel, 1974). Two conflicting problems arise. As the flow increases, the temperature of the thermistor decreases toward the ambient value, so the instrument becomes increasingly temperature sensitive and decreasingly speed sensitive. The obvious solution, increased self-heating, causes free convection at very low speeds and reduces the dependability of bridge-balancing in still water. To circumvent these problems, a more complex circuit was devised (LaBarbera and Vogel, 1976), and has been used by a large number of people in a wide variety of applications. Its main feature is a self-balancing bridge which keeps the thermistor at constant temperature and resistance. Consequently, the sensor need be heated only modestly. Increasing the flow increases the heat dissipation of the sensor; the resulting imbalance of the bridge is amplified and used to increase the current flowing through the bridge, offsetting the extra heat loss. A fringe benefit is that the thermistor will not overheat if it is removed from the water. The output of the feedback amplifier is itself amplified to provide a signal for meter or recorder.

A specific circuit for such a flowmeter is given in Figure A2.2. It is a somewhat simplifed version of the previously published circuit and is usable, as shown, between 0.5 and 50 cm s^{-1}. Parts cost (1980) about $50; the only specialized components are the thermistors, which must be ordered from the manufacturers (VECO, 128 Springfield Avenue, Springfield, NJ 07081; Fenwal Electronics, 63 Fountain Street, Framingham, MA 01701; Thermometrics, 808 US #1, Edison, NJ 08817).

A series of suggestions for construction follow, and a possible layout of components is given in Figure A2.3.

1. As shown, two speed ranges may be selected by use of a double-pole, double-throw switch. Few applications should require more, but if only one range is needed, then R4, SW2, and the 47K feedback resistor can be omitted.

2. Note that each feedback resistor has its own 5K offset potentiometer†: for each range added, both resistor and potentiometer are needed. Only one range need begin at zero speed. The offset potentiometer can set the

† In the original diagram (LaBarbera and Vogel, 1976), the 5K trimpot on the second 741 was shown incorrectly connected, for which we apologize. The slide wire (arrow) should connect to the 741 and the two fixed contacts to ground and +15V, respectively, in the same manner as the meter zeroing trimpot is connected.

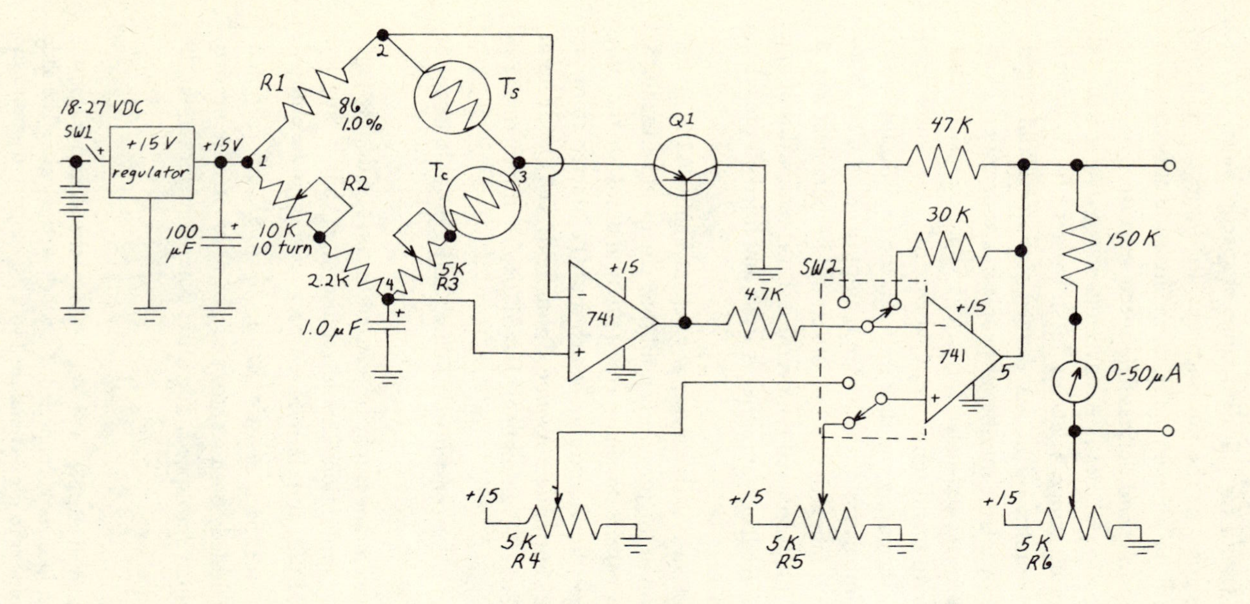

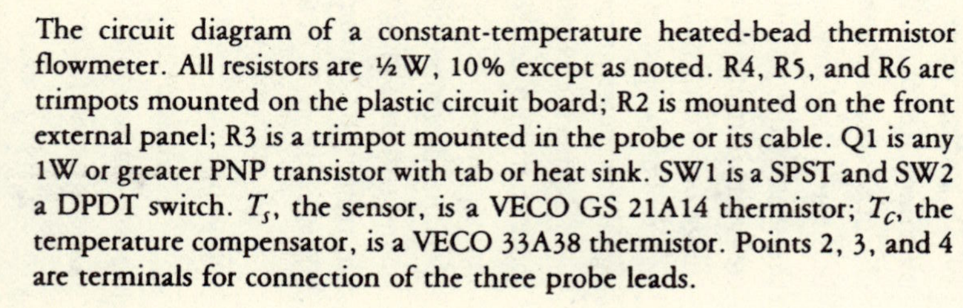
FIGURE A2.2

The circuit diagram of a constant-temperature heated-bead thermistor flowmeter. All resistors are ½ W, 10% except as noted. R4, R5, and R6 are trimpots mounted on the plastic circuit board; R2 is mounted on the front external panel; R3 is a trimpot mounted in the probe or its cable. Q1 is any 1W or greater PNP transistor with tab or heat sink. SW1 is a SPST and SW2 a DPDT switch. T_s, the sensor, is a VECO GS 21A14 thermistor; T_c, the temperature compensator, is a VECO 33A38 thermistor. Points 2, 3, and 4 are terminals for connection of the three probe leads.

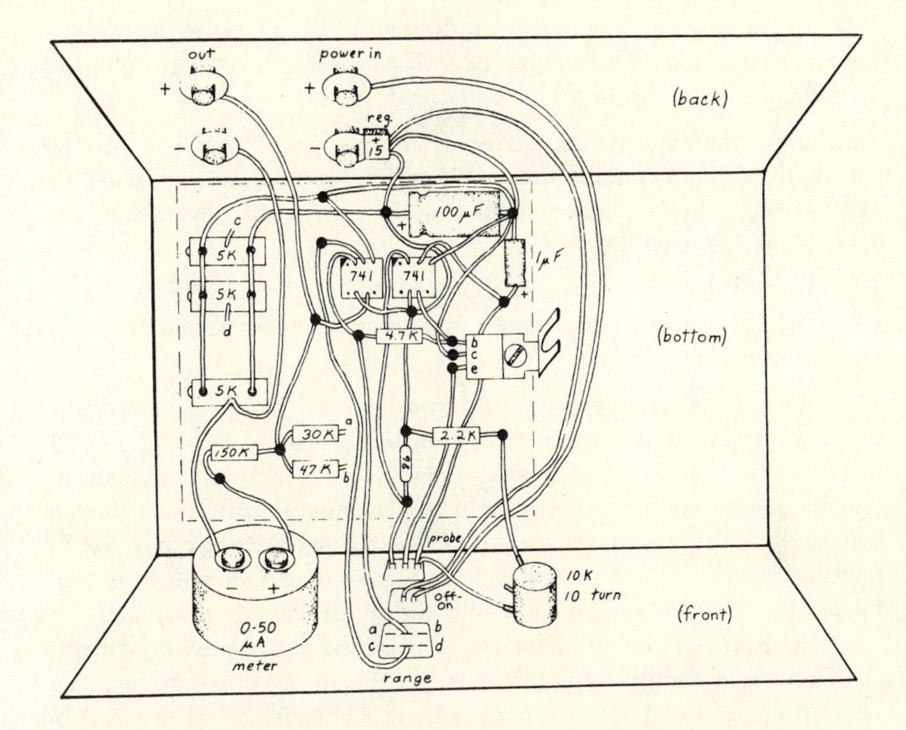

A possible layout for the circuit of Fig. A2.2. Most components mount on a perforated circuit board; the outer case is a 8.9 × 20 × 14.4-cm metal cabinet (Radio Shack, Inc.). The number one pin positions for the 741 IC's are marked; the IC's are shown in bottom view. Note the connections to base (*b*), collector (*c*), and emitter (*e*) of the transistor and the omitted wires leading to the DPDT range switch. In no case is component placement critical.

FIGURE A2.3

zero reading of the meter to correspond to some higher speed for the other range(s).

3. The sensitivity of a range varies directly, not inversely, with the value of the feedback resistor, so higher values correspond to narrower ranges of flow speed.

4. The circuit draws about 50 mA. Two small 9 V transistor radio batteries will power it but only for short periods, so larger batteries are generally handier.

5. The most costly component is the meter. Use of a less sensitive meter requires only that the 150K resistor be reduced; thus for a 0–50 mA meter, a 1K resistor is appropriate.

6. For routine operation at speeds above 30 to 50 cm s^{-1}, it may be useful to heat the thermistor further. The latter may be done by reducing the value of R1 to, say, 47 ohms and reducing the amplification by lowering the value of the 30K feedback resistor.

7. In the layout shown in Figure A2.3, space has been left in the box for either batteries, an AC power supply, a voltage to frequency converter, or a second flowmeter.

The original description gives instructions for making an especially rugged probe with a flexible tip. One can make a simpler probe by housing R3 in a small can near the point at which the cable plugs into the circuit box, so only the thermistors are in the probe itself. Thermistor leads can be painted with an insulating coating and then covered with a quick-setting epoxy. We've used lead lengths of up to 20 m with no problems—long leads reduce the need for waterproof circuit housings and other such distressing complications.

The initial balancing of the circuit is carried out in a series of steps: (1) With the probe fixed in still water, R2 should be set so the potential between points 1 and 3 is 4.5 V; (2) R4 and R5 should be set to give 3 V between point 5 and ground; and (3) R6 should be set to give a zero indication on the meter.

Temperature compensation is set with the aid of two still-water baths, one about 6° above and the other 6° below the anticipated typical ambient temperature. With the probe in one bath, R2 is adjusted to give 4.5 V between points 1 and 3. The probe is then transferred to the other bath and the magnitude and direction of the voltage change is noted. The value of R3 is altered, the probe returned to the first bath, R2 readjusted to give 4.5 V, and the procedure repeated until R3 reaches a value at which the same voltage is obtained in either bath. After temperature compensation of the first probe, R4, R5, and R6 should be set as described above and not changed again. Before use, each temperature-compensated probe is first fixed in still water, and the meter is nulled with R2 while set to its most sensitive range.

Two other versions of this flowmeter have proven useful. For better spatial and temporal resolution, a "micro" flowmeter can be made by using a VECO 31A70 thermistor which is about 0.5 instead of 1.5 mm in diameter. The maximum speed, however, is about 10 instead of 50 cm s^{-1}. Other changes: the temperature-compensating thermistor and R3 are replaced by a

4.7K fixed resistor; R1 is replaced by a 220 ohm 10% resistor; and the range resistors are 22K (0 to 10 cm s^{-1}) and 47K (0 to 2 cm s^{-1}).

For use in air, one needs to make different changes. A bead thermistor, Fenwal GB31J1 (or VECO 31A2) or smaller 1000 ohm unit, is used, T_c and R3 are again replaced by a 4.7K resistor, and the range resistors are 15K (0 to 5 m s^{-1}) and 39K (0 to 0.5 m s^{-1}). R1 remains at 86 ohms. A multipurpose circuit box can be made in which R1 and the range resistors can be altered without soldering, so all three versions of the flowmeter are available in a single unit.

For some years we routinely calibrated for both water and air flows by moving probes on a rotating beam through otherwise still fluid. With no place to leave the bulky whirling arm apparatus assembled, the system was quite a nuisance. More recently, we've used a few other schemes, the most convenient of which is based on the distribution of speeds in fully developed laminar flow in a circular pipe. You'll recall from Chapter 10 that flow along the axis is twice the average flow. Average flow can be determined with a stopwatch and graduate, so one needs only to locate a probe on the axis where flow is fully developed (equation 10.9) and the Reynolds number is under 2000. Possible set-ups for water and air are shown in Figure A2.4. The microflowmeter presented special problems since it is routinely used at speeds of about 1.0 mm s^{-1}. But it proved, finally, the easiest of all. We just mount rods on the tool holder of a three-foot metal lathe and use the rods to hold the probe in a trough alongside the lathe. The lead-screw drive of the lathe then moves the probe very slowly and steadily through the trough.

None of these thermistor flowmeters will work well at very high velocities, which perhaps is why they have been accorded little attention by engineers. The problem rests largely in the glass envelope around the actual bead; maintaining the bead at constant temperature does not necessarily keep the outer surface of the glass at constant temperature. At high flow rates, the temperature gradient occurs mainly between bead and outer surface rather than between surface and free stream, so changes in flow have little effect on the overall rate of heat transfer. It must be admitted, however, that we've put little effort into optimizing the designs given here for use at the highest practical velocities since our projects have rarely needed respose at very high speeds.

Another inexpensive electronic flowmeter for air has been described by Brumbly and Arbas (1979). They use the exposed, heated filaments of small incandescent lamps as sensors in an instrument which covers speeds up to 16 m s^{-1} with a response time of 2 ms. These are better specifications than those of our thermistor-based devices, but the associated circuitry is more complex and the geometry of the probe somewhat less versatile.

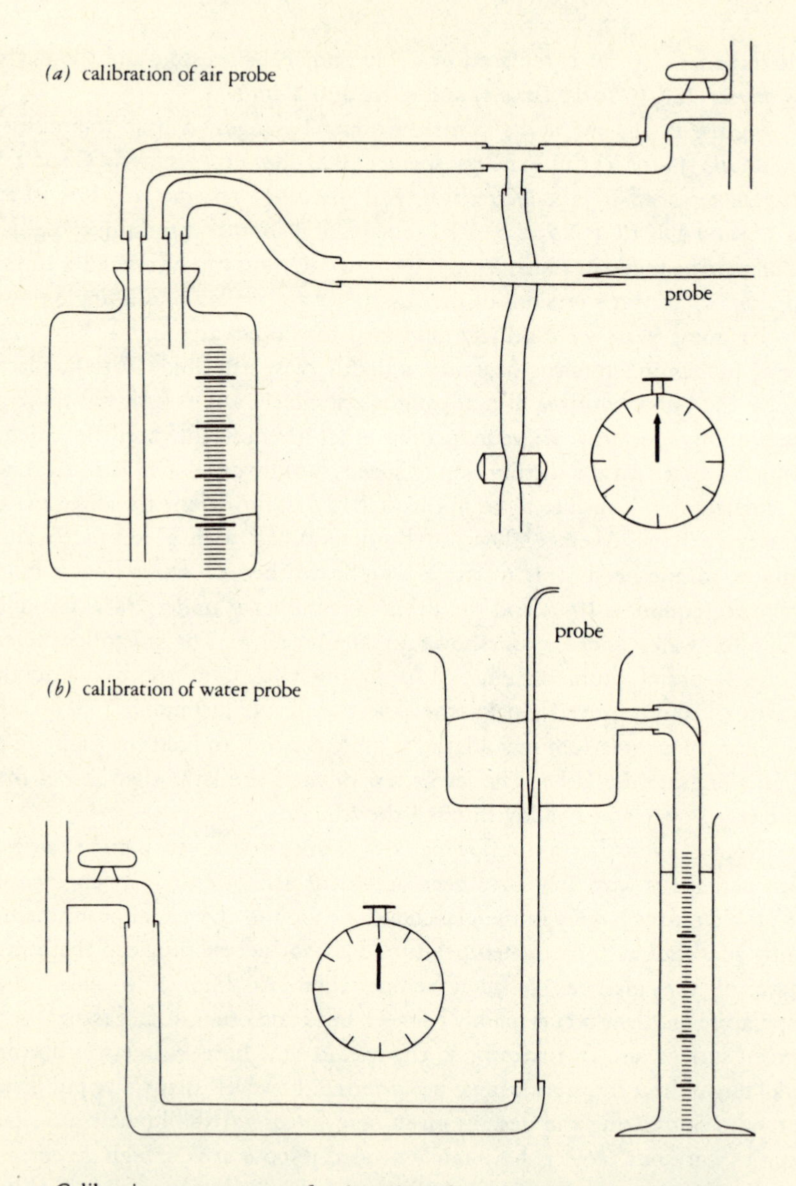

Calibration arrangements for thermistor flowmeters. In (*a*), water rising in a calibrated carboy forces air through a long pipe. A siphon is provided to drain the carboy. In (*b*) tap or temperature-conditioned water emerges from a long pipe at rates which are measured with the aid of an overflow and graduate.

FIGURE A2.4

force measurement

Measurement of forces, a frequent necessity, brings one into yet another realm of technology. Again, no single, simple recipe can be depended upon—the characteristics of organisms and models are too diverse and the range of the relevant forces too wide. And force *per se* is no easy thing to measure; work, or force times displacement, is what deflects an instrument. In practice, the displacement is either kept to a minimum, arranged so as not to change the geometry of flow around an object, or offset in some null-balancing system. There is, as well, the connection between object and measuring device to worry about. The most common arrangement is a stiff support of minimal thickness coming from the rear of the object. Such a support is termed a "sting," presumably even if used in determining the body drag of a hymenopteran. The drag or lift of any support has to be subtracted from the measured force, with some appropriate estimate of the interaction of support and test object.

For many purposes, arrangements of pivots and beams are easily improvised, either *de novo* or from disassembled old laboratory balances. The main addition which may be required is compensation for the jitteriness which seems an inevitable concomitant of forces on objects in fluids. A simple beam balance hunts excessively unless some extra restoring force is supplied (such as a substantial mass attached below the pivot point) or unless sufficient damping is available (such as a flat plate in oil as a dashpot). A beam balance can be used as a deflection indicator or (by moving weights about) as a null-balancing device. Large scale versions are called flight balances and are described in the books mentioned on wind tunnels. Although I haven't tried it, I suspect that many modern laboratory balances can be arranged to work as force measuring devices by the addition of appropriate strings, beams, and/or pulleys. The resulting direct, digital display of numbers proportional to force should be a fine convenience. Again, additional damping may be necessary.

An instrument suitable for measurement of forces down to 10^{-7} N can be made from the movement of an ordinary direct current microammeter (Vogel and Chapman, 1966). The load is applied to the needle of the meter (or an extension on it), and the deflection of the needle is offset by an electrical current. The current is proportional to the force applied by the load. Calibration is done with the needle horizontal and deflected by small objects of predetermined weight such as loops of fine wire. The polar diagram of a fruit fly wing (Figure 11.12, page 190) was derived from data obtained with such an instrument.

Transducers designed for recording isometric tension in muscles work well for flow forces of somewhat larger magnitude. Still another source of a coarse force transducer is a digital bathroom scale. The one I disemboweled contained a linear variable displacement transducer (LVDT) with its associated circuitry and a large spring. The spring could easily be changed to alter the effective range, and the device could be repackaged for field use without much trouble.

The most general solution to the problems of force measurement is undoubtedly the use of strain gauges. Strain, it should be stressed, is a displacement, so, to act as a stress- or force-measuring instrument, a strain gauge must be used with a stiff beam of reasonably constant elastic modulus. The cheapest and commonest strain gauges are thin metallic foil arrays which can be glued onto test objects or beams. As a strain gauge is stretched, its resistance changes very slightly from the unstressed value (often 120 ohms). With the advent of inexpensive, stable direct current amplifiers (operational amplifiers), it is a simple matter to use a strain gauge in a Wheatstone bridge and amplify the imbalance of the bridge resulting from stretch of the gauge. For measuring lift, drag, torque, or thrust on an object in a flow, it is most convenient to glue a gauge onto a strip of metal which is rigidly supported at one end and bears the sting at the other. The best metal strips seem to be "shim stock," thin strips of varying thicknesses used as so-called "feeler gauges." The thinner the strip, the more sensitive is the resulting force transducer. Often it is necessary to use a second, unstressed strain gauge in an adjacent arm of the bridge to compensate for temperature-induced drift. The fixed resistors in the bridge must be of sufficient precision and stability: 1% precision resistors work adequately. The gauge itself should not be subjected to more than 2 to 3 V or it may heat intolerably. Further information about strain gauges may be found in Perry and Lissner (1962) and Neubert (1967). Gauges, adhesives, and other supplies are available from William T. Bean, Inc. (18915 Grand River Avenue, Detroit, MI 48223). Strain gauges on shim stock are particularly mass-produceable and student-proof.

Of course, the matter of force measurement can sometimes be circumvented by measuring flow speeds upstream and downstream from an object and using the integration scheme presented in Chapter 4. But this indirect system for force measurement is probably best reserved for situations in which direct measurements are impractical—where, for example, an organism cannot be detached from its normal substratum.

list of symbols

(To approximate common practice as much as possible, certain symbols have been used for two different quantities; in context, no serious ambiguity should result.)

a	radius
AR	aspect ratio
a_s	Stokes radius
b	wing span
$[C]$	concentration
C	arbitrary constant
c	wing chord
C_D	drag coefficient
C_{DL}	local drag coefficient
C_{DVS}	viscous drag coefficient, referred to $(\text{surface})^{1/2}$
C_{DVV}	viscous drag coefficient, referred to $(\text{volume})^{1/3}$
C_L	lift coefficient
C_O	orifice coefficient
C_p	thermal capacity
D	drag
d	diameter
d	zero plane displacement
D_c	diffusion constant
Di	distance index
E	shear modulus
F	force
f	friction factor
Fr	Froude number
g	gravitational acceleration
Gr	Grashof number
H	total head of pressure
h	height, vertical distance
Hy	Hydra number
J	advance ratio
K	numerical constant
k	thermal conductivity
k	von Karman's constant
L	lift
L	the dimension *length*
L'	entrance length
l	length, characteristic length of an object
M	the dimension *mass*
m	mass
n	revolution rate, frequency
Nu	Nusselt number
P	power
p	pressure
Pe	Peclet number
Pr	Prandtl number
Q	volume flow rate

q_c	rate of convective heat transfer
q_m	rate of convective mass transfer
R	resistance
R	resultant force
r	incremental radius
Ra	Rayleigh number
Re	Reynolds number
Ri	Richardson number
S	area
Sc	Schmidt number
Sh	Sherwood number
St	Strouhal number
T	thrust
T	temperature
T	the dimension *time*
t	time
U	velocity
U	mean velocity
$U_{\max}$	peak velocity
V	volume
W	weight
We	Weber number
Wo	Womersley number
x	distance in direction of free-stream flow
y	distance normal to free-stream flow
z	distance normal to y and free-stream flow
z	distance upward from earth
z_0	roughness parameter
α	angle of attack
β	thermal expansion coefficient
Γ	circulation
δ	boundary layer thickness
ε	roughness height
Θ	the dimension *temperature*
θ	angular deformation
θ	glide angle
μ	dynamic viscosity
υ	kinematic viscosity
ρ	density
σ	surface tension
τ	shear stress
ϕ	wingbeat amplitude

bibliography and index of citations

(Asterisks indicate useful general sources; italic numbers in parentheses refer to citations in the text.)

Alcaraz, M., G.-A. Paffenhofer, and J.R. Strickler (1980) Catching the algae. In *The Evolution and Ecology of Zooplankton Communities,* W.C. Kerfoot, ed. Hanover, N.H.: University Press of New England and American Society of Limnology and Oceanography. (*263*)

Alexander, D.E. and J. Ghiold (1980) The functional significance of lunules in the sand dollar, *Mellita quinquiesperforata. Biol. Bull.* **159**: 561–70. (*51*)

*Alexander, R.M. (1968) *Animal Mechanics.* Seattle: University of Washington Press. (*118*)

Alexander, R.M. (1971) *Size and Shape.* London: Edward Arnold. (*94, 205*)

Alexander, R.M. (1977) Swimming. In *Mechanics and Energetics of Animal Locomotion,* R.M. Alexander and G. Goldspink, eds. pp. 222–48. London: Chapman and Hall. (*118, 217*)

Aleyev, Y.G. (1977) *Nekton.* Boston: Kluwer Boston. (*23, 117*)

Altman, P.L. and D.S. Dittmer, eds. (1971) *Biological Handbooks: Respiration and Circulation.* Bethesda, Md.: Federation of American Societies for Experimental Biology. (*22, 29*)

*Ambühl, H. (1959) Die Bedeutung der Stromüng als okölogisches Faktor. *Schweiz Z. Hydrol.* **21**: 133–264. (*142, 143, 145*)

Arnold, G.P. and D. Weihs (1978) The hydrodynamics of rheotaxis in the plaice (*Pleuronectes platessa* L.). *J. Exp. Biol.* **75**: 147–70. (*50*)

Baker, D.J. (1966) A technique for the precise measurement of small fluid velocities. *J. Fluid Mech.* **26**: 573–75. (*306*)

Balding, F.R. and G.L. Cunningham (1976) A comparison of heat transfer characteristics of simple and pinnate leaf models. *Bot. Gaz.* **137**: 65–74. (*282*)

Balsam, W.L. and S. Vogel (1973) Water movement in archaeocyathids: evidence and implications of passive flow in models. *J. Paleont.* **47**: 979–84. (*82*)

Bange, G.G.J. (1953) On the quantitative explanation of stomatal transpiration. *Acta Bot. Neerlandia* **2**: 255–96. (*149*)

Barnes, H. and H.T. Powell (1950) The development, general morphology, and subsequent elimination of barnacle populations, *Balanus crenatus* and *B. balanoides* after a heavy initial settlement. *J. Anim. Ecol.* **19**: 175–79. (*154*)

Bartholomew, G.A. and R.C. Lasiewski (1965) Heating and cooling rates, heart rate, and simulated diving in the Galapagos marine iguana. *Comp. Biochem. Physiol.* **16**: 575–82. (*22*)

*Bascom, W. (1980) *Waves and Beaches.* 2nd ed. Garden City, N.Y.: Anchor/Doubleday. (*4, 98*)

*Batchelor, G.K. (1967) *An Introduction to Fluid Dynamics.* Cambridge, U.K.: Cambridge University Press. (*19*)

* Berg, H.C. (1975) Bacterial locomotion. In *Swimming and Flying in Nature*, Vol. 1, T. Y.-T. Wu et al., eds. pp. 1-12. New York: Plenum Press. (*258*)

Berg, H.C. and R.A. Anderson (1973) Bacteria swim by rotating their flagellar filaments. *Nature* **245**: 380-82. (*259*)

* Beyer, W.H. (1978) *CRC Standard Mathematical Tables*. West Palm Beach, Fla.: CRC Press. (*247*)

Bidder, G.P. (1923) The relation of the form of a sponge to its currents. *Quart. J. Microsc. Soc.* **67**: 292-323. (*31, 155*)

Bilo, D. (1971) Flugbiophysik von Kleinvögeln. I. Kinematik und Aerodynamik des Flügelabschlages bein Haussperling (*Passer domesticus* L.). *Z. vergl. Physiol.* **71**: 382-454. (*205, 217*)

Bilo, D. (1972) Flugbiophysik von Kleinvögeln. II. Kinematik und Aerodynamik des Flügelaufschlages bein Haussperling (*Passer domesticus* L.). *Z. vergl. Physiol.* **76**: 426-37. (*205, 217*)

* Bird, R.B., W.E. Stewart, and N. Lightfoot (1960) *Transport Phenomena*. New York: John Wiley. (*285*)

Birkhoff, G. (1960) *Hydrodynamics*. 2nd ed. Princeton: Princeton University Press. (*275*)

Bishop, W. (1961) The development of tailless aircraft and flying wings. *J. Roy. Aero. Soc.* **65**: 799-806. (*221*)

* Blevins, R.D. (1977) *Flow-Induced Vibration*. New York: Van Nostrand Reinhold. (*269*)

Bold, H.C. and M.J. Wynne (1978) *Introduction to the Algae*. Englewood Cliffs, N.J.: Prentice-Hall. (*104*)

Bowerbank, J.S. (1864) *A Monograph of the British Spongiadae*. London: The Ray Society. (*30*)

Branch, G.M. and A.C. Marsh (1978) Tenacity and shell shape in six *Patella* species: adaptive features. *J. Exp. Mar. Biol. Ecol.* **34**: 111-30. (*147*)

Brennen, C. and H. Winet (1977) Fluid mechanics of propulsion by cilia and flagella. *Ann. Rev. Fluid Mech.* **9**: 339-98. (*246, 261*)

Brewer, R. and P.G. Merritt (1978) Wind throw and tree replacement in a climax beech-maple forest. *Oikos* **30**: 149-52. (*94*)

Brodie, H.J. (1955) Springboard dispersal operated by rain. *Can.J. Bot.* **33**: 156-67. (*159*)

Brodie, H.J. and P.H. Gregory (1953) The action of wind in the dispersal of spores from cup-shaped plant structures. *Can. J. Bot.* **31**: 402-10. (*160*)

Brown, S.C. (1977) Biomechanics of water pumping by *Chaetopterus variopedatus* Renier: kinetics and hydrodynamics. *Biol. Bull.* **153**: 121-32. (*181*)

Brumbley, D.R. and E.A. Arbas (1979) An inexpensive hot-wire anemometer suitable for behavioral research. *Comp. Biochem. Physiol.* **64A**: 449-50. (*315*)

Buller, A.H.R. (1933) *Researches on Fungi*. Vol. 5. London: Longmans. (*160*)

Buller, A.H.R. (1934) *Researches on Fungi*. Vol. 6. London: Longmans. (*159*)

Burrows, F.M. (1973) Calculations of the primary trajectories of plumed seeds in steady winds with variable convection. *New Phytol.* **72**: 647-64. (*254*)

Burrows, F.M. (1975) Wind-borne seed and fruit movement. *New Phytol.* **75**: 405-18. (*222*)

* Burton, A.C. (1972) *Physiology and Biophysics of the Circulation*. Chicago: Year Book Medical Publishers. (*29*)

Buss, L.W. (1979) Bryozoan overgrowth interaction: the interdependence of competition for food and space. *Nature* **281**: 475-77. (*155*)

Caborn, J.M. (1965) *Shelterbelts and Windbreaks*. London: Faber and Faber. (*139*)

Carey, F.G., J.M. Teal, J.W. Kanwisher, and K.D. Lawson (1971) Warm-bodied fish. *Amer. Zool.* **11**: 137-45. (*23*)

* Caro, C.G., T.J. Pedley, R.C. Schroter, and W.A. Seed (1978) *The Mechanics of the Circulation*. Oxford, U.K.: Oxford University Press. (*32, 174, 179, 186, 274, 285*)

Carstens, T. (1968) Wave forces on boundaries and submerged bodies. *Sarsia* **34**: 37-60. (*99*)

Chadwick, L.E. (1953) Aerodynamics and flight metabolism. In *Insect Physiology*, K.D. Roeder, ed. pp. 615-36. New York: John Wiley. (*234*)

Chamberlain, J.A. (1976) Flow patterns and drag coefficients of cephalopod shells. *Paleontol.* **19**: 539-63. (*118*)

Chamberlain, J.A. and R.R. Graus (1975) Water flow and hydromechanical adaptations of branched reef corals. *Bull. Mar. Sci.* **25**: 112-25. (*108*)

Chapman, D.C., R.H. Rand, and J.R. Cooke (1977) A hydrodynamical model of bordered pits in conifer tracheids. *J. Theor. Biol.* **67**: 11-24. (*53*)

Chapman, G. (1968) The hydraulic system of *Urechis caupo* Fisher and MacGinitie. *J. Exp. Biol.* **49**: 657-67. (*181*)

Charters, A.C., M. Neushul, and C. Barilotti (1969) The functional morphology of *Eisenia arborea*. *Int. Seaweed Symp.* **6**: 89-105. (*101, 102*)

Cheung, A.T.W. and H. Winet (1975) Flow velocity profile over a ciliated surface. In *Swimming and Flying in Nature*, Vol. 1, T.Y.-T. Wu et al., eds. pp. 223-34. New York: Plenum Press. (*179*)

Chien, S., S. Usami, R.J. Dellenback, and C.A. Bryant (1971) Comparative hemorheology: hematogolical implications of species differences in blood viscosity. *Biorheology* **8**: 35-57. (*22*)

* Clayton, B.R. and B.S. Massey (1967) Flow visualization in water: a review of techniques. *J. Sci. Instrum.* **44**: 2-11. (*303*)

Cone, C.D. (1962) The soaring flight of birds. *Sci. Amer.* **206** (4): 130-40. (*225*)

Cook, P.L. (1977) Colony-wide currents in living Bryozoa. *Cah. Biol. Mar.* **18**: 31-47. (*154*)

* Coull, B.C. and S.S. Bell (1979) Perspectives of marine meiofaunal ecology. In *Ecological Processes in Coastal and Marine Systems*, R.J. Livingston, ed. pp. 189-216. New York: Plenum Press. (*264*)

Cox, R.G. (1970) The motion of long, slender bodies in a viscous fluid. Part 1. General theory. *J. Fluid Mech.* **44**: 791-810. (*246*)

Crenshaw, D.G. (1980) How interstitial animals respond to viscous flows. Ph.D. Thesis, Duke University. (*36, 264*)

Crisp, D.J. (1955) The behavior of barnacle cyprids in relation to water movement over a surface. *J. Exp. Biol.* **32**: 569-90. (*162*)

Crisp, D.J. (1960) Factors influencing growth rate in *Balanus balanoides*. *J. Anim. Ecol.* **29**: 95-116. (*153*)

Csicàky, M.J. (1977) Body-gliding in the zebra finch. *Fortschr. Zool.* **24** (2/3): 275-86. (*214, 219*)

Cummins, K.W. (1964) Factors limiting the microdistribution of the caddisflies *Pycnopsyche lepida* (Hagen) and *Pycnopsyche guttifer* (Walker) in a Michigan stream (Trichoptera: Limnephilidae). *Ecol. Monogr.* **34**: 271-95. (*144*)

* Davis, L.B.,Jr. and R.C. Birkebak (1975) Convective energy transfer in fur. In *Perspectives of Biophysical Ecology*, D.M. Gates, ed. pp. 525-48. New York: Springer-Verlag. (*282*)

* Defant, A. (1961) *Physical Oceanography*, Vol. 1. New York: Pergamon Press. (*139*)

DeLaubenfels, M.W. (1947) Ecology of the sponges of a brackish water environment at Beaufort, N.C. *Ecol. Monogr.* **17**: 31-46. (*105*)

Delf, E.M. (1932) Experiments with the stipes of *Fucus* and *Laminaria*. *J. Exp. Biol.* **9**: 300-13. (*102*)

Dodds, G.S. and F.L. Hisaw (1924) Ecological studies of aquatic insects. I. Adaptations of mayfly nymphs to swift streams. *Ecology* **5**: 137-48. (*142, 143, 152*)

Dodds, G.S. and F.L. Hisaw (1925) Ecological studies on aquatic insects. III. Adaptations of caddisfly larvae to swift streams. *Ecology* **6**: 123-37. (*144*)

Doty, M.S. (1971) Measurement of water movement in reference to benthic algal growth. *Botanica Marina* **14**: 32-35. (99)

* Dowdell, R.B., ed. (1974) *Flow: Its Measurement and Control in Science and Industry*, Vol. 1, 3 parts. Pittsburgh: Instrument Society of America. (*303*)

DuBois, A.B., G.A. Cavagna, and R.S. Fox (1974) Pressure distribution on the body surface of swimming fish. *J. Exp. Biol.* **60**: 581-91. (*50*)

DuBois, A.B., G.A. Cavagna, and R.S. Fox (1975) The forces resisting locomotion in bluefish. In *Swimming and Flying in Nature*, Vol. 2, T.Y.-T. Wu et al., eds. pp. 541-51. New York: Plenum Press. (*50*)

Dury, G.H. (1969) Hydraulic geometry. In *Introduction to Fluvial Processes*, R.J. Chorley, ed. pp. 146-56. New York: Barnes and Noble. (*33*)

Eisenhart, C. (1968) Expression of the uncertainties of final results. *Science* **160**: 1201-1204. (*8*)

Ellington, C.P. (1975) Non-steady-state aerodynamics of the flight of *Encarsia formosa*. In *Swimming and Flying in Nature*, Vol. 2, T.Y.-T. Wu et al., eds. pp. 783-96. New York: Plenum Press. (*237*)

*Ellington, C.P. (1978) The aerodynamics of normal hovering flight: three approaches. In *Comparative Physiology: Water, Ions and Fluid Mechanics*, K. Schmidt-Nielsen, L. Bolis, and S.H.P. Maddrell, eds. pp. 327-45. Cambridge, U.K.: Cambridge University Press. (*230, 239*)

*Fauchald, K. and P.A. Jumars (1979) The diet of worms: a study of polychaete feeding guilds. *Oceanogr. Mar. Biol. Ann. Rev.* **17**: 193-284. (*156*)

Forster, G.F. (1977) Effect of leaf surface wax on deposition of airborne propagules. *Trans. Br. Mycol. Soc.* **68**: 245-50. (*161*)

*Foster-Smith, R.L. (1978) An analysis of water flow in tube-living animals. *J. Exp. Mar. Biol. Ecol.* **34**: 73-95. (*181*)

Fraser, A.I. (1962) Wind tunnel studies of the forces acting on the crowns of small trees. *Rep. Forest Res.* (U.K.) **1962**: 178-83. (*95, 97*)

*Gates, D.M. (1962) *Energy Exchange in the Biosphere.* New York: Harper and Row. (*278*)

*Geiger, R. (1965) *The Climate Near the Ground.* Cambridge, Mass.: Harvard University Press. (*137, 139*)

Gerard, V.A. and K.H. Mann (1979) Growth and reproduction of *Laminaria longicruris* populations exposed to different intensities of water movement. *J. Phycol.* **15**: 33-41. (*151*)

Gibo, D.L. and M.J. Pallett (1979) Soaring flight of monarch butterflies, *Danaus plexippus* (Lepidoptera: Danaidae), during the late summer migration in southern Ontario. *Can. J. Zool.* **57**: 1393-1401. (*220, 221, 225*)

Gies, J. (1963) *Bridges and Men.* Garden City, N.Y.: Doubleday. (*28*)

*Goldstein, S. (1938) *Modern Developments in Fluid Dynamics.* Reprint (2 vols.). New York: Dover Publications, 1965. (*4, 13, 60, 80, 114, 131, 206*)

*Gordon, J.E. (1978) *Structures, or Why Things Don't Fall Down.* Harmondsworth, Middlesex, England: Penguin Books. (*4, 88*)

Gosline, J.M. (1971) Connective tissue mechanics of *Metridium senile*. I. Structural and compositional aspects. *J. Exp. Biol.* **55**: 763-74. (99)

*Grace, J. (1977) *Plant Response to Wind.* London: Academic Press. (*138, 139, 149, 271*)

Grace, J. and M.A. Collins (1976) Spore liberation from leaves by wind. In *Microbiology of Aerial Plant Surfaces*, C.H. Dickinson and T.F. Preece, eds. pp. 185-98. London: Academic Press. (*158*)

Grace, J. and J. Wilson (1976) The boundary layer over a *Populus* leaf. *J. Exp. Bot.* **27**: 231-41. (*133*)

Grant, R.E. (1825) Observations of the structure and function of the sponge. *Edinburgh Phil. Journ.* **13**: 94-107, 333-46. (*30*)

Gray, J. (1936) Studies in animal locomotion. VI. The propulsive powers of the dolphin. *J. Exp. Biol.* **13**: 192-99. (*118*)

Gray, J. (1968) *Animal Locomotion.* London: Weidenfeld and Nicolson. (*118*)

*Greenewalt, C.H. (1962) Dimensional relationships for flying animals. *Smithson. Misc. Collns.* **144** (2): 1-46. (*205, 229*)

Gregory, P.H. (1971) The leaf as a spore trap. In *Ecology of Leaf Surface Microorganisms*, T.F. Preece and C.H. Dickenson, eds. pp. 239-43. London: Academic Press. (*161*)

*Gregory, P.H. (1973) *The microbiology of the atmosphere.* 2nd ed. Aylesbury, U.K.: Leonard Hill Books. (*160, 161, 254*)

Gregory, S.G. and J.A. Petty (1973) Valve action in bordered pits in conifers. *J. Exp. Bot.* **24**: 763-67. (*53*)

Guard, C.L. and D.E. Murrish (1975) Effects of temperature on the viscous behavior of blood from antarctic birds and mammals. *Comp. Biochem. Physiol.* **52A**: 287-90. (*22*)

Hairston, N.G. (1957) Observations on the behavior of *Draco volans* in the Philippines. *Copeia* **1957**: 262-65. (*223*)

* *Handbook of Chemistry and Physics.* 55th ed. (1974-5), R.C. Weast, ed. Cleveland, Ohio: CRC Press. (*19*)

* Happel, J. and H. Brenner (1965) *Low Reynolds Number Hydrodynamics.* Englewood Cliffs, N.J.: Prentice-Hall. (*243, 247, 248, 250, 260*)

Harper, J.L., P.H. Lovell, and K.G. Moore (1970) The shapes and sizes of seeds. *Ann. Rev. Ecol. System.* **1**: 327-56. (*253*)

Harrod, J.J. (1965) Effect of current speed on the cephalic fans of the larva of *Simulium. Hydrobiologia* **26**: 8-12. (*144*)

Hartshorn, G.S. (1978) Tree falls and tropical forest dynamics. In *Tropical Trees as Living Systems,* P.B. Tomlinson and M.H. Zimmeman, eds. pp. 617-38. Cambridge, U.K.: Cambridge University Press. (*97*)

Haslam, S.M. (1978) *River Plants: The Macrophytic Vegetation of Watercourses.* Cambridge, U.K.: Cambridge University Press. (*151*)

Hebert, P.D.N. (1978) The adaptive significance of cyclomorphosis in *Daphnia:* more possibilities. *Freshwater Biol.* **8**: 313-20. (*24*)

Hersh, G.L. (1960) A method for the study of the water currents of invertebrate ciliary filter feeders. *Veliger* **2**: 77-83. (*305*)

Hertel, H. (1966) *Structure, Form and Movement.* New York: Reinhold. (*144, 221, 236*)

Herzog, H.O. (1925) More facts about the Flettner rotor ship. *Sci. Amer.* **132**: 82-83. (*194*)

Higdon, J.J.L. and S. Corrsin (1978) Induced drag of a bird flock. *Amer. Nat.* **112**: 727-44. (*210*)

Hildebrandt, J. and A.C. Young (1965) Anatomy and physics of respiration. In *Physiology and Biophysics,* T.C. Ruch and H.D. Patton, eds. pp. 733-60. Philadelphia: Saunders. (*181*)

Hinds, T.E., F.G. Hawksworth, and W.J. McGinnies (1963) Seed dispersal in *Arceuthobium:* a photographic study. *Science* **140**: 1236-38. (*160*)

Hirst, J.M. and O.J. Stedman (1971) Patterns of spore dispersal in crops. In *Ecology of Leaf Surface Micro-organisms,* T.F. Preece and C.H. Dickenson, eds. pp. 229-37. London: Academic Press. (*161*)

* Hoar, W.S. and D.J. Randall, eds. (1978) *Fish Physiology, Vol. 7, Locomotion.* New York: Academic Press. (*217*)

Hocking, B. (1953) The intrinsic range and speed of flight of insects. *Trans. Roy. Ent. Soc. Lond.* **104**: 223-345. (*214*)

* Hoerner, S.F. (1965) *Fluid-Dynamic Drag.* S.F. Hoerner, 2 King Lane, Greenbriar, Brick Town, N.J. 08723. (*4, 88, 89, 93, 100, 101, 108, 113, 114, 194*)

Holbo, H.R., T.C. Corbett, and P.J. Horton (1980) Aeromechanical behavior of selected Douglas-fir. *Agric. Meteorol.* **21**: 81-91. (*97, 271*)

* Holwill, (1977) Low Reynolds number undulatory propulsion of organisms of different sizes. In *Scale Effects in Animal Locomotion,* T.J. Pedley, ed. pp. 233-42. London: Academic Press. (*258*)

Horn, H.S. (1971) *The Adaptive Geometry of Trees.* Princeton: Princeton University Press. (*97*)

Horsfield, K. and G. Cumming (1967) Angles of branching and diameters at branches in the human bronchial tree. *Bull. Math. Biophysics* **29**: 245-59. (*184*)

Howland, B., G.S. Springer, and M.G. Hill (1966) Use of electrochemiluminescence in visualizing separated flows. *J. Fluid. Mech.* **24**: 697-704. (*306*)

Hubbard, J.A.E.B. (1974) Scleractinian coral behavior in a calibrated current experiment: an index to their distribution pattern. *Proc. 2nd Int. Coral Reef Symp.* **2**: 107-26. (*107, 108*)

Hughes, R.G. (1975) The distribution of epizoites on the hydroid *Nemertesia antennina* (L.). *J. Mar. Biol. Assoc. U.K.* **55**: 275-94. (*153*)

* Hutchinson, G.E. (1957) *A Treatise on Limnology,* Vol. 1. New York: John Wiley. (*255, 277*)

* Hutchinson, G.E. (1967) *A Treatise on Limnology,* Vol. 2. New York: John Wiley. (*22, 24, 243, 249*)

* Hynes, H.B.N. (1970) *The Ecology of Running Waters.* Liverpool, U.K.: Liverpool University Press. (*33*)

Ingold, C.T. (1953) *Dispersal in Fungi.* Okford, U.K.: Clarendon Press. (*253*)

*Ingold, C.T. (1965) *Spore Liberation.* Oxford, U.K.: Clarendon Press. (*160*)

*Ippen, A.T., ed. (1966) *Estuary and Coastline Hydrodynamics.* New York: McGraw-Hill. (*98*)

Jaag, O. and H. Ambühl (1964) The effect of the current on the composition of biocoenoses in flowing water streams. *Int. Conf. Wat. Pollut. Res. Lond.* Oxford, U.K.: Pergamon Press. pp. 31–49. (*36*)

Jaeger, E.C. (1930) *A Dictionary of Greek and Latin Combining Forms used in Zoological Names.* Springfield, Ill.: Charles C. Thomas. (*141*)

*Jaffrin, M.Y. and A.H. Shapiro (1971) Peristaltic pumping. *Ann. Rev. Fluid Mechanics* **7**: 213–47. (*187*)

*Jahn, T.L. and J.J. Votta (1972) Locomotion of protozoa. *Ann. Rev. Fluid Mech.* **4**: 93–116. (*258*)

Jefferies, R.P.S. and P. Minton (1965) The mode of life of two Jurassic species of 'Posidonia' (Bivalva). *Paleontology* **8**: 156–85. (*81*)

Jeje, A.A. and M.H. Zimmerman (1979) Resistance to water flow in xylem vessels. *J. Exp. Bot.* **30**: 817–28. (*183*)

Jensen, M. (1954) *Shelter Effect.* Copenhagen: Danish Technical Press. (*139*)

*Jensen, M. (1956) Biology and physics of locust flight. III. The aerodynamics of locust flight. *Phil. Trans. Roy. Soc. Lond.* **B 239**: 511–52. (*209, 212, 213, 214, 217, 220, 221, 229*)

Jobin, W.R. and A.T. Ippen (1964) Ecological design of irrigation canals for snail control. *Science* **145**: 1324–26. (*87*)

*Johnson, S.E. II (1975) Microclimate and energy flow in the marine rocky intertidal. In *Perspectives of Biophysical Ecology,* D.M. Gates, ed. pp. 559–87. New York: Springer-Verlag. (*282*)

Jokiel, P.L. (1978) Effects of water movement on reef corals. *J. Exp. Mar. Biol. Ecol.* **35**: 87–97. (*108*)

Jones, W.E. and A. Demetropoulos (1968) Exposure to wave action: measurements of an important ecological parameter on rocky shores in Anglesey. *J. Exp. Mar. Biol. Ecol.* **2**: 46–63. (*100, 101*)

*Jørgensen, C.B. (1966) *Biology of Suspension Feeding.* Oxford, U.K.: Pergamon Press. (*152, 181, 262*)

*Jørgensen, C.B. (1975) Comparative physiology of suspension feeding. *Ann. Rev. Physiol.* **37**: 57–79. (*104, 105, 156, 262*)

Kingsolver, J. (1980) Thermal and hydric aspects of environmental heterogeneity in the pitcher plant mosquito. *Ecol. Monogr.* **49**: 357–76. (*285*)

Kirby-Smith, W.W. (1972) Growth of the bay scallop: the influence of experimental water currents. *J. Exp. Mar. Biol. Ecol.* **8**: 7–18. (*153*)

Knight-Jones, E.W. and J. Moyse (1961) Intraspecific competition in sedentary marine animals. *Symp. Soc. Exp. Biol.* **15**: 72–95. (*154*)

Knoerr, K.R. and L.W. Gay (1965) Tree leaf energy balance. *Ecology* **46**: 17–24. (*282*)

Koehl, M.A.R. (1976) Mechanical design in sea anemones. In *Coelenterate Ecology and Behavior,* G.O. Mackie, ed. pp. 23–31. New York: Plenum Press. (*106*)

Koehl, M.A.R. (1977*a*) Effects of sea anemones on the flow forces they encounter. *J. Exp. Biol.* **69**: 87–105. (*106*)

Koehl, M.A.R. (1977*b*) Water flow and the morphology of zoanthid colonies. *Proc. 3rd. Int. Coral Reef Symp.* **1**: 437–44. (*107*)

Koehl, M.A.R. and S.A. Wainwright (1977) Mechanical adaptations of a giant kelp. *Limnol. Oceanogr.* **22**: 1067–71. (*102*)

Kokshaysky, N.V. (1979) Tracing the wake of a flying bird. *Nature* **279**: 146–48. (*36*)

Kramer, M.O. (1965) Hydrodynamics of the dolphin. *Adv. Hydrosci.* **2**: 111–30. (*118*)

Kramer, P.J. (1959) Transpiration and the water economy of plants. In *Plant Physiology,* Vol. 2, F.C. Steward, ed. pp. 607–726. New York: Academic Press. (*30*)

Kramer, P.J. and T.T. Koslowski (1960) *Physiology of Trees.* New York: McGraw-Hill. (*30*)

Kreith, F. (1973) *Principles of Heat Transfer.* 3rd ed. New York: Harper and Row. (*278*)

Kristiansen, U.R. and O.K.O. Petterson (1978) Experiments on the noise heard by human beings when exposed to atmospheric winds. *J. Sound Vib.* **58**: 285–92. (*75*)

Kuethe, A.M. (1975) On the mechanics of flight of small insects. In *Swimming and Flying in*

Nature, Vol. 2, T.Y.-T. Wu et al., eds. pp. 803-13. New York: Plenum Press. (*215*)

Kunkel, W.B. (1948) Magnitude and character of errors produced by shape factors in Stokes' law estimates of particle radius. *J. Appl. Phys.* **19**: 1056-58. (*252*)

LaBarbera, M. (1978) Particle capture by a Pacific brittle star: experimental test of the aerosol suspension feeding model. *Science* **201**: 1147-49. (*263*)

LaBarbera, M. and S. Vogel (1976) An inexpensive thermistor flowmeter for aquatic biology. *Limnol. Oceanogr.* **21**: 750-56. (*311*)

*Lamb, H. (1932) *Hydrodynamics.* 6th ed. Reprint. New York: Dover Publications, 1945. (*276*)

Lange, W.H., Jr. (1956) Aquatic Lepidoptera. In *Aquatic Insects of California,* R.L. Usinger, ed. pp. 271-88. Berkeley: Univ. of California Press. (*52*)

Langmuir, I. (1938) The speed of the deer fly. *Science* **87**: 233-34. (*15*)

Leech, H.B. and H.P. Chandler (1956) Aquatic Coleoptera. In *Aquatic Insects of California,* R.L. Usinger, ed. pp. 293-371. Berkeley: Univ. of California Press. (*52*)

Leopold, L.B. and T. Maddock, Jr. (1953) The hydraulic geometry of stream channels and some physiographic implications. *U.S. Geol. Survey Professional Paper 252.* (*33*)

Leversee, G.J. (1976) Flow and feeding in fan-shaped colonies of the gorgonian coral, *Leptogorgia. Biol. Bull.* **151**: 344-56. (*105*)

*Lewis, J.R. (1964) *The Ecology of Rocky Shores.* London: The English Universities Press. (*98*)

*Leyton, L. (1975) *Fluid Behavior in Biological Systems.* Oxford, U.K.: Clarendon Press. (*4, 118, 130, 135, 139, 264, 278, 285*)

*Lighthill, J. (1969) Hydromechanics of aquatic animal propulsion. *Ann. Rev. Fluid Mech.* **1**: 413-46. (*217*)

*Lighthill, J. (1972) *Physiological Fluid Mechanics.* New York: Springer-Verlag. (*187*)

*Lighthill, J. (1975) Aerodynamic aspects of animal flight. In *Swimming and Flying in Nature,* Vol. 2, T.Y.-T. Wu et al., eds. pp. 423-91. New York: Plenum Press. (*274*)

Lissaman, P.B.S. and C. Shollenberger (1970) Formation flight of birds. *Science* **168**: 1003-1005. (*210*)

*Lowry, W.P. (1967) *Weather and Life: An Introduction to Biometerorology.* New York: Academic Press. (*138, 139*)

Lund, J.W.G. (1959) Buoyancy in relation to the ecology of freshwater phytoplankton. *Brit. Phycol. Bull.* (*Phycol. Bull.*) **7**: 1-17. (*255*)

Lundegårdh, H. (1966) *Plant Physiology.* New York: American Elsevier Publishing Co. (*30, 32*)

Machin, J. (1964*a*) The evaporation of water from *Helix aspera.* II. Measurement of air flow and the diffusion of water vapor. *J. Exp. Biol.* **41**: 771-81. (*149*)

Machin, J. (1964*b*) The evaporation of water from *Helix aspera.* III. The application of evaporation formulae. *J. Exp. Biol.* **41**: 783-92. (*149*)

Mathieson, A.C., E. Tveter, M. Daly, and J. Howard (1977). Marine algal ecology in a New Hampshire tidal rapid. *Botanica Marina* **20**: 277-90. (*98*)

Maull, D.J. and P.W. Bearman (1964) The measurement of the drag of bluff bodies by the wake traverse method. *J. Roy. Aero. Soc.* **68**: 843. (*60*)

Mayhead, G.J. (1973*a*) Some drag coefficients for British forest trees derived from wind tunnel studies. *Agric. Meteorol.* **12**: 123-30. (*95*)

Mayhead, G.J. (1973*b*) Sway periods of forest trees. *Scot. For.* **27**: 19-23. (*97*)

McCutchen, C.W. (1977) The spinning rotation of ash and tulip tree samaras. *Science* **197**: 691-92. (*194, 222, 236*)

McGahan, J. (1973) Gliding flight of the Andean condor in nature. *J. Exp. Biol.* **58**: 225-37. (*205, 210, 220*)

McIntire, C.D. (1966) Some effects of current velocity on periphyton communities in laboratory streams. *Hydrobiologia* **27**: 559-70. (*161*)

McMahon, T.A. (1975) The mechanical design of trees. *Sci. Amer.* **233** (1): 93-102. (*97*)

McNown, J.S. and J. Malaika (1950) Effects of particle shape on settling velocity at low Reynolds numbers. *Trans. Amer. Geophys. Union.* **31**: 74-82. (*249*)

*Mechtly, E.A. (1973) *The International System of Units: Physical Constants and Conversion Factors.* Second revision. Washington, D.C.: National Aeronautics and Space Administration SP-7012. (*7*)

*Meidner, H. and T.A. Mansfield (1968) *Physiology of Stomata.* London: McGraw-Hill. (*149*)

*Merritt, R.W. and K.W. Cummins (1978) *An Introduction to the Aquatic Insects of North America.* Dubuque, Iowa: Kendall-Hunt Publishing Co. (*142, 263*)

*Merzkirch, W. (1974) *Flow Visualization.* New York: Academic Press. (*303*)

Miller, P.L. (1966) The supply of oxygen to the active flight muscles of some large beetles. *J. Exp. Biol.* **45**: 285-304. (*54*)

Minton, R. and R.P.S. Jefferies (1966) Hydrodynamic simulation of small, passively sinking biological objects. *Nature* **209**: 829-30. (*81*)

*Mises, R. von (1945) *Theory of Flight.* Reprint. New York: Dover Publications, 1959. (*4, 113, 204, 228, 234*)

*Monteith, J.L. (1973) *Principles of Environmental Physics.* London: Edward Arnold. (*138, 139, 149, 285*)

Moore, I.J. (1964) Effects of water currents on fresh water snails *Stagnicola plaustris* and *Physa propinqua. Ecology* **45**: 558-64. (*101*)

*Mortimer, C.H. (1974) Lake hydrodynamics. *Mitt. Internat. Verein. Limnol.* **20**: 124-97. (*227*)

Munk, W.H. and G.A. Riley (1952) Absorption of nutrients by aquatic plants. *J. Mar. Res.* **11**: 215-40. (*255*)

Murdock, G.R. and S. Vogel (1978) Hydrodynamic induction of water flow in a keyhole limpet (Gastropoda, Fissurellidae). *Comp. Biochem. Physiol.* **61A**: 227-31. (*54*)

Muus, B.J. (1968) A field method for measuring "exposure" by means of plaster balls. *Sarsia* **34**: 61-68. (99)

Nachtigall, W. (1964) Zur Aerodynamic des Coleopterenflugs: wirken die Elytren als Tragflügel? *Verh. Deut. Zool. Ges.* (*Kiel*) **58**: 319-26. (*120, 209, 214*)

Nachtigall, W. (1966) Die Kinematik der Schlagflügelbewegungen von Dipteren: Methodische und analytische Grundlagen zur Biophysik des Insektenflugs. *Zeitschr. Vergl. Physiol.* **52**: 155-211. (*217*)

*Nachtigall, W. (1974) Locomotion: mechanics and hydrodynamics of swimming in aquatic insects. In *The Physiology of Insects,* 2nd ed., M. Rockstein, ed. **Vol. 3**: 381-432. New York: Academic Press. (*117, 215, 256*)

Nachtigall, W. (1977*a*) Swimming mechanics and energetics of locomotion of variously sized water beetles—Dytiscidae, body length 2 to 35 mm. In *Scale Effects in Animal Locomotion.* T.J. Pedley, ed. pp. 269-83. London: Academic Press. (*117*)

Nachtigall, W. (1977*b*) Die aerodynamische Polare des *Tipula*-Flügels und eine Einrichtung zur halbautomatischen Polarenaufnahme. *Fortschr. Zool.* **24**: (2/3): 347-52. (*209, 213*)

Nachtigall, W. (1979*a*) Gleitflug des Flugbeutlers *Petaurus breviceps papuanus.* II. Filmanalysen zur Einstellung von Gleitbahn und Rumpf sowie zur Steurung des Gleitflugs. *J. Comp. Physiol.* **133**: 89-95. (*209, 211, 220*)

Nachtigall, W. (1979*b*) Gleitflug des Flugbeutlers *Petaurus breviceps papuanus* (Thomas). III. Modelmessungen zum Einfluss des Fell besatzes auf Umströmung und Luftkrafterzeugung. *J. Comp. Physiol.* **133**: 339-49. (*212*)

Nachtigall, W. (1979*c*) Der Taubenflügel in Gleitflugstellung: Geometrische Kenngrössen der Flügelprofile und Luftkrafterzeugung. *J. für Ornithol.* **120**: 30-40. (*219*)

Nachtigall, W., R. Grosch, and T. Schultze-Westrum (1974) Gleitflug des Flugbeutlers *Petaurus breviceps papuanus* (Thomas). I. Flugverhalten und Flugsteuerung. *J. Comp. Physiol.* **92**: 105-15. (*209*)

Nachtigall, W. and B. Kempf (1971) Vergleichende Untersuchung zur Flugbiologischen Funktion des Daumenfittich (Alula spuria) bei Vögeln. I. Der Daumenfittich als Hochauftriebserzeuger. *Z. Vergl. Physiol.* **71**: 326-41. (*204, 209*)

National Bureau of Standards (US) (1977) *The International System of Units (SI).* Washington, D.C.: U.S. Government Printing Office NBS SP-330. (7)

Needham, J.G., J.R. Traver, and Y. Hsu (1935) *The Biology of Mayflies.* Ithaca, N.Y.: Comstock Publishing Co. (*142*)

*Neubert, H.K.P. (1967) *Strain Gauges.* London: Macmillan. (*318*)

Newman, B.G., S.B. Savage, and D. Schouella (1977) Model tests on a wing section of an *Aeschna* dragonfly. In *Scale Effects in Animal Locomotion,* T.J. Pedley, ed. pp. 445-77. London: Academic Press. (*206*)

Newman, J.N. and T.Y. Wu (1975) Hydromechanical aspects of fish swimming. In *Swimming and Flying in Nature*, Vol. 2, T.Y. Wu et al., eds. pp. 615–34. New York: Plenum Press. *(118)*

Nielsen, A. (1950) The torrential invertebrate fauna. *Oikos* **2**: 176–96. *(52, 142, 144, 146)*

Nobel, P.S. (1974) Boundary layers of air adjacent to cylinders: estimation of effective thickness and measurement on plant material. *Pl. Physiol.* **54**: 177–81. *(133, 149)*

Nobel, P.S. (1975) Effective thickness and resistance of the air boundary layer adjacent to spherical plant parts. *J. Exp. Bot.* **26**: 120–30. *(133, 149)*

* Norberg, R.A. (1973) Autorotation, self-stability, and structure of single-winged fruits and seeds (samaras) with comparative remarks on animal flight. *Biol. Rev.* **48**: 561–96. *(236, 253)*

Norberg, R.A. (1975) Hovering flight of the dragonfly *Aeschna juncea* L., kinematics and aerodynamics. In *Swimming and Flying in Nature*, Vol. 2, T.Y.-T. Wu et al., eds. pp. 763–81. New York: Plenum Press. *(239)*

Norberg, U.M. (1975) Hovering flight in the pied flycatcher (*Ficedula hypoleuca*). In *Swimming and Flying in Nature*, Vol. 2, T.Y.-T. Wu et al., eds. pp. 869–81. New York: Plenum Press. *(238)*

Norberg, U.M. (1976) Aerodynamics of hovering flight in the long-eared bat, *Plecotus auritus*. *J. Exp. Biol.* **65**: 459–70. *(239)*

Norton, T.A. (1973) Oriented growth of *Membranipora membranacea* (L.) on the thallus of *Saccorhiza polyschides* (Lightf.) Batt. *J. Exp. Mar. Biol. Ecol.* **13**: 91–95. *(155)*

Oliver, H.R. and G.J. Mayhead (1974) Wind measurements in a pine forest during a destructive gale. *Forestry* **47**: 185–94. *(97, 137, 138, 139)*

Oliver, J.A. (1951) "Gliding" in amphibians and reptiles, with a remark on arboreal adaptation in the lizard *Anolis carolinensis* Voight. *Amer. Nat.* **85**: 171–76. *(222)*

O'Neill, P.L. (1978) Hydrodynamic analysis of feeding in sand dollars. *Oecologia* **34**: 157–74. *(207, 208)*

* Ower, E. and F.C. Pankhurst (1977) *Measurement of Air Flow*. Oxford, U.K.: Pergamon Press. *(303)*

Palmer, E. and G. Weddell (1964) The relationship between structure, innervation, and function of the skin of the bottlenose dolphin (*Tursiops truncatus*). *Proc. Zool. Soc. Lond.* **143**: 553–68. *(23)*

* Pankhurst, R.C. and D.W. Holder (1952) *Wind-Tunnel Technique*. London: Sir Isaac Pitman & Sons. *(44, 291, 303)*

Parkhurst, D.F., P.R. Duncan, D.M. Gates, and F. Kreith (1968) Convection heat transfer from broad leaves of plants. *J. Heat Transfer* **90**: 71–76. *(278)*

* Parkhurst, D.F. and O.L. Loucks (1972) Optimal leaf size in relation to environment. *J. Ecol.* **60**: 505–37. *(282)*

Parkhurst, D.F. and G.I. Pearman (1974) Convective heat transfer from a semi-infinite flat plate to periodic flow at various angles of incidence. *Agric. Meteorol.* **13**: 383–93. *(283)*

Parlange, J.-Y. and P.E. Waggoner (1972) Boundary layer resistance and temperature distribution on still and flapping leaves. II. Field experiments. *Pl. Physiol.* **50**: 60–63. *(133)*

Parry, D.A. (1949) The structure of whale blubber, and a discussion of its thermal properties. *Quart. J. Microscop. Sci.* **90**: 13–26. *(23)*

* Pedley, T. J. (1977) Pulmonary fluid dynamics. *Ann. Rev. Fluid Mech.* **9**: 229–74. *(32, 181, 184)*

* Pennak, R.W. (1978) *Fresh water Invertebrates of the United States*. 2nd ed. New York: John Wiley. *(144)*

Pennington, R.H. (1965) *Introductory Computer Methods and Numerical Analysis*. 2nd ed. London: Collier-MacMillan. *(175)*

Pennycuick, C. J. (1960) Gliding flight of the fulmer petrel. *J. Exp. Biol.* **37**: 330–38. *(210)*

Pennycuick, C. J. (1968) A wind-tunnel study of gliding flight in the pigeon *Columba livia*. *J. Exp. Biol.* **49**: 509–26. *(112, 119, 205, 210)*

Pennycuick, C. J. (1971) Control of gliding angle in Rüppell's griffon vulture *Gyps ruppellii*. *J. Exp. Biol.* **55**: 39–46. *(119, 210, 220)*

* Pennycuick, C. J. (1972) *Animal Flight*. London: Edward Arnold. *(86, 223)*

*Pennycuick, C. J. (1974) *Handy Matrices of Unit Conversion Factors for Biology and Mechanics.* London: Edward Arnold. (7)

Perrier, E.R., A. Aston, and G.F. Arkin (1973) Wind flow characteristics on a soybean leaf compared with a leaf model. *Pl. Physiol.* **28**: 106-12. (*133*)

*Perry, C.C. and H.R. Lissner (1962) *The Strain Gage Primer.* New York: McGraw-Hill. (*318*)

Philipson, G.N. and B.H.S. Moorhouse (1974) Observations on ventilatory and net-spinning activities of the genus *Hydropsyche* Pictet (Trichoptera, Hydropsychidae) under experimental conditions *Freshwater Biol.* **4**: 525-33. (*263*)

Plate, E.J. (1970) The aerodynamics of shelterbelts. *Agric. Meteorol.* **8**: 203-22. (*139*)

Plesset, M.S., C.G. Whipple, and H. Winet (1975) Analysis of the steady state of the bioconvection in swarms of swimming micro-organisms. In *Swimming and Flying in Nature,* Vol. 1, T.Y.-T. Wu et al., eds. pp. 339-60. New York: Plenum Press. (*255*)

*Pope, A. and J.J. Harper (1966) *Low-Speed Wind Tunnel Testing.* New York: John Wiley. (*288, 291*)

*Porter, W.P. and D.M. Gates (1969) Thermodynamic equilibria of animals with environment. *Ecol. Monogr.* **39**: 227-44. (*278, 282*)

*Prandtl, L. and O.G. Tietjens (1934) *Applied Hydro- and Aeromechanics.* Reprint. New York: Dover Publications, 1957. (*4, 44, 55, 60, 167, 174, 271, 276*)

Prange, H.D. and K. Schmidt-Nielsen (1970) The metabolic cost of swimming in ducks. *J. Exp. Biol.* **53**: 763-77. (*86*)

*Purcell, E.M. (1977) Life at low Reynolds number. *Amer. J. Physics* **45**: 3-11. (*242, 243*)

Rabel, H. von (1965) *Modellflug Profile.* Munich, Germany. (*206*)

Ramsay, J.A., C.G.B. Butler, and J.H. Sang (1938) The humidity gradient at the surface of a transpiring leaf. *J. Exp. Biol.* **15**: 255-65. (*149*)

Randall, R.H. and L.G. Eldredge (1977) Effects of typhoon Pamela on the coral reefs of Guam. *Proc. 3rd Int. Coral Reef Symp.* **2**: 525-32. (*108*)

Rayleigh, Lord (1883) The soaring of birds. *Nature* **27**: 534-35. (*225*)

Rayner, J.M.V. (1977) The intermittent flight of birds. In *Scale Effects in Animal Locomotion,* T. J. Pedley, ed. pp. 437-43. London: Academic Press. (*214*)

Rayner, J.M.V. (1979) A new approach to animal flight mechanics. *J. Exp. Biol.* **80**: 17-54. (*239*)

Rees, C. J.C. (1975) Aerodynamic properties of an insect wing section and a smooth airfoil compared. *Nature* **258**: 141-42. (*206*)

Reiswig, H.M. (1971*a*) *In situ* pumping activities of tropical demospongiae. *Mar. Biol.* **9**: 38-50. (*105*)

Reiswig, H.M. (1971*b*) Particle feeding in natural populations of three marine demosponges. *Biol. Bull.* **141**: 568-91. (*262*)

Reiswig, H.M. (1974) Water transport, respiration, and energetics of three tropical marine sponges. *J. Exp. Mar. Biol. Ecol.* **14**: 231-49. (*30*)

Reiswig, H.M. (1975*a*) The aquiferous systems of three marine demospongiae. *J. Morph.* **145**: 493-502. (*31*)

Reiswig, H.M. (1975*b*) Bacteria as food for temperate-water marine sponges. *Can. J. Zool.* **53**: 582-89. (*262*)

Resh, V.H. and J.O. Solem (1978) Phylogenetic relationships and evolutionary adaptations of aquatic insects. In *An Introduction to the Aquatic Insects of North America,* R.W. Merritt and K.W. Cummins, eds. pp. 33-42. Dubuque, Iowa: Kendall-Hunt Publishing Co. (*142, 144*)

Reynolds, O. (1883) An experimental investigation of the circumstances which determine whether the motion of water shall be direct or sinuous, and of the law of resistance in parallel channels. *Trans. Roy. Soc. Lond.* **174**: 935-82. (*39*)

*Riedl, R. J. (1971*a*) Water movement. In *Marine Ecology,* Vol. 1, pt. 2, O. Kinne, ed. pp. 1085-88, 1124-56. London: Wiley-Interscience. (*98, 101, 104*)

Riedl, R. J. (1971*b*) How much water passes through sandy beaches? *Int. Revue Geo. Hydrobiol.* **56**: 923-46. (*264*)

Riedl, R. J. and H. Forstner (1968) Wasserbewegung im Mikrobereich des Benthos. *Sarsia* **34**: 163-88. (*106*)

* Riedl, R. J. and R. Machan (1972) Hydrodynamic patterns of lotic sands and their bioclimatological implications. *Mar. Biol.* **13**: 179-209. (*263*)

Roffey, J. (1963) Observations on gliding in the desert locust. *Anim. Behav.* **11**: 359-66. (*225*)

Roos, F.W. and W.W. Willmarth (1969) Hydrogen bubble flow visualization at low Reynolds numbers. *AIAA Journ.* **7**: 1635-37. (*306*)

* Rouse, H. (1938) *Fluid Mechanics for Hydraulic Engineers.* Reprint. New York: Dover Publications, 1961. (*4, 130, 184*)

* Rubenstein, D.I. and M.A.R. Koehl (1977) The mechanisms of filter feeding: some theoretical considerations. *Amer. Natur.* **111**: 981-94. (*22, 156, 262*)

Salisbury, E. (1976) Seed output and the efficacy of dispersal by wind. *Proc. Roy. Soc.* **B 192**: 323-29. (*254*)

* Schmidt-Nielsen, K. (1969) The neglected interface: the biology of water as a liquid-gas system. *Q. Rev. Biophys.* **2**: 283-304. (*149*)

* Schmidt-Nielsen, K. (1972) *How Animals Work.* Cambridge, U.K.: Cambridge University Press. (*178*)

* Schmidt-Nielsen, K. (1979) *Animal Physiology.* 2nd ed. Cambridge, U.K.: Cambridge University Press. (*282*)

Schmitz, F.W. (1960) *Aerodynamik des Flugmodells.* Duisberg, Germany: Carl Lange Verlag. (*206*)

Schraub, F.A., S.J. Kline, J. Henry, P.W. Runstadler, Jr., and A. Littell (1965) Use of hydrogen bubbles for quantitative determination of time-dependent velocity fields in low-speed water flows. *J. Basic Engin.* (*Trans. ASME*): June, pp. 429-44. (*306*)

Schrodter, H. (1960) Dispersal by air and water: the flight and landing. In *Plant Pathology,* Vol. 3, J.G. Horsfall and A.E. Dimont, eds. pp. 170-227. New York: Academic Press. (*254*)

Schuepp, P.H. (1972) Studies of forced convection heat and mass transfer of fluttering realistic leaf models. *Boundary Layer Meteorol.* **2**: 263-74. (*272*)

Schumacher, G. J. and H.A. Whitford (1965) Respiration and P^{32} uptake in various species of freshwater algae as affected by a current. *J. Phycol.* **1**: 78-80. (*151*)

Schwenke, H. (1971) Water movement—plants. In *Marine Ecology,* Vol 1, pt. 2, O. Kinne, ed. pp. 1091-1121. London: Wiley-Interscience. (*98, 102*)

Scorer, R.S. (1978) *Environmental Aerodynamics.* Chichester, U.K.: Ellis Horwood. (*277*)

Sculthorpe, C.D. (1967) *The Biology of Aquatic Vascular Plants.* London: Edward Arnold. (*104, 151*)

Seeley, L.E., R.L. Hummel, and J.W. Smith (1975) Experimental velocity profiles in laminar flow around spheres at intermediate Reynolds numbers. *J. Fluid Mech.* **68**: 591-608. (*93, 112*)

* Shapiro, A.H. (1961) *Shape and Flow.* Garden City, New York: Doubleday.(*4, 80*)

Sheldon, J.C. and F.M. Burrows (1973) The dispersal effectiveness of the achene-pappus units of selected compositae in steady winds with convection. *New Phytol.* **72**: 665-75. (*253*)

* Simiu, E. and R.H. Scanlan (1978) *Wind Effects on Structures.* New York: John Wiley. (*269*)

Simons, D.B. (1969) Open channel flow. In *Introduction to Fluvial Processes,* R. J. Chorley, ed. pp. 124-45. New York: Barnes & Noble. (*140*)

* Slatyer, R.O. (1967) *Plant-Water Relationships.* New York: Academic Press. (*149*)

* Sleigh, M.A. (1978) Fluid propulsion by cilia and flagella. In *Comparative Physiology: Water, Ions, and Fluid Mechanics,* K. Schmidt-Nielsen, L. Bolis, and S.H.P. Maddrell, eds. pp. 255-66. Cambridge, U.K.: Cambridge University Press. (*260*)

Sleigh, M.A. and J.R. Blake (1977) Methods of ciliary propulsion and their size limitations. In *Scale Effects in Animal Locomotion,* T. J. Pedley, ed. pp. 243-56. London: Academic Press. (*261*)

* Smayda, T. J. (1970) The suspension and sinking of phytoplankton in the sea. *Oceanogr. Mar. Biol. Ann. Rev.* **8**: 354-414. (*254, 255*)

Smayda, T. J. (1974) Some experiments on the sinking characteristics of two freshwater diatoms. *Limnol. Oceanogr.* **19**: 628-35. (*255*)

Smith, A.P. (1972) Buttressing of tropical trees: a descriptive model and new hypotheses. *Amer. Natur.* **106**: 32-46. (*97*)

* Smith, I.R. (1975) Turbulence in Lakes and Rivers. *Freshwater Biol. Assoc. Sci. Publ.* **29**: 1-79. (*140*)

Smith, J. Maynard (1952) The importance of the nervous system in the evolution of animal flight. *Evolution* **6**: 127-29. (*221*)

Sotavalta, O. (1953) Recordings of high wing-stroke and thoracic vibration frequency in some midges. *Biol. Bull.* **104**: 439-44. (*229*)

Sreeramulu, T. (1962) Aerial dissemination of barley loose smut (*Ustilago nuda*). *Trans. Br. Mycol. Soc.* **45**: 373-84. (*159*)

Stanley, E.N. and R.C. Batten (1968) Viscosity of sea water at high pressures and moderate temperatures. *Nav. Ship Res. Dev. Center Rept.* No. 2827. (*18*)

Stanley, E.N. and R.C. Batten (1969) Viscosity of sea water at moderate temperatures and pressures. *J. Geophys. Res.* **74**: 3415. (*18*)

Steinman, D.B. (1955) Suspension bridges: the aerodynamic problem and its solution. In *Science in Progress*, 9th ser., G.A. Baitsell, ed. pp. 241-91. New Haven: Yale University Press. (*270, 272*)

Steinmann, P. (1907) Die Tierwelt der Gebirgsbäche. *Ann. Biol. Lacust. Bruxelles.* **2**: 20-150. (quoted by Nielsen, Ambühl) (*142*)

Stoker, M.G.P. (1973) Role of diffusion boundary layer in contact inhibition of growth. *Nature* **246**: 200-202. (*152*)

* Streeter, V.L. and E.B. Wylie (1975) *Fluid Mechanics.* 7th ed. New York: McGraw-Hill. (*4*)

Strickler, J.R. (1975*a*) Intra- and interspecific information flow among planktonic copepods: receptors. *Verh. Internat. Verein. Limnol.* **19**: 2951-58. (*257*)

* Strickler, J.R. (1975*b*) Swimming of planktonic *Cyclops* species (Copepoda, Crustacea): pattern, movements and their control. In *Swimming and Flying in Nature*, Vol. 2, T.Y.-T. Wu et al., eds. pp. 599-613. New York: Plenum Press. (*258*)

Strickler, J.R. (1977) Observation of swimming performances of planktonic copepods. *Limnol. Oceanogr.* **22**: 165-70. (*256*)

Strickler, J.R. and S. Twombly (1975) Reynolds number, diapause, and predatory copepods. *Verh. Internat. Verein. Limnol.* **19**: 2943-50. (*257*)

Stride, G.O. (1955) On the respiration of an African beetle, *Potomodytes tuberosus* Hinton. *Ann. Entom. Soc. Amer.* **48**: 344-51. (*51*)

Stuart, A.M. (1958) The efficiency of adaptive structures in the nymph of *Rhithrogena semicolorata* (Curtis) (Ephemeroptera). *J. Exp. Biol.* **35**: 27-38. (*144*)

* Sutton, O.G. (1949) *The Science of Flight.* Harmondsworth, Middlesex, England: Penguin Books, Ltd. (*4*)

* Sverdrup, H.U., M.W. Johnson, and R.H. Fleming (1942) *The Oceans: Their Physics, Chemistry, and General Biology.* Englewood Cliffs, N.J.: Prentice-Hall. (*18, 22*)

Swedmark, B. (1964) The interstitial fauna of marine sands. *Biol. Rev.* **39**: 1-42. (*265*)

Taylor, G.I. (1951) Analysis of the swimming of microscopic organisms. *Proc. Roy. Soc. Lond.* **A 211**: 225-39. (*259*)

Thom, A. and P. Swart (1940) The forces on an aerofoil at very low speeds. *J. Roy. Aero. Soc.* **44**: 761-70. (*78, 215*)

Thom, A.S. (1968) The exchange of momentum, mass, and heat between an artificial leaf and airflow in a wind tunnel. *Q. J. Roy. Meterol. Soc.* **94**: 44-55. (*97, 285*)

Thomas, D. (1958) Tree crown deformation as an index of exposure intensity. *Forestry* **31**: 121-31. (*98*)

Thomas, T.M. (1973) Tree deformation by wind in Wales. *Weather* **28**: 46-58. (*98*)

* Thompson, D'Arcy W. (1942) *On Growth and Form.* 2nd ed. Cambridge, U.K.: Cambridge University Press. (*215*)

Titman, D. (1975) A fluorometric technique for measuring sinking rates of freshwater plankton. *Limnol. Oceanogr.* **20**: 869-76. (*255*)

Tolbert, W.W. (1977) Aerial dispersal behavior of two orb weaving spiders. *Psyche* **84**: 13-27. (*159*)

Tracy, C.R. and P.R. Sotherland (1979) Boundary layers of bird eggs: do they ever constitute a significant barrier to water loss? *Physiol. Zool.* **52**: 63-66. (*149*)

Tucker, V.A. (1973) Bird metabolism during flight: evaluation of a theory. *J. Exp. Biol.* **58**: 689–709. (*119*)

*Tucker, V.A. and G.C. Parrott (1970) Aerodynamics of gliding flight in a falcon and other birds. *J. Exp. Biol.* **52**: 345–67. (*112, 115, 210, 219, 220, 221, 293*)

Vahl, O. (1971) Growth and density of *Patina pellucida* (L.) (Gastropoda: Prosobranchia) on *Laminaria hyperborea* (Gunneries) from western Norway. *Ophelia* **9**: 31–50. (*147*)

Vahl, O. (1972) On the position of *Patina pellucida* (L.) (Gastropoda) on the frond of *Laminaria hyperborea. Ophelia* **10**: 1–9. (*145*)

Van der Pijl, L. (1972) *Principles of Dispersal in Higher Plants.* 2nd ed. New York: Springer-Verlag. (*159, 254*)

Vermeij, G.J. (1973) Morphological patterns in high-intertidal gastropods: adaptive strategies and their limitations. *Mar. Biol.* **20**: 319–46. (*282*)

Vogel, S. (1962) A possible role of the boundary layer in insect flight. *Nature* **193**: 1201–1202. (*130, 162*)

Vogel, S. (1966) Flight in *Drosophila.* I. Flight performance of tethered flies. *J. Exp. Biol.* **44**: 567–78. (*7, 122, 229*)

Vogel, S. (1967*a*) Flight in *Drosophila.* II. Variations in stroke parameters and wing contour. *J. Exp. Biol.* **46**: 383–92. (*230, 238*)

Vogel, S. (1967*b*) Flight in *Drosophila.* III. Aerodynamic characteristics of fly wings and wing models. *J. Exp. Biol.* **46**: 431–43. (*83, 206, 209, 213, 214*)

*Vogel, S. (1969) Low-speed wind tunnels for biological investigations. In *Experiments in Physiology and Biochemistry,* Vol. 2, G.A. Kerkut, ed. pp. 295–325. London: Academic Press. (*122, 291, 310*)

Vogel, S. (1970) Convective cooling at low airspeeds and the shape of broad leaves. *J. Exp. Bot.* **21**: 91–101. (*97, 282*)

Vogel, S. (1974) Current-induced flow through the sponge, *Halichondria. Biol. Bull.* **147**: 443–56. (*311*)

Vogel, S. (1976) Flows in organisms induced by movements of the external medium. In *Scale Effects in Animal Locomotion,* T. J. Pedley, ed. pp. 285–97. London: Academic Press. (*55*)

Vogel, S. (1977) Current-induced flow through sponges *in situ. Proc. Nat. Acad. Sci. USA* **74**: 2069–71. (*54*)

Vogel, S. (1978*a*) Organisms that capture currents. *Sci. Amer.* **239** (2): pp. 128–39. (*54, 153, 233*)

Vogel, S. (1978*b*) Evidence for one-way valves in the water flow system of sponges. *J. Exp. Biol.* **76**: 137–48. (*105*)

Vogel, S. and W.L. Bretz (1972) Interfacial organisms: passive ventilation in the velocity gradients near surfaces. *Science* **175**: 210–11. (*54*)

Vogel, S. and R.D. Chapman (1966) Force measurements with d'Arsonval galvanometers. *Rev. Scient. Instrum.* **37**: 520. (*317*)

Vogel, S., C.P. Ellington, Jr., and D.C. Kilgore, Jr. (1973) Wind-induced ventilation of the burrow of the prairie dog, *Cynomys ludovicianus. J. Comp. Physiol.* **84**: 1–14. (*54, 82*)

Vogel, S. and N. Feder (1966) Visualization of low-speed flow using suspended plastic particles. *Nature* **209**: 186–87. (*37, 83, 305*)

*Vogel, S. and M. LaBarbera (1978) Simple flow tanks for research and teaching. *Bioscience* **28**: 638–43. (*297*)

*Von Karman, T. (1954) *Aerodynamics.* New York: McGraw-Hill. (*4, 86*)

Vosburgh, F. (1977) The response to drag of the reef coral *Acropora reticulata. Proc. 3rd Int. Coral Reef Symp.* **1**: 477–82. (*82, 107*)

*Wainwright, S.A., W.D. Biggs, J.D. Currey, and J.M. Gosline (1976) *Mechanical Design in Organisms.* New York: John Wiley. (*7, 101, 107*)

Wainwright, S.A. and J.R. Dillon (1969) On the orientation of sea fans (genus *Gorgonia*). *Biol. Bull.* **136**: 130–40. (*87, 105*)

*Wallace, J.B. and R.W. Merritt (1980) Filter-feeding ecology of aquatic insects. *Ann. Rev. Entomol.* **25**: 103–32. (*48, 156, 262, 263*)

Wallace, J.B., J.R. Webster, and W.R. Woodall (1977) The role of filter-feeders in flowing waters. *Arch. Hydrobiol.* **79**: 506–32. (*144, 263*)

Walne, P.R. (1972) The influence of current speed, body size and water temperature on the filtration rate of five species of bivalves. *J. Mar. Biol. Assoc. U.K.* **52:** 345–74. (*153*)

Walters, V. (1962) Body form and swimming performance of the scombroid fishes. *Amer. Zool.* **2:** 143–49. (*23*)

Warburton, K. (1976) Shell form, behavior, and tolerance to water movement in the limpet, *Patina pellucida* (L.) (Gastropoda: Prosobranchia). *J. Exp. Mar. Biol. Ecol.* **23:** 307–28. (*145, 147*)

*Webb, P.W. (1975) Hydrodynamics and energetics of fish propulsion. *Bull. Fish. Res. Bd. Canada* **190:** 1–158. (*23, 115, 118, 217*)

Webb, P.W. (1979) Mechanics of escape responses in crayfish (*Orconectes virilis*). *J. Exp. Biol.* **79:** 245–63. (*276*)

Weihs, D. (1974) Energetic advantages of burst swimming of fish. *J. Theor. Biol.* **48:** 215–29. (*219*)

Weihs, D. (1975) Some hydrodynamical aspects of fish schooling. In *Swimming and Flying in Nature,* Vol. 2, T.Y.-T. Wu et al., eds. pp. 703–18. New York: Plenum Press. (*210*)

Weir, J.S. (1973) Air flow, evaporation and mineral accumulation in mounds of *Macrotermes subhyalinus* (Rambur). *J. Anim. Ecol.* **42:** 509–20. (*54*)

Weis-Fogh, T. (1956) Biology and physics of locust flight. II. Flight performance of the desert locust (*Schistocerca gregaria*). *Phil. Trans. Soc. Lond.* **B 239:** 459–510. (*120*)

*Weis-Fogh, T. (1973) Quick estimates of flight fitness in hovering animals, including novel mechanisms for lift production. *J. Exp. Biol.* **59:** 169–230. (*230, 237*)

*Weis-Fogh, T. (1975*a*) Flapping flight in birds and insects, conventional and novel mechanisms. In *Swimming and Flying in Nature,* Vol. 2, T.Y.-T. Wu et al., eds. pp. 729–62. New York: Plenum Press. (*237*)

*Weis-Fogh, T. (1975*b*) Unusual mechanisms for the generation of lift in flying animals. *Sci. Amer.* **233** (5): 81–87. (*237*)

*Weis-Fogh, T. and R.M. Alexander (1977) The sustained power output from striated muscle. In *Scale Effects in Animal Locomotion,* T. J. Pedley, ed. pp. 511–25. London: Academic Press. (*118, 230*)

Weis-Fogh, T. and W.B. Amos (1972) Evidence for a new mechanism of cell motility. *Nature* **236:** 301–304. (*156*)

Westlake, D.F. (1967) Some effects of low-velocity currents on the metabolism of aquatic macrophytes. *J. Exp. Bot.* **18:** 187–205. (*151*)

White, C.M. (1946) The drag of cylinders in fluids at low speeds. *Proc. Roy. Soc. Lond.* **A 186:** 472–79. (*73, 250*)

*White, F.M. (1974) *Viscous Fluid Flow.* New York: McGraw-Hill. (*243, 244, 246*)

Wilbur, H.M. (1976) Life history evolution of seven milkweeds of the genus Asclepias. *J. Ecol.* **64:** 223–40. (*253*)

Withers, P.C. (1979) Aerodynamics and hydrodynamics of the 'hovering' flight of Wilson's storm petrel. *J. Exp. Biol.* **80:** 83–91. (*211*)

Withers, P.C. and P.L. (O'Neill) Timko (1977) The significance of ground effect to the aerodynamic cost of flight and energetics of the black skimmer (*Rhyncops nigra*). *J. Exp. Biol.* **70:** 13–26. (*210*)

Womersley, J.R. (1955) Method for the calculation of velocity, rate of flow, and viscous drag when the pressure gradient is known. *J. Physiol.* **127:** 553–63. (*274*)

*Wu. T.Y.-T., C.J. Brokaw, and C. Brennen, eds. (1975) *Swimming and Flying in Nature.* 2 vols. New York: Plenum Press. (*261*)

Yoshino, M.M. (1973) Wind-shaped trees in the subalpine zone in Japan. *Arctic and Alpine Res.* **5:** A115–26. (*98*)

Zaret, R.E. and W.C. Kerfoot (1980) The shape and swimming technique of *Bosmina longirostris. Limnol. Oceanogr.* **25:** 126–33. (*247, 256*)

Zeleny, J. and L.W. McKeehan (1909) An experimental determination of the terminal velocity of fall of small spheres in air. *Science* **29:** 469. (*83*)

Zimmermann, M.H. (1971) Transport in the xylem. In *Trees: Structure and Function,* M.H. Zimmermann and C.L. Brown, eds. pp. 169–220. New York: Springer-Verlag. (*182, 183*)

subject index

(Abbreviations: BL, boundary layer; *Re,* Reynolds number)

index

index

index

index